PROSPECTIVE EVALUATION OF APPLIED ENERGY RESEARCH AND DEVELOPMENT AT DOE (PHASE TWO)

Committee on Prospective Benefits of DOE's Energy Efficiency and Fossil Energy R&D Programs (Phase Two)

Board on Energy and Environmental Systems
Division on Engineering and Physical Sciences

NATIONAL RESEARCH COUNCIL
OF THE NATIONAL ACADEMIES

THE NATIONAL ACADEMIES PRESS
Washington, D.C.
www.nap.edu

THE NATIONAL ACADEMIES PRESS   500 Fifth Street, N.W.   Washington, DC 20001

NOTICE: The project that is the subject of this report was approved by the Governing Board of the National Research Council, whose members are drawn from the councils of the National Academy of Sciences, the National Academy of Engineering, and the Institute of Medicine. The members of the committee responsible for the report were chosen for their special competences and with regard for appropriate balance.

This report and the study on which it is based were supported by Contract No. DE-AT01-05EE13073. Any opinions, findings, conclusions, or recommendations expressed in this publication are those of the authors and do not necessarily reflect the views of the organizations or agencies that provided support for the project.

International Standard Book Number-13: 978-0-309-10467-8
International Standard Book Number-10: 0-309-10467-X

Available in limited supply from:
Board on Energy and Environmental Systems
National Research Council
500 Fifth Street, N.W.
Keck W934
Washington, DC 20001
202-334-3344

Additional copies available for sale from:
The National Academies Press
500 Fifth Street, N.W.
Lockbox 285
Washington, DC 20055
800-624-6242 or 202-334-3313 (in the
   Washington metropolitan area)
http://www.nap.edu

Copyright 2007 by the National Academy of Sciences. All rights reserved.

Printed in the United States of America

THE NATIONAL ACADEMIES
Advisers to the Nation on Science, Engineering, and Medicine

The **National Academy of Sciences** is a private, nonprofit, self-perpetuating society of distinguished scholars engaged in scientific and engineering research, dedicated to the furtherance of science and technology and to their use for the general welfare. Upon the authority of the charter granted to it by the Congress in 1863, the Academy has a mandate that requires it to advise the federal government on scientific and technical matters. Dr. Ralph J. Cicerone is president of the National Academy of Sciences.

The **National Academy of Engineering** was established in 1964, under the charter of the National Academy of Sciences, as a parallel organization of outstanding engineers. It is autonomous in its administration and in the selection of its members, sharing with the National Academy of Sciences the responsibility for advising the federal government. The National Academy of Engineering also sponsors engineering programs aimed at meeting national needs, encourages education and research, and recognizes the superior achievements of engineers. Dr. Charles M. Vest is president of the National Academy of Engineering.

The **Institute of Medicine** was established in 1970 by the National Academy of Sciences to secure the services of eminent members of appropriate professions in the examination of policy matters pertaining to the health of the public. The Institute acts under the responsibility given to the National Academy of Sciences by its congressional charter to be an adviser to the federal government and, upon its own initiative, to identify issues of medical care, research, and education. Dr. Harvey V. Fineberg is president of the Institute of Medicine.

The **National Research Council** was organized by the National Academy of Sciences in 1916 to associate the broad community of science and technology with the Academy's purposes of furthering knowledge and advising the federal government. Functioning in accordance with general policies determined by the Academy, the Council has become the principal operating agency of both the National Academy of Sciences and the National Academy of Engineering in providing services to the government, the public, and the scientific and engineering communities. The Council is administered jointly by both Academies and the Institute of Medicine. Dr. Ralph J. Cicerone and Dr. Charles M. Vest are chair and vice chair, respectively, of the National Research Council.

www.national-academies.org

COMMITTEE ON PROSPECTIVE BENEFITS OF DOE'S ENERGY EFFICIENCY AND FOSSIL ENERGY R&D PROGRAMS (PHASE TWO)

MAXINE L. SAVITZ, NAE,[1] *Chair*, Honeywell, Inc. (retired), Los Angeles, California
LINDA R. COHEN, University of California, Irvine
JAMES CORMAN, Energy Alternatives Studies, Inc., Schenectady, New York
PAUL A. DeCOTIS, New York State Energy Research and Development Authority, Albany
RAMON L. ESPINO, University of Virginia, Charlottesville
ROBERT W. FRI, Resources for the Future, Washington, D.C.
W. MICHAEL HANEMANN, University of California, Berkeley
WESLEY L. HARRIS, NAE, Massachusetts Institute of Technology, Cambridge
MARTHA A. KREBS, California Energy Commission, Sacramento
LESTER B. LAVE, IOM,[2] Carnegie Mellon University, Pittsburgh, Pennsylvania
RICHARD G. NEWELL, Resources for the Future, Washington, D.C.
JACK S. SIEGEL, Energy Resources International, Inc., Washington, D.C.
JAMES E. SMITH, Duke University, Durham, North Carolina
TERRY SURLES, University of Hawaii, Manoa
JAMES L. SWEENEY, Stanford University, California
MICHAEL TELSON, University of California, Washington, D.C.

Project Staff

Board on Energy and Environmental Systems (BEES)

MARTIN OFFUTT, Study Director
ALAN CRANE, Senior Program Officer
JAMES J. ZUCCHETTO, Director, BEES
PANOLA GOLSON, Program Associate
DANA CAINES, Financial Associate
JENNIFER BUTLER, Financial Assistant

Transportation Research Board

JILL WILSON, Senior Program Officer

Consultant

KAREN JENNI, Insight Decisions

---

[1] NAE, member, National Academy of Engineering.
[2] IOM, member, Institute of Medicine.

BOARD ON ENERGY AND ENVIRONMENTAL SYSTEMS

DOUGLAS M. CHAPIN, NAE,[1] *Chair*, MPR Associates, Alexandria, Virginia
ROBERT W. FRI, *Vice Chair*, Resources for the Future, Washington, D.C.
RAKESH AGRAWAL, NAE, Purdue University, West Lafayette, Indiana
ALLEN J. BARD, NAS,[2] University of Texas, Austin
MARILYN BROWN, Georgia Institute of Technology, Atlanta
PHILIP R. CLARK, NAE, GPU Nuclear Corporation (retired), Boonton, New Jersey
MICHAEL CORRADINI, NAE, University of Wisconsin, Madison
E. LINN DRAPER, JR., NAE, American Electric Power, Inc., Austin, Texas
CHARLES GOODMAN, Southern Company, Birmingham, Alabama
DAVID G. HAWKINS, Natural Resources Defense Council, Washington, D.C.
DAVID K. OWENS, Edison Electric Institute, Washington, D.C.
WILLIAM F. POWERS, NAE, Ford Motor Company (retired), Ann Arbor, Michigan
TONY PROPHET, Hewlett-Packard Company, Cupertino, California
MICHAEL P. RAMAGE, NAE, ExxonMobil Research and Engineering Company (retired), Moorestown, New Jersey
MAXINE L. SAVITZ, NAE, Honeywell, Inc. (retired), Los Angeles, California
PHILIP R. SHARP, Resources for the Future, Washington, D.C.
SCOTT W. TINKER, University of Texas, Austin

Staff

JAMES J. ZUCCHETTO, Director
DUNCAN BROWN, Senior Program Officer (part time)
ALAN CRANE, Senior Program Officer
MARTIN OFFUTT, Senior Program Officer
DANA CAINES, Financial Associate
PANOLA GOLSON, Program Associate
JENNIFER BUTLER, Financial Assistant

---

[1]NAE, member, National Academy of Engineering.
[2]NAS, member, National Academy of Sciences.

Acknowledgments

The Committee on Prospective Benefits of DOE's Energy Efficiency and Fossil Energy R&D Programs (Phase Two) wishes to acknowledge and thank the many individuals who contributed time and effort to this National Research Council (NRC) study. The presentations at committee meetings and ongoing dialogue with interested participants from within the administration and outside experts provided valuable information and insight. In particular, the Department of Energy's Office of Energy Efficiency and Renewable Energy, the Office of Fossil Energy, and the Office of the Undersecretary for Energy and Environment supplied extensive data and analysis for the study. Their valuable information on and insight into advanced technologies and development initiatives assisted the committee in formulating the recommendations included in this report. Ongoing dialogue with the Office of Management and Budget, the Office of Science and Technology Policy, and congressional staff provided valuable guidance for the selection of case studies and presentation of results.

The chair particularly wants to acknowledge the participation of the expert panels, who performed the analysis with professionalism and insight in a timely and informative way. The Phase Two panels exhibited the same high quality as those of Phase One.

All of the committee members served as chairs or members of the expert panels and as members of the methodology and process subcommittees, displaying both expertise and leadership. The chair would like to thank William Fisher, University of Texas, Austin, who, although not a member of the full committee, chaired the panel for the Natural Gas Exploration and Production Program.

In particular, the chair is very grateful for the work and support of the staff of the NRC. The staff included Martin Offutt, James Zucchetto, Alan Crane, and Jill Wilson. Martin Offutt did an excellent job in assembling six expert panels in a relatively short period of time, staffing the work of three of them, and helping to drive the production of this report. This study differs from many other NRC projects and has placed additional demands on the expertise of the staff. Panola Golson did an excellent job of supporting the six panels and the committee throughout. Karen Jenni, the consultant, assisted the panels in understanding the methodology and developing the decision trees in a manner consistent across all of the panels.

Lastly, the chair would like to thank Robert Fri, who stepped in over the last few months to assist the chair in getting the report completed.

This report has been reviewed in draft form by individuals chosen for their diverse perspectives and technical expertise, in accordance with procedures approved by the NRC's Report Review Committee. The purpose of this independent review is to provide candid and critical comments that will assist the institution in making its published report as sound as possible and to ensure that the report meets institutional standards for objectivity, evidence, and responsiveness to the study charge. The review comments and draft manuscript remain confidential to protect the integrity of the deliberative process. We wish to thank the following individuals for their review of this report:

William Agnew, NAE, General Motors (retired),
David Archer, NAE, Westinghouse (retired),
William Banholzer, NAE, Dow Chemical,
Joseph Cordes, George Washington University,
Roland Horne, NAE, Stanford University,
Trevor Jones, NAE, Biomec, Inc.,
James J. Markowsky, NAE, American Electric Power (retired),
Dexter Peach, General Accounting Office (retired),
Tim Pinkston, Southern Company Generation,
Edward Rubin, Carnegie Mellon University,
Rosalie Ruegg, Independent Consultant,

Robert Shaw, Arete Corporation,
Jack Wise, NAE, Mobil Research and Development Corporation (retired),
Jim Wolf, Independent Consultant, and
Frankie Wood-Black, Philips Petroleum Company.

Although the reviewers listed above have provided many constructive comments and suggestions, they were not asked to endorse the conclusions or recommendations nor did they see the final draft of the report before its release. The review of this report was overseen by John Ahearne, NAE, Sigma Xi, and Lawrence Papay, NAE, Science Applications International Corporation (retired). Appointed by the National Research Council, they were responsible for making certain that an independent examination of this report was carried out in accordance with institutional procedures and that all review comments were carefully considered. Responsibility for the final content of this report rests entirely with the authoring committee and the institution.

Contents

SUMMARY 1

1 INTRODUCTION 8
Background, 8
Applied Energy Research and Development at DOE, 9
Evaluating the Federal Investment in Energy Research and Development, 10
The Current Study, 13

2 RESULTS OF APPLYING THE METHODOLOGY 14
Case Study Selection and Execution, 14
Statement of Task, 14
Advice to Users, 15
Results of Case Studies, 16

3 METHODOLOGY FOR PROSPECTIVE EVALUATION OF 40
DEPARTMENT OF ENERGY PROGRAMS
Valuation of Environmental and Security Benefits, 40
Methodological Considerations in Measuring Benefits, 47

4 EXPERT PANEL PROCESS 52
Introduction, 52
Expert Panel Composition, 52
Role of the Panel Chair, the Consultant(s), and the Oversight Committee, 53
Panel Activities and Process, 54
Interactions with DOE and Information Request, 56
Duration and Frequency of the Expert Panel Reviews, 57
Assessment of Activities by Non-DOE Entities, 57
General Issues, 57
Quality Assurance, 58
Full-Scale Implementation of the Methodology, 58

5 CONCLUSIONS AND RECOMMENDATIONS 60
Introduction, 60
Priorities Identified for Phase Three of This Project, 60
The Process for Obtaining Results, 61
Using the Results, 62
Estimating National Security and Environmental Benefits, 62
Continuity: Institutionalizing the Evaluation Process, 63

REFERENCES 64

APPENDIXES

A PART Assessment Questions 69
B Committee Biographies 71
C Statement of Task 75
D Letter Report 77
E Committee and Panel Activities 92
F Guidance on Prospective Benefits Evaluation 95
G Information to Be Requested of the Department of Energy 107
H Report of the Panel on DOE's Integrated Gasification Combined Cycle Technology R&D Program 111
I Report of the Panel on DOE's Carbon Sequestration Program 132
J Report of the Panel on DOE's Natural Gas Exploration and Production R&D Program 152
K Report of the Panel on DOE's Distributed Energy Resources Program 171
L Report of the Panel on DOE's Light-Duty Vehicle Hybrid Technology R&D Program 187
M Report of the Panel on DOE's Chemical Industrial Technologies Program 208

Tables and Figures

TABLES

S-1   Benefits of Three EERE R&D Programs, 6
S-2   Benefits of Three FE R&D Programs, 7

1-1   Summary of Congressional Appropriations Oversight of EERE Programs, 10
1-2   Stated Goals of GPRA and PART, 11

3-1   Estimates of the Social Damage Costs of Air Emissions, 43
3-2   Estimates of Pollution Abatement Costs, 43

H-1   Baseline and 2010 Goals for Total IGCC System as Given in DOE's Advanced IGCC Research, 112
H-2   Evolutionary Improvements Due to DOE Advanced IGCC Research, 113
H-3   Evolutionary Improvements Due to DOE Advanced Gasification Research, 113
H-4   Revolutionary or Long-Term Improvements Due to DOE Advanced Gasification Research, 113
H-5   Costs for 500-MW Power Plants Using a Range of Technologies Without Carbon Capture and Storage, 118
H-6   Annual Additions of IGCC Capacity Used to Calculate Program Benefits, 131

I-1   Top-Level Carbon Sequestration Roadmap, 134

J-1   DOE's Performance Targets for Expanding Domestic ERR by 50 Tcf Through 2015, 153
J-2   Panel Assessments of Technical and Market Risks for Each Subprogram, 156

L-1   Panel Estimates of Fuel Economy Improvements Relative to Conventional Vehicles, 197
L-2   Panel Estimates of Fuel Economy Improvement for Vehicles with Specified Technical Improvements, 197

FIGURES

S-1 Results matrix, 3

2-1 Findings for DOE's gasification technologies R&D, 17
2-2 Findings for DOE's carbon sequestration program, 20
2-3 Findings for DOE's natural gas exploration and production program, 23
2-4 Findings for DOE's distributed energy resources program, 29
2-5 Findings for DOE's light-duty vehicle hybrid technology R&D, 32
2-6 Findings for DOE's chemical industrial technologies program, 37

F-1 Results matrix for evaluating benefits and costs prospectively, 96
F-2 Decision tree, 99
F-3 Example of decision tree applied to advanced lighting program, 100
F-4 Template for presenting panel results, 104-105

G-1 Three-page program assessment summary (PAS) form, to be completed by DOE, 108-110

H-1 Decision tree representing the panel's assessment of the likely technical outcomes of IGCC R&D, 120
H-2 Estimated COE for IGCC under different technical success assumptions for the AEO Reference Case global scenario, 121
H-3 Cumulative amount of IGCC built under three different technical success assumptions and two different global scenarios, 122
H-4 Cumulative distribution on the net present value of IGCC research under the AEO Reference Case scenario, assuming IGCC replaces PC, 124
H-5 Results matrix of the Panel on DOE's Integrated Gasification Combined Cycle Technology R&D Program, 126

I-1 Funding requirements for DOE's carbon sequestration program, 135
I-2 Decision tree used by carbon sequestration panel, 139
I-3 Summary of probability assessment results, 140
I-4 Panel assessment of sequestration risks, 141
I-5 Effect of carbon tax and incremental cost of CCS on COE for IGCC, 143
I-6 Comparison of COE with competing technologies, 145
I-7 Impact of DOE program funding on cumulative builds of IGCC with CCS, through 2025, 146
I-8 Cumulative distribution on the NPV of the economic benefits of carbon sequestration research, with and without the DOE program, 147
I-9 Results matrix of the Panel on DOE's Carbon Sequestration Program, 148

J-1 Decision tree for the existing fields subprogram, 155
J-2 Decision tree for the drilling, completion, and stimulation subprogram, 158
J-3 Decision tree for the advanced diagnostics and imaging subprogram, 159
J-4 Decision tree for the Deep Trek subprogram, 161
J-5 Estimated increase in domestic natural gas attributed to the program by year, 163
J-6 Results matrix of the Panel on DOE's Natural Gas Exploration and Production Program, 164
J-7 Demand curve and two supply curves for natural gas, 169
J-8 Change in total surplus from the change in the supply curve, 170

K-1   Decision tree for combined heat and power program, 175
K-2   Assessment of technical success with DOE funding and high electricity prices, 176
K-3   Assessment of technical success for the AEO Reference Case scenario, 176
K-4   Assessment of technical success for the High Oil and Gas Prices scenario, 177
K-5   Assessment of technical success for the Carbon Constrained scenario, 177
K-6   Assessment of technical success for the Electricity Constrained scenario, 178
K-7   Assessment of market success, 179
K-8   Panel's estimate of the amount of CHP added with and without the DOE program for each scenario, 180
K-9   Uncertainty around estimated CHP additions for the Electricity Constrained scenario, 181
K-10 Results matrix of the Panel on DOE's Distributed Energy Resources Program, 182

L-1   Decision tree representing the panel's evaluation of the batteries program, 194
L-2   Decision tree representing the panel's evaluation of the lightweighting research program, 195
L-3   Decision tree representing the panel's evaluation of the advanced combustion engines program, 196
L-4   Fraction of new vehicles purchased that are conventional and hybrid electric, for two HEV market scenarios, 199
L-5   Fraction of total vehicle miles driven by year by vehicle type, for two HEV market scenarios, 200
L-6   Range and likelihood of discounted total consumer expenditures for vehicles and gas between 2006 and 2050, assuming the Low HEV market scenario and Reference Case prices, 201
L-7   Results matrix of the Panel on DOE's Light-Duty Vehicle Hybrid Technology R&D Program, 202

M-1   Probability of technical success, probability of market success, and value of benefits if the project is successful, 210
M-2   Uncertainty surrounding estimates of program benefits, 213
M-3   Implied decision tree representing the panel's evaluation of the ITP–Chemicals portfolio, 214
M-4   Results matrix of the Panel on DOE's Industrial Technologies Program–Chemicals, 215
M-5   Sample project information sheet, 217

Summary

OVERVIEW

In recent years, federal oversight of public expenditures has sought to integrate performance and budgeting. Notably, the Government Performance and Results Act (GPRA) was passed in 1993 "in response to questions about the value and effectiveness of federal programs" (GAO, 1997, p. 11). GPRA and other mandates have led agencies to develop indicators of program performance and program outcomes. The development of indicators has been watched with keen interest by Congress, which asked the National Research Council (NRC) for a series of reports using quantitative indicators to evaluate the effectiveness of applied energy R&D. The first such report[1] took a retrospective view of the first 23 years of R&D programs sponsored by the Department of Energy (DOE) on fossil energy and energy efficiency.[2] That report found that DOE-sponsored research had netted large commercial successes—such as advanced refrigerator compressors, electronic lighting ballasts, and emission control technology for flue gas desulfurization (NRC, 2001). Other programs, however, were judged to have been costly failures in which large R&D expenditures did not result in a commercial energy technology (NRC, 2001). A follow-up NRC committee was assigned the task of adapting the retrospective methodology to the assessment of the future payoff of continuing programs (NRC, 2005a). The present report continues the NRC's investigation of R&D outcome indicators and applies the benefits evaluation methodology to six DOE R&D activities. The report further defines indicators for environmental and security benefits and refines the evaluation process based on the experience with the case studies.

Evaluating the outcome of R&D expenditures requires an analysis of program costs and benefits. Doing so is not a trivial matter. First, the analysis of costs and benefits must reflect the full range of public benefits that are envisioned, accounting for environmental and energy security impacts as well as economic impacts. Second, the analysis must consider how likely the research is to succeed and how valuable the research will be if successful. Finally, the analysis must consider what might happen if the government does not support the project: Would some non-DOE entity undertake it or an equivalent activity that would produce some or all of the benefits of government involvement? The process and methodology developed by the Phase One committee, summarized in Appendix F, are designed to address these challenges.

Pursuant to GPRA, DOE submits its own analysis of program costs and benefits in an annual report to Congress accompanying the President's Budget Request. Additional analyses are required of DOE and all federal agencies by the Office of Management and Budget (OMB), the executive agency that formulates and administers the federal budget. The President's Management Agenda (PMA) (OMB, 2001) sets forth nine agency-specific reforms and five government-wide goals (OMB, 2005); a set of R&D investment criteria was spelled out in 2003, implementing one of the agency-specific goals of the PMA (Marburger and Daniels, 2003). Also in 2003, OMB inaugurated the Program Assessment Rating Tool (PART) (OMB, 2003, pp. 47-53) to assess and improve program performance.

The principal responsibility at DOE for developing applied energy technologies resides in three offices—the Office of Energy Efficiency and Renewable Energy (EERE), the Office of Fossil Energy (FE), and the Office of Nuclear Energy Science and Technology (NE).[3] Of their combined budget authority of $2.4 billion in FY05, the three offices devoted approximately $1.3 billion to R&D. The R&D pro-

---

[1] Requested by Congress in the conference report of the Consolidated Appropriations Act for fiscal year (FY) 2000 (House Report 106-479, p. 493. November 18, 1999. Washington, D.C.: U.S. Government Printing Office).

[2] These programs include only those that were at the time under the jurisdiction of the U.S. House Appropriations Subcommittee on the Interior and Related Agencies.

[3] The Office of Electricity Delivery and Energy Reliability, created in 2002, could be considered a fourth applied energy program.

grams are complemented by policy measures such as tax incentives to encourage early adoption of advanced technologies by consumers, efficiency standards for household appliances, and production tax credits for certain renewable energy sources (EIA AEO, 2006).

The programs evaluated by the NRC were limited to R&D within FE and to the portion of EERE's R&D devoted to energy conservation.[4] Research within FE has traditionally been divided between the Office of Coal and Power Systems (CPS) and the Office of Oil and Natural Gas. CPS administers a suite of clean coal R&D, which has the goal of ensuring the generation of clean, reliable, and affordable electricity from coal. The Office of Oil and Natural Gas supports research and policy options to ensure clean, reliable, and affordable supplies of oil and natural gas for American consumers.[5] The energy conservation portion of EERE's R&D is related to technologies for the efficient end-use of fuels and electricity in vehicles, in industrial processes, and within building envelopes.

Two general activities have been the focus of this Phase Two study: refining the methodology developed in Phase One and applying it to additional R&D projects. The committee improved the methodology for estimating environmental benefits (from, for example, reduced emissions) and estimating national security benefits (from, for example, reduced oil imports). In parallel, the committee selected six R&D activities to be the subject of case studies, which were carried out by separate expert panels appointed by the NRC. The activities selected for review included three within EERE—the Chemicals subprogram of the Industrial Technologies Program, the Distributed Energy Resources Program, and the activities related to light-duty hybrid electric vehicles within the Vehicles Technologies Program—and three within FE—the Integrated Gasification Combined Cycle subprogram, the Carbon Sequestration Program, and the Exploration and Production activities of the Natural Gas Technologies Program. This Phase Two study shows that the basic analytical structure, using decision trees, works well and can be implemented with the appropriate panels.

HOW PROGRAM BENEFITS ARE EVALUATED

Introduction

The primary effects of DOE's programs are seen to be these: (1) they reduce technical risk, (2) they reduce market risk, and (3) they accelerate the introduction of the technology into the marketplace. The methodology developed by the Phase One committee used expert panels to review the DOE R&D program and estimate the expected economic, environmental, and energy security benefits of the program in three different global economic scenarios, with the results summarized in a matrix such as that shown in Figure S-1 (see Appendix F for generalized definitions of economic, environmental, and energy security benefits). The expert panel evaluation process is facilitated by a decision analysis consultant, and the panels construct simple decision trees to describe the main technical and market uncertainties associated with the program and the impact of DOE support on the probability of various technical and market outcomes. The decision trees used by all the panels assessed changes in technical and market risks. The acceleration effect was considered separately by each panel. In some cases, acceleration increases the likelihood that a project will attain the program goals of completion by a critical date, which is then accounted for in the assessment of technical risk. In other cases, the panels accounted for acceleration in their benefit calculations, which assume that if the technology is ultimately developed in the absence of the government program the net benefits accrue only for a limited time.

The calculations based on the decision trees allow the benefits of each R&D project to be estimated for combinations of outcomes—technical and market—in each of these global economic scenarios. These scenario- and outcome-specific results would typically be estimated using a simple spreadsheet model in conjunction with more sophisticated models such as the Energy Information Administration's (EIA's) National Energy Modeling System (NEMS).[6] For example, NEMS might be used to estimate prices and demand for various energy sources in a particular global economic scenario. A spreadsheet model might be used to estimate the demand for the particular technology of interest given different levels of effectiveness and/or costs.

The overall benefit of the DOE R&D program is given as the difference between the expected benefits with DOE support and the expected benefits without DOE support, where the expected benefits are given as a probability-weighted average of the benefits in particular technical and market outcomes. To ensure consistency across the expert panels, the process calls for the use of common scenarios and assumptions across evaluations and an oversight committee that provides guidance to the panels. This kind of panel-based probabilistic assessment of R&D programs is common in many industries, in particular, the pharmaceutical industry (Sharpe and Keelin, 1998).

---

[4]The remainder of EERE's R&D was devoted to energy supply from renewable resources. Beginning with FY06, funding for all FE and EERE R&D programs was consolidated into one appropriations account subject to the jurisdiction of the House Appropriations Subcommittee on Energy and Water Development and Related Agencies (CRS, 2005).

[5]Available at <http://www.fossil.energy.gov/programs/oilgas/index.html>.

[6]NEMS is a computer-based, energy-economy system for modeling U.S. energy markets that projects the production, imports, conversion, consumption, and prices of energy, subject to assumptions about macroeconomic and financial factors, world energy markets, resource availability and costs, behavioral and technological choice criteria, cost and performance characteristics of energy technologies, and demographics.

Program Name:

Program Goals:

Year Goals Achieved:

Costs:

Current Funding Cycle:

Expected Cost to Completion:

		Global Scenario		
		Reference Case	High Oil and Gas Prices	Carbon Constrained
Program Risks	Technical Risk			
	Market Risks			
Expected Program Benefits	Economic Benefits			
	Environmental Benefits			
	Security Benefits			

FIGURE S-1 Results matrix.

Methodology

Public programs often have social benefits that are not valued by markets. Assessing the value of such benefits is inherently difficult, involves ambiguity, and, even as an academic matter, a range of possible answers. For the DOE programs, two broad classes of benefits have this characteristic: the environmental benefits of energy technology and the security benefits of energy savings or energy alternatives. These program attributes are in general critical components of the benefits package—indeed, if a program can be justified simply on a market benefits basis, the rationale for government participation might be open to question.

Environmental Benefits

Although there are a host of land, water, and, perhaps, public health impacts to consider, the evaluation of benefits in monetary units related to reducing criteria air pollutants is both the primary environmental benefit identified in the present study and the class of benefits for which valuation methods are most advanced.

Recommendation 1: Panels should apply valuations in monetary units to criteria air pollutant emissions in the results matrix but not to other types of pollutant emissions.

The valuations used should be the allowance price forecasts for the future period.

Energy Security Benefits

Electricity. While the complex relationship between electricity supply and security is becoming clearer, analysts are a long way from having methods for valuing reductions in security threats contributed by technologies such as distributed generation.[7]

Recommendation 2: Panels conducting prospective benefits assessments should describe reductions in threats to energy security related to electricity supply as physical quantities of oil and gas.

Oil and Gas. Increases in U.S. oil and gas consumption and imports may impose incremental costs that are not fully reflected in the market price. For oil, several social cost components have been estimated in various studies, together comprising the so-called "oil premium." In principle, similar estimates could be made for natural gas, but the committee is unaware of any such research.

Recommendation 3: Panels should describe energy security benefits related to reduced oil and natural gas consumption quantitatively in the benefits matrix as physical quantities of oil and gas. The time pattern of the oil consumption impacts should be made explicit, along with an assessment of the probable state of the oil market during those future times.

Conclusions

The committee also reached the following conclusions in regard to the methodology:

• The committee endorses the decision tree framework for use in estimating the benefits of DOE's applied research programs. However, panels must take care and follow the guidance of a decision analyst (the consultant) to understand how to assign probabilities and how to specify the government role clearly.
• The global scenarios developed by the committee for all panels proved to be a valuable tool for characterizing and quantifying the benefits of the DOE R&D programs. However, the Phase Two experience shows that panels sometimes needed to clarify aspects of a scenario to address issues that were important to the specific R&D program.
• The NEMS model is important for providing baseline energy prices and demands, but using it to estimate the prices and demands of all different program outcomes is unlikely to yield refinements that can affect the estimated benefits in a meaningful way.
• The success or failure of competing or complementary technologies can significantly affect the value of a DOE applied research project. Although Phase Two shed light on this issue and provided some methodological guidance for dealing with it, more work is required to describe a method for estimating the benefits of DOE's overall portfolio made up of separate programs or subprojects.
• The panels were generally successful in implementing the committee's methodology, but the Phase Two experience did highlight some process issues that should be considered in future studies.

Process

The experience with the six expert panels and case studies has led the committee to make several recommendations on the process. In particular, the committee found that the commitment and the technology background of the panel members determined the quality of the assessment of a program.

Recommendation 4: Panel composition and level of expertise must be critically considered during the selection process. If a panel concludes that certain skills are not possessed by its members, it should consider adding members or using an outside expert to brief it.

The leadership role of the panel chair cannot be overemphasized. For a panel to succeed, the chair had to take the lead in interacting with DOE to ensure that the best possible information was available to the panel before its first meeting. Moreover, the panels where the chairs spent significant time ensuring that all panel members were fully familiar with the process and methodology produced the best assessments.

Recommendation 5: The panel chair should spend a fair amount of time outside the actual panel meetings working with the DOE program managers, DOE management, and the independent consultant(s).

The primary responsibility of the consultant(s) was to maintain consistency across the panels in applying the methodology and facilitating the analysis. This included structuring the decision trees, facilitating the assignment of probabilities to technical and market outcomes, and providing guidance for and assistance in modeling benefits. Phase

---

[7]The Distributed Energy Resources Panel noted that security pertains to (1) safety from terrorist attacks, (2) insensitivity to energy disruptions caused by reductions in oil imports (or other imports), and (3) customer protection from the effects of disruptions to utility electric service. A unique benefit of distributed energy production technologies is to decentralize power production and locate it at or close to loads, hence providing benefits of types (1) and (3).

Two made use of the same consultant for all of the panels, and the arrangement worked very well.

Recommendation 6: Depending on the number of programs being evaluated and the panels' schedules, it might be necessary to have more than one consultant. One consultant should focus on the decision tree development and probability assessment and the other on the modeling of benefits.

The timeliness and quality of information provided by DOE to the various panels played a critical role in facilitating the deliberations and conclusions of the panel and impacted the quality and utility of its evaluation. Completion of panel evaluations was contingent on the panel's receiving synoptic information and benefits calculations inputs.

Recommendation 7: Since the usefulness of the benefits estimates depends on the quality and timeliness of information available to the panels, DOE management should give its full support for providing the necessary information. DOE at all levels should buy into this process because it is useful for managing and assessing its programs. If this commitment is not clear, the committee should explore all avenues for gaining DOE support.

Quality control continued to be important in ensuring the consistency, and therefore the utility, of the panel evaluations.

Recommendation 8: An oversight committee should apply the quality control process to several elements of the study process, including ensuring appropriate panel membership and composition, orienting the panel chair and consultant, monitoring the panel's progress, monitoring information received from DOE for adequacy and consistency, and reviewing and revising the process itself.

RESULTS OF BENEFITS EVALUATIONS

Approach

The expert panels evaluated benefits using the process and methodology summarized above and described in detail in Chapters 3 and 4 of the Phase One report (NRC, 2005a): That process and the associated methodology are summarized here in Appendix F, which was provided to panelists before their first meeting. The improvements to the methodology described in the above section on methodology were suggested by the committee in parallel with the case studies and thus did not alter the approach taken by the expert panels. However, future case studies will make use of these suggestions.

Results of Six Case Studies

As noted above, the committee selected six case studies to test the proposed methodology and guide refinements and extensions, three within EERE and three within FE. The results of the benefits evaluation for EERE activities are summarized in Table S-1. The three EERE activities had combined annual funding of $115 million in FY05 out of a total of $768 million spent on R&D by EERE. The results of the benefits evaluation for FE activities are summarized in Table S-2. The three FE activities had combined annual funding of $105 million in FY05 out of a total of $561 million spent on R&D by FE. The completion costs are cumulative quantities, calculated assuming the program receives level (constant) funding at the FY05 amount, starting in FY06 through the year in which major goals are achieved.

Tables S-1 and S-2 display program benefits as calculated in each of three standard scenarios that were used in all six case studies: (1) a reference scenario; (2) a high oil and gas prices scenario; and (3) a carbon-constrained scenario. The reference scenario is based on the EIA Reference Case from its *Annual Energy Outlook 2005*, with oil and gas prices ranging from $34 per barrel and $5.3 per thousand cubic feet (Mcf), respectively, in 2005 to $30 per barrel and $4.8 per Mcf, respectively, in 2025. In contrast, the High Oil and Gas Prices scenario has oil and gas prices ranging from $68 per barrel and $10.6 per thousand cubic feet (Mcf), respectively, in 2005, to $78 per barrel and $9.7 per Mcf, respectively, in 2025. The Carbon Constrained scenario assumes a carbon price of $100 per ton of carbon emissions, equivalent to $27 per ton of carbon dioxide. A fourth scenario is invoked as needed to calculate benefits in a future state of the world that would appeal to the unique performance characteristics of the technology under consideration. Economic benefits are measured in dollars, environmental benefits in terms of the physical quantities of avoided emissions, and security benefits as physical quantities of reduced resource consumption (some of which would be reflected in reduced imports). All benefits are computed net of savings that would have been realized in the absence of the DOE program—that is, as a result of R&D by U.S. industry or foreign entities. The results summarized in Tables S-1 and S-2 are described in more detail in Chapter 1 and Appendixes H-M.

When reviewing these results it is important to bear in mind that these studies were conducted to test the proposed methodology and guide refinements and extensions. The committee believes that the estimated benefits are useful indications of program benefits, but the reader should be aware that some panels expressed concern about the reliability of their estimates because they lacked good information about some aspects of the program.

TABLE S-1 Benefits of Three EERE R&D Programs

Program	Program Completion Costs (Assuming Level Funding)	Economic Benefits (Cumulative Net Savings)[a]	Environmental Benefits (Cumulative Reduction in Emissions)	Security Benefits (Cumulative Reduction in Resource Consumption)
Industrial Technologies Program—Chemicals	$75 M through 2015	Ref: $534 M High O&G: $950 M CC: $550 M	All scenarios[b] 24,700 MT CO 15,000 MT SO_2 22,600 MT NO_x 280 MT PM 540 MT VOCs 2.87 MMTCE	All scenarios Natural gas: 89 Bcf Petroleum: 1.3 million bbl
Distributed Energy Resources[c]	$205 M through 2015	Ref: $57 M High O&G: $46 M CC: $64 M Other: $83 M[e]	Unknown[d]	Ref: 10 TBtu of primary energy High O&G: 8 TBtu CC: 11 TBtu Other: 15 TBtu[e]
Light-Duty Vehicle Hybrid Technologies[f]	$567 M through 2012	Ref: $5.9 B to $7.2 B[g] High O&G: $27.5 B to $28.2 B CC: $7.3 B to $8.5 B[g]	Ref: 28 MMTCE High O&G: 51 MMTCE CC: 32 MMTCE	Ref: 219-224 M bbl gasoline High O&G: 398-405 M bbl gasoline CC: 248-252 M bbl gasoline

NOTES: High O&G, High Oil and Gas Prices scenario; Ref, Reference Case from EIA's *Annual Energy Outlook 2005* (EIA, 2005b); CC, Carbon Constrained scenario; Other, fourth scenario added by panel; MMTCE, million metric tons carbon equivalent; TBtu, trillion British thermal units; Bcf, billion cubic feet; bbl, barrels; GHG, greenhouse gases; MT, metric tons; M, million; B, billion; CO, carbon monoxide; SO_2, sulfur dioxide; NO_x, oxides of nitrogen; VOCs, volatile organic compounds; PM, particulate matter.

[a]For ITP–Chemicals program, benefits are cumulative through 2030. For the DER program, benefits are cumulative through 2025. For LDV Hybrid Technologies program, benefits are cumulative through 2050. Economic benefits have been discounted at 3 percent annually.

[b]The chemical industries panel concluded that the scenarios would produce insignificantly small changes in the volumes of oil and gas saved; therefore, the physical quantities reported for environmental and security benefits are the same for all scenarios. Economic benefits differ because prices differ from one scenario to the next.

[c]Includes only the end-use system integration and interface activity of the DER program.

[d]DER program can improve or worsen the environment depending on location, fuel used or displaced, and the technology deployed. The DER panel was not confident about assigning environmental benefits to the DER program as modeled and presented by DOE.

[e]The "other" scenario for distributed energy assumed constrained electricity transmission.

[f]Includes the following elements of the vehicles technology program: Hybrid and Electric Propulsion, excluding projects related to heavy vehicles; Advanced Combustion R&D, limited to the combustion and emission control R&D activity; and Materials Technology, excluding projects related to heavy vehicles and excluding the high-temperature materials laboratory activity.

[g]The two values correspond to two different market scenarios—high and low—for hybrid vehicle penetration.

ADVICE TO USERS OF THE BENEFITS EVALUATION RESULTS

The committee has developed some insights into the methodological strengths and weaknesses of the proposed process. These insights, recorded in the form of recommendations below, may assist decision makers with interpreting and applying the results of the analysis.

Policy measures that have nothing to do with research can have an effect on when and whether the benefits of some programs will be realized. For example, the benefits of carbon capture and sequestration depend on the size and timing of a carbon tax (or equivalent policy intervention in the market). The scenarios are a valuable tool for characterizing and quantifying the benefits of the DOE R&D program.

Recommendation 9: Decision makers should consider the impact of other policy measures—that is, policies not related to research—in all domains of action (federal, state, and international, say) when considering the results of prospective benefits evaluations. Having a common set of scenarios is useful in general, although additional scenarios may be called for in some cases. While defining the scenarios more completely would be helpful for interpreting the outcomes of the analysis, at the same time it is essential to preserve flexibility by keeping the scenarios as broad as possible.

The panel evaluations permit calculation of a benefit-cost ratio. However, the benefit-cost ratio is not the correct metric for optimizing the allocation of additional resources among the programs in a portfolio.

Recommendation 10: To allocate resources, DOE should know the marginal benefit of a budget increase on a program-by-program basis. To calculate the marginal benefit, the decision tree should be examined to identify the outcomes that

TABLE S-2 Benefits of Three FE R&D Programs

Program	Program Completion Costs (Assuming Level Funding)	Economic Benefits (Cumulative Net Savings)[a]	Environmental Benefits (Cumulative Reduction in Emissions)	Security Benefits (Cumulative Reduction in Consumption or Importation of Natural Gas)[b]
Natural Gas Exploration and Production[c]	$140 M through 2015	Ref: $220 M High O&G: $590 M CC: $300 M	Not quantified[d]	Ref: 1.2 Tcf High O&G: 0.6 Tcf CC: 1.2 Tcf
Integrated Gasification Combined Cycle Technology	$750 M through 2020	Ref: $6.4 B to $7.8 B High O&G: $7 B to $47 B	Ref: −90 to 30 MMTCE High O&G: 34 to 36 MMTCE	Ref: up to 4.5 Tcf natural gas High O&G: up to 3.6 Tcf natural gas
Carbon Sequestration	$875 M through 2020	CC: $3.5 B Other:[e] $3.9 B	CC and other:[e] Net environmental benefits are likely zero.[f]	CC and other:[e] Net security benefits are likely zero.[g]

NOTES: High O&G, High Oil and Gas Prices scenario; Ref, Reference Case from Energy Information Administration's *Annual Energy Outlook 2005* (EIA, 2005b); Other, fourth scenario added by panel; CC, Carbon Constrained scenario, which assumes a $100 tax per ton of carbon emissions; MMTCE, million metric tons carbon equivalent; Tcf, trillion cubic feet; quad, quadrillion British thermal units; MT, metric tons; M, million; B, billion.

[a]For the Natural Gas Exploration and Production program, benefits are cumulative through 2025. For IGCC and Carbon Sequestration, benefits accrue over the 20-year book life of new plants built through 2025. Economic benefits have been discounted at 3 percent annually.

[b]For the natural gas technology program, security benefits are expressed as offsets of imported natural gas. For the IGCC program, benefits are reduced U.S. consumption.

[c]Includes the exploration and production activities only.

[d]Modest environmental benefits could result from a reduction in disturbed area and, if fuel switching from coal occurs, from lower carbon emissions.

[e]The "other" scenario devised for the carbon sequestration program assumes a $300 tax per ton of carbon emissions.

[f]Other technologies that could be deployed in a CC regime would reduce emissions by the same amount as those of the DOE program, albeit at higher cost.

[g]The least-cost alternatives to the technology under development in the DOE program would be a combination of nuclear generation and renewable electricity that also would reduce natural gas consumption. The costs of other technologies that would reduce carbon emissions were taken from DOE forecasts.

would be most sensitive to changes in budget levels. In the Phase One study, for example, the lighting program proved to be highly budget-dependent. When such sensitivities exist, the decision tree can be re-estimated for a different budget level, using the committee's methodology. The marginal benefit associated with the change in budget level is the difference between the net benefits of the two calculations.

The methodology estimates benefits for each of three scenarios that describe future states of the world, but it does not attempt to combine the three sets of benefits into a single set. As a result, the decision maker must weigh the alternative scenarios in arriving at judgments about the benefits of the overall research portfolio. Presumably, the portfolio would contain a balance of projects that would produce acceptable results across the range of scenarios outcome.

The methodology estimates public benefits in three areas—economic, environmental, and energy security. While these three types of benefits reflect DOE's strategic goals (DOE, 2005a), the committee recognizes that other kinds of benefits may be important in evaluating some projects. For example, markets demand that automobile manufacturers produce cars that not only meet the fuel economy objective that is important to DOE but also satisfy several other needs as well. An example of another need might be employment.

Recommendation 11: If benefits in areas other than economic, environmental, or energy security are found to occur, they should be noted in the text accompanying the results matrix. However, the matrix should stay focused on the three main types of benefits to facilitate comparisons across programs.

1

Introduction

BACKGROUND

Since its inception in 1977 from an amalgam of federal authorities, the U.S. Department of Energy (DOE) has administered numerous programs aimed at developing applied energy technologies. The better portion of the annual expenditures dedicated to such technologies is spent on research and development (R&D) and is administered by three DOE offices—the Office of Energy Efficiency and Renewable Energy (EERE), the Office of Fossil Energy (FE), and the Office of Nuclear Energy Science and Technology (NE).[1] These three offices received approximately $2.6 billion total for fiscal year (FY) 2006, approximately $1.5 billion of which is being devoted to R&D. The R&D programs are complemented by policy measures, such as tax incentives to encourage early adoption of advanced technologies by consumers, efficiency standards for household appliances, and production tax credits for a variety of energy production technologies, including certain renewable energy sources (EIA, 2006).

In recent years, federal oversight of public expenditures has emphasized the integration of performance and budgeting. Notably, the Government Performance and Results Act (GPRA) was passed in 1993 "in response to questions about the value and effectiveness of federal programs" (GAO, 1997, p. 11). GPRA and other mandates have led agencies to develop indicators of program performance and program outcomes. The development of indicators has been watched with keen interest by Congress, which has requested of the National Research Council (NRC) a series of reports using quantitative indicators to evaluate the effectiveness of applied energy R&D. The first such report[2] took a retrospective view of the first 23 years of DOE R&D programs on fossil energy and energy efficiency.[3] The report found that DOE-sponsored research had netted large commercial successes, such as advanced refrigerator compressors, electronic lighting ballasts, and emission control technology for flue gas desulfurization (NRC, 2001). However, some programs were judged to be costly failures in which large R&D expenditures did not result in a commercial energy technology (NRC, 2001). A follow-up NRC committee was assigned the task of adapting the methodology to the assessment of the future payoff of continuing programs (NRC, 2005a).

Evaluating the outcome of R&D expenditures requires an analysis of program costs and benefits. Doing so is not a trivial matter. First, the analysis of costs and benefits must reflect the full range of public benefits that are envisioned, accounting for environmental and energy security impacts as well as economic effects. Second, the analysis must consider how likely the research is to succeed and how valuable the research will be if successful. Finally, the analysis must consider what might happen if the government did not support the project: Would some non-DOE entity undertake it or an equivalent activity that would produce some or all of the benefits of government involvement?

The present report continues to investigate the development and use of R&D outcome indicators and applies the benefits evaluation methodology to six DOE R&D activities. It provides further definition for the development of indicators for environmental and security benefits and refines the evaluation process based on its experience with the six DOE R&D case studies.

---

[1] The Office of Electricity Delivery and Energy Reliability, created in 2002, could be considered a fourth applied energy program.

[2] Requested by Congress in the conference report of the Consolidated Appropriations Act for FY00 (House Report 106-479, p. 493. November 18, 1999).

[3] These programs include only those that were at the time under the jurisdiction of the U.S. House Appropriations Subcommittee on the Interior and Related Agencies.

APPLIED ENERGY RESEARCH AND DEVELOPMENT AT DOE

The DOE divides its mission according to four strategic goals—one each dealing with defense, science, environment, and energy[4] (DOE, 2005a)—and seven general goals. The energy strategic goal is supported by the general goal of energy security.[5] The general goal underscores the role of technology development:

> Improve energy security by developing technologies that foster a diverse supply of reliable, affordable and environmentally sound energy by providing for reliable delivery of energy, guarding against energy emergencies, exploring advanced technologies that make fundamental improvement in our mix of energy options, and improving energy efficiency. (DOE, 2005a, p. 20)

The three applied energy offices implement this goal using R&D programs, intergovernmental grants, reserves of fossil fuels, and appliance efficiency standards. Within EERE, R&D programs can be considered in two more or less distinct groups. The first consists of R&D directed at biomass and biorefinery systems R&D, geothermal technology, hydrogen technology, hydropower, solar energy, and wind energy technologies. Program activities range from basic research at universities and national laboratories to cost-shared applied research, development, and field validation in partnership with the private sector (OMB, 2005). The second group within EERE R&D consists of energy conservation R&D relating to the efficient end use of fuels and electricity in vehicles, in industrial processes and manufacturing, and in building envelopes. In general, this second group of energy conservation EERE R&D programs has the objective of achieving an output of a good or service with less energy input[6]—that is, achieving greater energy efficiency. The FY06 appropriation for EERE was $1,173 million,[7,8] the R&D component of which totaled $770 million, including $151 million for congressionally mandated R&D activities.[9]

Research in FE has traditionally been divided between the Office of Coal and Power Systems (CPS) and the Office of Oil and Natural Gas. CPS administers a suite of clean coal R&D, which aims at ensuring the generation of clean, reliable, affordable electricity from coal. One large planned demonstration project is FutureGen, which has as its aim the development of a zero-atmospheric-emissions power plant. Additional programs focus on all the key technologies needed for FutureGen—carbon sequestration, membrane technologies for oxygen and hydrogen separation, advanced turbines, fuel cells, technologies related to gasifiers for coal-to-hydrogen conversion, and others. The Clean Coal Power Initiative emphasizes the development of pollution controls and efficiency improvements for existing plants (DOE, 2006a). The Office of Oil and Natural Gas supports research and policy options to ensure clean, reliable, and affordable supplies of oil and natural gas for American consumers.[10] The FY06 appropriation for FE was $841 million, with $592 million of it allotted for fossil energy R&D.[11]

NE supports innovative applications of nuclear technology and administers four principal R&D programs. The Generation IV Nuclear Energy Systems Initiative and the Nuclear Hydrogen Initiative, which according to the information DOE provided on its budget for FY07 (DOE, 2006b) will develop a new generation of nuclear technologies to produce cost-effective electricity and commercial quantities of economic hydrogen to support the development of a transportation infrastructure. Nuclear Power 2010 is a demonstration program applicable to untested licensing and other regulatory processes. The Advanced Fuel Cycle Initiative aims to improve the efficiency and proliferation resistance of the nuclear fuel cycle.

The scope of programs subject to the NRC studies in this series has been limited to fossil energy R&D and the energy conservation portion of EERE's R&D. Prior to FY06, funding for this group of programs fell under the jurisdiction of the Interior and Related Agencies appropriation subcommittees owing to the programs' origins in the Department of Interior; the balance of EERE's R&D funding, on energy efficiency, was included on a separate appropriations bill. However, in FY06, the House Committee on Appropriations restructured its subcommittees' jurisdictions and reduced them in number from 13 to 10. Funding for all FE and EERE R&D programs was consolidated into one account subject to the jurisdiction of the House Appropriations Subcommittee on Energy and Water Development and Related Agencies

---

[4]The energy strategic goal is "[t]o protect our national security by promoting a diverse energy supply and delivery of reliable, affordable, and environmentally sound energy" (DOE, 2005, p. 7).

[5]The final level in the goal cascade is the GPRA unit defining a major activity that aligns resources with a unique program goal. Some examples of GPRA units are hydrogen technology, solar energy, and wind energy.

[6]Examples of this include, in the automotive industry, vehicles that obtain more miles per gallon (mpg); in the aluminum industry, production of more pounds of aluminum per Btu of energy; in lighting, more lumens per watt (NRC, 2001).

[7]*Budget of the United States Government, Fiscal Year 2007—Appendix*, p. 390.

[8]DOE. *FY2007 Control Table by Organization*, p. 5. Available at <http://www.cfo.doe.gov/budget/07budget/Content/orgcontrol.pdf>.

[9]DOE. *Department of Energy 2007 Budget Request: Fossil Energy and Other*. Available at <http://www.cfo.doe.gov/budget/07budget/Content/Volumes/Vol_7_FE.pdf>.

[10]Available at <http://www.fossil.energy.gov/programs/oilgas/index.html>.

[11]The balance of budget authority within the Office of Fossil Energy went for the Strategic Petroleum Reserve and Northeast Home Heating Oil Reserve ($164 million), a rescission of uncommitted balances for clean coal technology (–$20 million), and for other purposes.

TABLE 1-1 Summary of Congressional Appropriations Oversight of EERE Programs

Subaccount	FY05 and Earlier		FY06	
			EWD	
	Interior Bill: Energy Conservation Account	EWD: Energy Supply Account	Energy Supply and Conservation Account	Electricity Delivery and Energy Reliability Account
Vehicle technologies	X		X	
Fuel cell technologies	X		(1)	
Hydrogen technology		X	X	
Distributed energy	X			X
Building technologies	X		X	
Industrial technologies	X		X	
Biomass and biorefinery systems R&D	X		(2)	
Solar energy		X	X	
Wind energy		X	X	
Hydropower		X	X	
Geothermal technology		X	X	
Biomass and biorefinery systems R&D		X	X	

NOTES: (1) Merged with the hydrogen technology subaccount; (2) merged with the identically named biomass and biorefinery systems R&D subaccount on the energy and water development bill. FY, fiscal year; EWD, Energy and Water Development Bill.

(CRS, 2005).[12] Four of the program funding line items that made up the pre-FY06 accounts were merged into two. In another instance—that of distributed energy—a program was moved out of EERE and into the appropriation account corresponding to the DOE Office of Electric Transmission and Distribution. (See Table 1-1 for a listing of programs and legislated funding sources in FY05 and earlier years and in FY06 and beyond.)

EVALUATING THE FEDERAL INVESTMENT IN ENERGY RESEARCH AND DEVELOPMENT

Federal Oversight

GPRA requires agencies to annually submit or update a strategic plan, an annual performance plan, and an annual performance report (GAO, 2005). Program costs and benefits are analyzed and submitted, pursuant to GPRA, in an annual report to Congress with the President's budget. GPRA and other mandates have led agencies to develop indicators of program performance and program outcomes. Additional analyses are required of agencies by the Office of Management and Budget (OMB), the executive agency that formulates and administers the federal budget and evaluates program effectiveness. The President's Management Agenda (PMA) (OMB, 2001) sets forth nine agency-specific reforms and five government-wide goals (OMB, 2005); a set of R&D investment criteria was spelled out in 2003, implementing one of the agency-specific goals of the PMA (Marburger and Daniels, 2003). Also in 2003, OMB inaugurated the Program Assessment Rating Tool (PART) (OMB, 2003) to assess and improve program performance. The Government Accountability Office—a congressional agency—regularly audits, examines, and evaluates government programs (GAO, 2005). A few of the aforementioned assessments, and their relationship to one another, are discussed below.

Reporting Requirements and Management Systems

GPRA and PART require DOE to report on the performance of its R&D programs. The OMB guidance document for PART (OMB, 2005) states that "PART is a vehicle for achieving the goals of GPRA." To encourage compatibility between the two systems, the PART guidance states that

> The PART . . . strengthens and reinforces performance measurement under GPRA by encouraging careful development of performance measures according to the outcome-oriented standards of the law and by requiring that agency goals be appropriately ambitious. Therefore, performance measures included in GPRA plans and reports and those developed or revised through the PART process must be consistent.

> The PART should help develop and identify meaningful performance measures to support GPRA reporting. . . . *[T]he measures used for GPRA should be the same as those included in the PART. In all cases, performance measures included in GPRA plans and reports should meet the standards of the*

---

[12]The name of the combined account is Energy Supply and Conservation. Page 97 of House Report 109-275 explains it thus: "Energy Conservation programs previously funded by the Interior and Related Agencies Appropriations Act are now funded by the Energy Supply and Conservation appropriation, and are combined with energy efficiency activities in the Energy Efficiency and Renewable Energy account." See also OMB (2005, p. 391).

INTRODUCTION

TABLE 1-2 Stated Goals of GPRA and PART

Program	GPRA Objective	PART Objective
Sequestration/ integrated gasification combined cycle (IGCC)	Create public/private partnerships to develop technology to ensure continued electricity generation and hydrogen production from the extensive U.S. fossil fuel resource, including control technologies to permit reasonable cost compliance with emerging regulations and, ultimately, by 2015, near-zero atmospheric emission plants (including carbon) that are fuel-flexible and capable of multi-product output and energy efficiencies over 60 percent with coal.	Long-term goals are 50% efficient coal-based power generation (IGCC) in 2010; CO_2 capture at 10% increase in cost of electricity by 2012; 90% reduction in mercury emissions at less than 75% current cost by 2012; $1,000/kW capital cost for IGCC technology in 2010.
Natural gas	Provide technology and policy options capable of ensuring abundant, reliable, and environmentally sound gas supplies.	Long-term goals are (1) by 2015, develop technologies to expand the 2002 domestic gas economically recoverable resource base by 100 Tcf; (2) develop technologies that will by 2025 add 20 Tcf of technically recoverable resources of natural gas from methane hydrates; (3) by 2013, reduce the cost of hydrogen production from natural gas by 25% from current baseline of $5.54/MMBtu (steam reforming of methane at natural gas price of $3.15/MMBtu).
Distributed energy	Develop and facilitate market adoption of a diverse array of cost-competitive, integrated distributed generation and thermal energy technologies in homes, businesses, industry, communities, and electricity companies, increasing the efficiency of electricity generation, delivery, and use, improving electricity reliability, and reducing environmental impacts.	The program has two long-term goals that capture most of the activities supported in each of the two subprograms. One subprogram focuses on the development of next-generation distributed energy technologies (e.g., microturbines, reciprocating engines, industrial gas turbines, thermally activated cooling and humidity control devices, combined heat and power systems) that are cleaner and more reliable, fuel efficient, fuel flexible, and affordable than existing equipment. The second subprogram concentrates on the development of technologies, tools, and techniques to enable prospective users of distributed energy systems—regardless of the type of technology—to evaluate benefits, install, operate, control, and maintain those systems in an optimized manner to meet the needs of their facilities and business operations and those of the electric power and natural gas utilities to which the systems are interconnected.
Vehicle technologies	Develop technologies that enable cars and trucks to become highly efficient, through improved power technologies and cleaner domestic fuels, and to be cost- and performance-competitive.	The program has established three long-term performance measures. These measures are focused on program outputs (not process-oriented) that could enable significant oil savings. However, success is dependant on the rate and level of market penetration, which are strongly affected by other policy instruments (such as fuel economy standards) and market conditions (such as fuel prices). The program also tracks progress against additional goals with its industry partners from the FreedomCAR and 21st Century Truck partnerships.
Industrial technologies	Partner with our most energy-intensive industries in strategic planning and energy-specific research, development, and demonstration to develop the technologies needed to use energy efficiently in their industrial processes and cost-effectively generate much of the energy they consume. The result of these activities will save feedstock and process energy, create domestic supply, improve the environmental performance of industry, and help America's economic competitiveness.	Not available.

PART – they must be outcome oriented and have ambitious targets. (OMB, 2005, p. 3, italics added)

In principle, therefore, PART and GPRA should use the same benefits estimates. However, it is difficult to tell whether this is being done because the DOE Performance and Accountability Report (DOE, 2005a) only presents a general statement of objectives and an assessment of whether a series of annual program targets was reached. Moreover, the PART Assessment Report[13] does not consistently state specific long-term goals for each program. Table 1-2 contains the goal statements in the 2005 GPRA and PART reports in

---

[13] Available at <http://www.whitehouse.gov/omb/budget/fy2006/pma/energy.pdf>.

the five technologies that the NRC evaluated in the current study.[14]

GPRA requires a statement of expected program results. EERE develops this statement through its benefits estimation process. That statement describes the relation between GPRA and benefits assessment as follows:

> EERE develops these benefits projections annually to help meet the requirements of the Government Performance and Results Act (GPRA) of 1993 and the President's Management Agenda (PMA). GPRA requires Federal Government agencies to develop and report on output and outcome measures for each program. This analysis helps meet GPRA requirements by identifying the potential outcomes and benefits of realizing EERE program goals (outputs). *The benefits estimates do not reflect the risk of realizing these goals, which is being addressed separately.* (DOE, 2005b, italics added)

This benefits assessment process is the same one that EERE discussed with the NRC Phase Two committee. Note that the GPRA benefits are not discounted for risk.

PART does not report benefits estimates per se but rather asks a series of questions about the program as a basis for assessing its performance and management. The specific questions are listed in Appendix A. The relationship between PART and the present study depends on whether the NRC committee's methodology helps to provide answers for the questions in PART. In particular, a credible benefits estimate is persuasive in answering PART questions on program purpose and design.

Analysis Within the Department of Energy

DOE has made important strides in developing tools for assessing the likely benefits of its R&D programs. Of special note is that elements of DOE—particularly the fossil energy and energy efficiency programs—have worked together toward common methodologies and approaches. In this regard, the efforts of DOE to improve and standardize its own estimates of benefits contributed to the committee's ability to fully understand the programs and their anticipated impacts. At the undersecretary level, DOE maintains the R&D Council to handle the PMA, including R&D investment criteria, portfolio analysis, and budgeting. Reporting to the R&D Council are the Science and Technology (S&T) Integration Working Group, which facilitates compliance with administration and DOE requirements through the use of common methods and procedures and provides portfolio level decision support briefings, and the S&T Laboratory Working Group, which enhances the alignment of the R&D agenda with national energy goals.[15]

National Research Council Studies

The conference report of the Consolidated Appropriations Act[16] for FY00 requested the NRC to assess the benefits and costs of DOE's R&D programs in fossil energy and energy efficiency, a task agreed on by the House and Senate subcommittees then having jurisdiction over funding of these programs. The retrospective study—*Energy Research at DOE: Was It Worth It?* (NRC, 2001)—reported that, in the aggregate, the benefits of federal applied energy R&D exceeded the costs but observed that the DOE portfolio included both striking successes and expensive failures. As important, the study noted that the methodologies by which DOE had calculated the benefits of its programs varied considerably, making comparisons of program benefits difficult.

Congress subsequently provided funds for "a continuing annual review by the [National] Academy [of Sciences] of programs . . . to measure the relative benefits expected to be achieved and to inform decision making on what programs should be continued, expanded, scaled-back, or eliminated."[17] The methodology suggested by the Phase One committee in its report, *Prospective Evaluation of Applied Energy Research and Development at DOE (Phase One): A First Look Forward* (NRC, 2005a), used expert panels to review the DOE R&D program. The panels estimated the expected benefits[18] in three categories—economic, environmental, and energy security—and in three global economic scenarios and summarized the results in a matrix. The panels constructed simple decision trees to describe the key technical and market uncertainties associated with the program and to evaluate the impact DOE support would have on the probability of various technical and market outcomes. To estimate benefits, simple spreadsheet models were deployed in conjunction with more sophisticated models (such as the Energy Information Administration's (EIA's) National Energy Modeling System (NEMS).[19] The overall benefit of the DOE R&D program was given as the difference between the expected benefits with DOE support and the expected ben-

---

[14]Table 1-2 also shows that the programs reported in PART and GPRA are generally larger than the specific technologies that the Phase Two study is evaluating. However, the programs in PART and GPRA appear to be the same.

[15]John Sullivan, Deputy Under Secretary for Energy, Science, and the Environment, "Energy, Science, and Environment R&D Council," Presentation to the committee on October 26, 2005.

[16]House Report 106-479, November 18, 1999.

[17]House Report 107-564, July 11, 2002.

[18]The expected benefit is a single quantity that incorporates information about the various possible outcomes and their respective probabilities and levels of benefits (NRC, 2005a).

[19]NEMS is a computer-based, energy-economy system for modeling U.S. energy markets that projects the production, imports, conversion, consumption, and prices of energy, subject to assumptions about macroeconomic and financial factors, world energy markets, resource availability and costs, behavioral and technological choice criteria, cost and performance characteristics of energy technologies, and demographics.

efits without DOE support. The process ensured consistency through the use of common scenarios and assumptions by each panel and the guidance on process and methodology provided by the parent committee, which also provided oversight.

THE CURRENT STUDY

The current NRC study continues the work of the first prospective benefits study, which evaluated the stream of benefits resulting from applied energy programs. It has the objective of testing, refining, and extending the proposed methodology. To conduct this study, the NRC formed the Committee on Prospective Benefits of DOE's Energy Efficiency and Fossil Energy Programs (Phase Two) (see biographies of committee members in Appendix B). The task assigned to the committee for Phase Two was as follows:

> The work of the Phase 2 committee will be supported by several panels that will be separately appointed by the NRC to apply the methodology developed in Phase 1 and evaluate the prospective benefits of individual EE and FE programs/projects. Since a methodology will have been developed in Phase 1, it is expected that a greater number of panels can be formed in Phase 2 and more time and resources can be devoted to evaluating prospective benefits. It is proposed that approximately 6 panels will be appointed by the NRC.

(The complete statement of task for Phase Two can be found in Appendix C.) Funds were appropriated in FY04 for Phase Two[20] and FY05 for Phase Three.[21]

To obtain feedback from stakeholders on its proposed methodology and its then-pending selection of DOE programs for further case study, the committee held a workshop on July 14, 2005, in Washington, D.C. Topics of discussion included the methodology developed in Phase One, the suggested revisions thereto, and the choice of case studies. Soon after the workshop, the committee selected six programs for case study and informed DOE of its choices. The programs selected comprised three in EE, including the Industrial Technologies Program (ITP)–Chemicals, the Distributed Energy Resources (DER) program, and the Light-Duty Vehicles Hybrid Technologies program, and three in FE, including the IGCC program, the Carbon Sequestration program, and the Natural Gas Exploration and Production program.

The NRC formed six panels of experts to conduct the case studies. Each panel held two meetings between September and December 2005. The panels carried out their evaluations following the methodology of Phase One with some modifications suggested by the Phase Two committee. The panels wrote short reports summarizing their findings and calculations and submitted them to the committee.

In a letter report issued December 14, 2005 (see Appendix D), the committee discussed the principal comments made during the workshop, the case studies it intended to perform in Phase Two, and the changes to the process and methodology that had occurred since Phase One. The committee maintained its dialogue with stakeholders throughout the study, including interfacing with the DOE R&D Council structure (see above) and ensuring frequent briefings by DOE to the committee and panels.

In addition, the committee refined its calculation of environmental and security benefits and added enhancements to the methodology based on its experience with the six panels. Activities undertaken by the committee during Phase Two are listed in Appendix E.

The balance of the current report is organized as follows: Chapter 2 presents the results of the six case studies. Chapter 3 discusses the methodology—the analytical framework for calculating quantitative program outcome indicators—in particular the way the framework has changed since the completion of the Phase One report (NRC, 2005a). Chapter 4 discusses the expert panel process—the sequence of tasks and events needed to apply the methodology to a specific program—and the changes to the process that have occurred since the Phase One study. Chapter 5 presents the conclusions and recommendations of the study. Appendix F contains material from Chapter 3 of the committee's Phase One report to provide guidance to the panels on the methodology. Appendix G describes the kinds of information that DOE should provide and suggests forms it can fill out to submit that information. Appendixes H to M are the reports prepared by the six panels.

---

[20]House Report 108-330, p. 83 and p. 163.
[21]House Report 108-792, p. 1032, and House Report 108-542, p. 125.

2

Results of Applying the Methodology

CASE STUDY SELECTION AND EXECUTION

The statement of task for the study called for case studies of the prospective benefits of approximately six DOE programs on energy efficiency and fossil energy R&D. Early in the study, during the open session of its July 14, 2005, meeting, the committee on prospective benefits held discussions with federal stakeholders on its pending selection of case studies. Based on that input, the committee finalized its choices and on July 22, 2005, sent letters to DOE—one each to the Office of Fossil Energy (FE) and the Office of Energy Efficiency and Renewable Energy (EERE)—informing them of the six selected programs (see Appendix D), which were as follows:

- FE programs and activities:
  —Integrated Gasification Combined Cycle (IGCC);
  —Carbon Sequestration; and
  —Natural gas technologies (exploration and production).
- EERE programs and activities:
  —Distributed Energy Resources (DER) program (end-use system integration and interface);
  —Vehicle technologies program (hybrid and electric propulsion, advanced combustion R&D and materials technology—excluding projects related to heavy-duty vehicles); and
  —Industrial Technologies Program–Chemicals.

In a few cases, the scope of the evaluation was limited to a subset of activities within a program. This is indicated in parentheses, where applicable.

The NRC formed six panels to conduct the case studies. The chairs of five of the six panels were members of the committee on prospective benefits. (A list of panel activities is included as Appendix E.) Each panel held two meetings between September and December 2005. A description of the evaluation process and methodology (see Appendix F) was distributed to panelists before the first panel meetings. After discussing this methodology early in the first meeting, the panel heard presentations delivered by DOE staff on their programs' FY05 budgets, objectives, R&D portfolios, and estimated future benefits; the latter had been calculated by DOE assuming program technical goals would be met. The panels used this and other information available to them, along with their own knowledge of the subjects, to construct a decision tree showing the timing and likelihood of the technical success and market acceptance of the technology or technologies supported by the program. The panels next calculated the benefits—economic, environmental, and security—associated with each level of technical and market success in three global economic scenarios. The overall benefit of the DOE R&D program is given as the difference between the expected benefits with DOE support and the expected benefits without DOE support. To ensure consistency, the panels employed the process recommending common scenarios and assumptions across evaluations, and they were aided by the committee, which served in an oversight capacity. A decision analysis consultant assisted with the preparation of decision trees and benefits calculations. The panels wrote short reports summarizing their findings and calculations and submitted them to the committee.

STATEMENT OF TASK

The statements of task for the six panels were nearly identical. Here is the statement of task for the Panel on Light-Duty Vehicle Hybrid Technology:

> The Panel on Light-Duty Vehicle Hybrid Technology will apply the methodology developed by the Committee on Prospective Benefits of Energy Efficiency and Fossil Energy R&D Programs to assess the potential benefits of DOE's R&D activities that are focused on hybrid electric vehicle technologies using more efficient internal combustion engine

power trains for light-duty vehicles. The program has R&D activities in combustion, lightweight materials for vehicle components and structures, electrical and power electronic systems, emission control technologies, and fuels. The panel will be composed of about 8 members who will have experience with hybrid vehicle technologies, combustion engine technology, emission controls, the automotive market and sector, economics, and vehicle engineering and marketing. Panel members may have expertise in more than one area.

The panel will apply the committee's methodology by first assessing the probability of success of meeting the DOE's time frame and targets for technology development, consider alternative paths of development with and without DOE funding, review DOE estimates of the economic, environmental, and national security benefits of its program on light-duty vehicle technology, and estimate the benefits under the alternative future states of the world (scenarios) that have been specified by the committee, namely, a base case EIA scenario, a high oil/gas price scenario, and a carbon-constrained scenario. Consideration of technical and market risk will be a key element of this assessment.

It is expected that the panel will hold two 2-day meetings and will write a short 10-page report on its assessment of the benefits of this program and fill in the benefits matrix and explanation of it as outlined in the committee's report, *Prospective Evaluation of Applied Energy Research and Development at DOE (Phase One): A First Look Forward*.

ADVICE TO USERS

The benefits methodology presented in this report is intended to provide consistent information that is useful to decision makers who are considering what programs should be continued, expanded, scaled back, or eliminated. The committee's charge did not extend to reaching conclusions or making recommendations about such decisions for the six programs that it reviewed for this Phase Two project. However, the committee has developed some insights that may assist decision makers with interpreting and applying the results of the analysis. These insights are recorded in the next section, "Commonly Encountered Issues."

Although the committee has developed a process whereby all the panels evaluating DOE programs will use a common methodology and has suggested a procedure whereby all evaluations will be performed on a consistent basis, it is important that the decision makers understand the assumptions that the panelists had to use in assessing the programs. This understanding is critical to giving them the confidence to use the results as they make decisions about DOE programs and as they recommend funding.

Commonly Encountered Issues

The committee's review of the six panel reports of Phase Two, along with the experience from the two earlier benefits evaluations, noted that several issues of interpretation came up frequently. Decision makers may thus encounter these same issues as they apply the results of the methodology presented here.

1. A clear understanding of the reason for government-funded research is fundamental to the entire evaluation. For this reason, the program summaries produced in Phase Two require an explicit statement of the reason for government action. The government must demonstrate on the one hand the importance of a technology and the compelling need for it, and, on the other hand, that it will not be achieved in a timely fashion without the government's participation. Readers are encouraged to consider this statement of justification carefully to verify that it accurately expresses the basic reason for the program.

2. Several of the expert panels concluded that the probability of meeting the technical goals of a DOE program on the stated timetable was very small. While it is important for a decision maker to know this, it is also frequently true that a program can produce significant benefits at lower levels of technical achievement or over longer times. The committee recommends that users investigate these trade-offs carefully.

3. The budgets of some programs may be inconsistent with the program goals. In some cases, this inconsistency is the reason that the program's stated goals are not likely to be met, as noted above. But in other cases the issue is that budgets have been declining for several years; the combined heat and power (CHP) program and the ITP–Chemicals subprogram in Phase Two study exhibited this characteristic. In the latter case, the benefits of the subprogram may be overambitious because the analysis assumes that the goals will be accomplished when the reality is that declining budgets, relative to those planned, make that unlikely. In these cases, decision makers should recognize that realizing the estimated benefit may require increasing (or at least stabilizing) the budget. At some point, declining budgets probably reach a threshold below which the program is not viable.[1]

4. The benefits of some programs are strongly influenced by nonresearch policy decisions. For example, the benefits of carbon capture and sequestration depend on the size and timing of the carbon tax (or equivalent policy intervention in the market). In this case, a delay in the imposition of the tax, by delaying an incentive to the private sector, would accelerate technology development attributable to government funding, although not on a one-for-one basis. Imposition of appliance efficiency standards might have a similar effect.

---

[1] At some point, however, a small program to maintain a federal presence might be justifiable. For example, the Panel on DOE's Distributed Energy Resources Program believes that the federal program is helpful in encouraging states to fund similar programs.

While not directly evaluated in this study, the social welfare impact of such policies could be considerable.

5. Programs having very large benefits might alter the supply and demand characteristics of the entire energy system. For example, a successful deployment of IGCC technology appears to be accompanied by a large role for DER in order to meet peak electricity demand. This, of course, would affect the benefits calculation for DER. NEMS is particularly useful in understanding these changes in the overall energy system.[2]

Using the Benefits Analysis for Decision Making

The expected net benefits calculated according to the methodology provide a useful basis for comparing programs. However, other elements of the analysis will contribute to the decision-making process. During its discussion of the results of the Phase Two work, the committee identified three such considerations that it believes deserve comment:

- The decision maker will often want to know how best to spend incremental resources (or how to minimize the impact on benefits of a reduction in resources). It is not the case, however, that incremental resources should be spent on programs with the highest benefit-cost ratio. What counts most in making this decision is finding the greatest incremental benefit produced by the additional resources. The methodology does not calculate marginal benefits and so is not directly useful in making these incremental decisions. However, the committee suggests that the decision tree analysis would be of considerable use in identifying actions most likely to increase the expected benefit. A recommended method for calculating marginal benefits appears in Chapter 5.
- The methodology presents benefits for each of three scenarios that describe future states of the world, but it does not attempt to combine the three sets of benefits into a single set. As a result, the decision maker must weigh the alternative scenarios in arriving at judgments about the benefits of the overall research portfolio. Ideally, a well-designed portfolio would contain a balance of projects that would produce acceptable results across the range of outcomes for different scenarios.
- The methodology estimates public benefits in three areas—economic, environmental, and energy security. While these three types of benefits reflect DOE's strategic goals (DOE, 2005a), the committee recognizes that other kinds of benefits may be important in evaluating some projects. For example, markets will demand that automobile manufacturers produce cars that not only meet the fuel economy of importance to DOE but have several additional attributes as well. Another example might be employment impacts. Because including such benefits could affect the comparability among programs, the committee recommends that they be considered apart from the results matrix developed by the methodology or that, if included, they should be part of the text of a panel report. The panel should also discuss noteworthy assumptions it has made in calculating benefits, particularly if the results are sensitive to that assumption or the assumptions itself is novel or interesting.

RESULTS OF CASE STUDIES

The six expert panels that evaluated DOE applied energy programs estimated prospective benefits—economic, environmental, and security—using quantitative outcome indicators. In some cases, security and environmental benefits had to be characterized qualitatively. These estimates were facilitated by an assessment of the key technical and market risk faced by the technology being developed. Each set of results is reported in a two-page format accompanied by a decision tree. The latter illustrates how the expert panel assessed the possible future outcomes of the R&D program in terms of the timing, probability, and levels of technical performance and market acceptance of the technology. The goals and history of the program are summarized, and the expert panels have offered further observations about the program. The expert panels have contemplated outcomes that might occur with and without federal funding and taken the difference between the outcomes as the net expected value of the program.

The findings of the six panels are presented in summary fashion in Figures 2-1 through 2-6:

- Figure 2-1, integrated gasification combined cycle technology, pages 17-19,
- Figure 2-2, carbon sequestration, pages 20-22,
- Figure 2-3, natural gas exploration and production, pages 23-28,
- Figure 2-4, distributed energy resources, pages 29-31,
- Figure 2-5, light-duty hybrid vehicles, pages 32-36, and
- Figure 2-6, Industrial Technologies Program—Chemicals, pages 37-39.

The complete panel reports from which this information was derived are found in Appendixes H through M.

---

[2]NEMS determines capacity additions in its electricity market module as follows: "Capacity expansion is determined by the least cost mix of all costs, including capital, O&M, and fuel." And further: "Fossil-fired steam and nuclear plant retirements are calculated endogenously within the model. Plants are retired if the market price of electricity is not sufficient to support continued operation." (EIA, 2003, pp.43-46)

Program Name: Gasification technologies program[a]

Program Goals and Timing: By 2010, complete R&D for advanced gasification combined-cycle technologies that can produce electricity from coal at 45%-50% efficiency at a capital cost of $1,000/kw or less. By 2020 develop zero-emissions plants (including carbon) that are fuel-flexible and capable of multiproduct output and efficiencies of >60% with coal.

Program Costs: $750 million through 2020

Industry and Foreign Government Funding: No information is available. There is no doubt that significant funding is being provided by several governments (e.g., Japan, European Union) and industry, especially IGCC equipment suppliers.

Key Complementary/Interdependent DOE Programs: Turbine programs, carbon sequestration

		Global Scenario[b]	
		AEO Reference Case	High Oil and Gas Prices
Program Risks	Technical Risks	Technical success was defined by the cost and efficiency of future IGCC plants. The panel identified four levels of technical success: $1,400/kW and 39% efficiency; $1,265/kW and 42% efficiency; $1,135/kW and 45% efficiency; $1,040/kW and 48 percent efficiency. Estimated probability of achieving the highest level of technical success was 10% with the DOE program, 0% without the DOE program.	
	Market Risks	Technologies that will compete with IGCC are natural gas generation and advanced PC generation, including USC-PC. Primary market competition depends on the price of natural gas, on the progress of research in other coal-fired generation technologies, and on the results of other research programs such as fuel cells and distributed generation.	
Expected Program Benefits	Economic Benefits	Economic benefits depend on the next-best alternative technology, which depends in part on the results of other DOE R&D programs.[c]	
		If IGCC displaces PC, $6.4 billion at 3% $2.4 billion at 7% If IGCC displaces NGCC, $7.8 billion at 3% $3 billion at 7%	If IGCC displaces PC, $7 billion at 3% $2.7 billion at 7% If IGCC displaces NGCC, $47 billion at 3% $18 billion at 7%
	Environmental Benefits	Environmental impacts depend on the next-best alternative technology, which depends on the results of other DOE R&D programs.	
		If IGCC displaces PC, carbon emissions decrease by 30 million t If IGCC displaces NGCC, carbon emissions increase by 90 million t	If IGCC displaces PC, carbon emissions decrease by 34 million t If IGCC displaces NGCC, carbon emissions decrease by 36 million t[d]
	Security Benefits	If IGCC plants are built instead of NGCC plants, there are security benefits related to the decrease in consumption of natural gas. The total amount of natural gas displaced by IGCC, assuming a 20-year plant life, is as follows:	
		If IGCC displaces NGCC, natural gas consumption is reduced by 5 quad	If IGCC displaces NGCC, natural gas consumption is reduced by 4 quad

NOTE: /kW, per kilowatt; IGCC, integrated gasification combined cycle; NGCC, natural gas combined cycle; PC, pulverized coal; quad, quadrillion British thermal units; and USC-PC, ultrasupercritical pulverized coal.

[a]For the purposes of this assessment, the panel defined the ICGCC research program as coal to electric power systems. It did not include IGCC/fuel cell configurations or carbon capture and sequestration technology options since these were being analyzed by other panels.

[b]The panel determined that the benefits of the IGCC research program in the Carbon Constrained scenario depend critically on the success of carbon sequestration research. Specifically, in the Carbon Constrained scenario there would be no market for IGCC plants unless the carbon emissions can be captured and sequestered. A separate panel evaluated the carbon sequestration program.

[c]Economic benefit in this table is calculated as the difference in the discounted expected value with and without the DOE program of all IGCC plants built between 2006 and 2025. Economic benefits are discounted at both 3% and 7% real, while environmental and security benefits are discounted at 3%.

[d]In the High Oil and Gas Prices scenario, almost the same amount of IGCC is expected to be built with or without the program. The reduction in carbon emissions results from the increased efficiency of IGCC.

FIGURE 2-1 Findings for DOE's gasification technologies R&D.

Comments and observations. The panel made three primary observations about the program. First, the panel believes DOE's estimates of the impact and benefits of its program are overly optimistic in two ways: in terms of the technical achievement of the program, and in terms of the time it will take from completion of the R&D to realize the technical and projected financial improvements in the market. Second, the benefits that are ultimately realized from DOE's gasification R&D depend on many factors outside the program's control, including, potentially, the success of fuel cell and distributed generation research. And, finally, the program is nevertheless expected to result in very significant benefits over the long run—in the panel's analysis the financial and technical improvements in IGCC technologies are not fully realized until post-2020, but the expected economic benefits of the program are several billion dollars under every different market condition considered.

Technical risks. Technical risks were identified for eight specific components of the program that were evaluated and are described in more detail in the main text of the report. These risks generally relate to the possibility the research may not reach its goals, that the results may not scale up to commercial application, and that the capital costs and/or parasitic energy requirement will be excessive. To achieve DOE's target levels of performance for cost and efficiency improvements, or nearly those levels, almost all components of the R&D program must succeed. However, it is not necessary for all components to succeed for the program to yield technical and cost improvements—significant improvements in IGCC systems may result from the program if a subset of technical activities is successful.

Market risks. Market factors that will affect the competitiveness of IGCC in the future include competing fuels (natural gas), the regulatory environment (particularly with regard to carbon emissions), and competing technologies (specifically, other advanced coal systems). The impact of natural gas prices and carbon regulations is addressed by the three global scenarios. In examining DOE's FY04 benefits analysis report and the results of its NEMS analyses, it appears that the success of R&D in other areas (fuels cells and distributed generation) affects what the next-best alternative to IGCC is in the future. Accordingly, the panel developed two estimates of benefits, depending on what the next-best technology turns out to be. It notes that the benefits of the program are significant regardless of the next-best technology.

Benefits. The economic, environmental, and security benefits are shown in the summary matrix.

FIGURE 2-1 Continued

RESULTS OF APPLYING THE METHODOLOGY

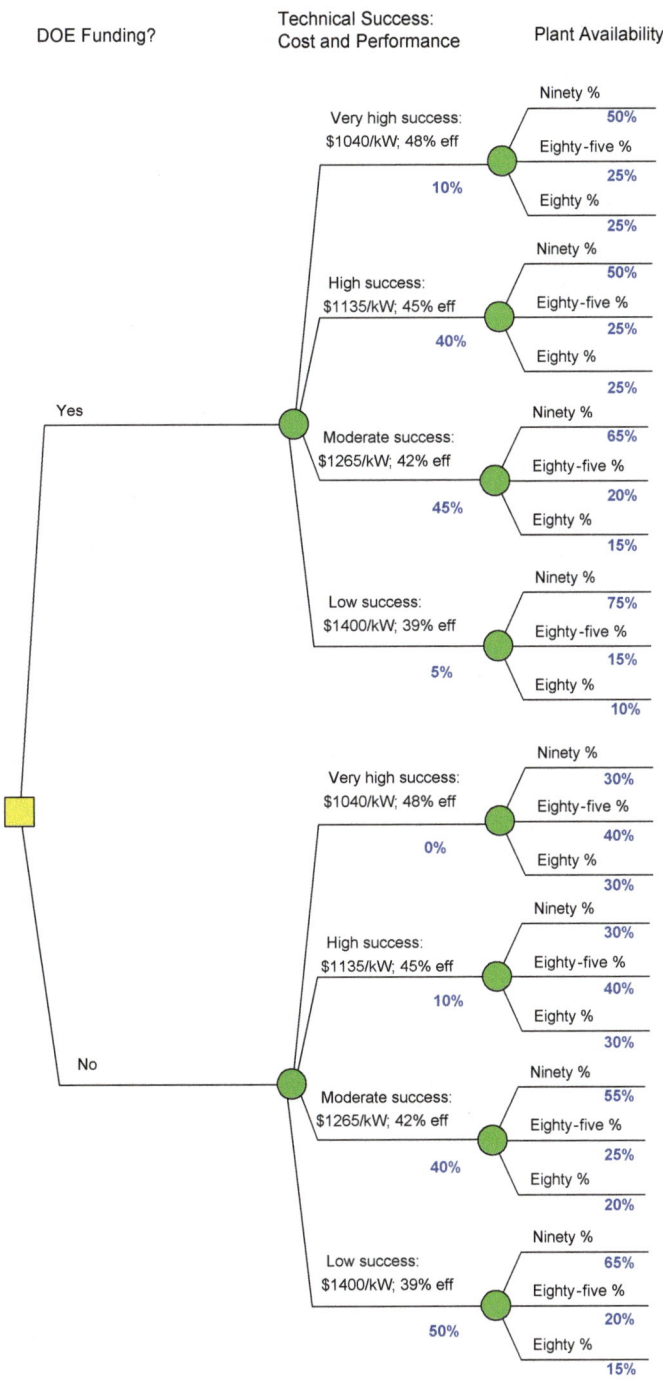

FIGURE 2-1 Continued

Program Name: Carbon sequestration
Program Goals: Achieve 90% CO_2 capture with 99% storage permanence at no more than 10% increase in the cost of energy services by 2012
Year Goals Expected to Be Achieved: 2012
Program Costs:
> Current (FY05) Funding: $44 million. Additional funding in the amount of $100 million over the next 4 years will go to the regional partnerships
> Proposed Year (FY06) Funding: $66 million
> Expected cost to completion: program costs through 2020 are expected to be $875 million

Key Complementary/Interdependent DOE Programs: IGCC

		Global Scenario[a]	
		Carbon Constrained	
		$100/ton Carbon Tax	$300/ton Carbon Tax
Program Risks	Technical Risks	Estimated as the probability of achieving specified impacts on the cost of electricity for IGCC plants with sequestration compared with those same plants without sequestration. Average of the panel assessments for the increase in cost of electricity (COE) for sequestration at three different times were as follows: 　　　　　2012　2017　2022 w/DOE program　33%　24%　18% w/o program　35%　28%　20%	2012　2017　2022 w/DOE program　30%　20%　15% w/o program　32%　24%　16%
	Market Risks	Estimated as the probability large-scale carbon sequestration would be allowed by both the regulators and the public. Probabilities were assigned for the "with DOE program" case and the "without program" case and were assumed to be the same in the two global scenarios considered: with DOE program, 77%; without program, 66%.	
Program Benefits	Expected Economic Benefits[b]	$3.5 billion at 3% Range: $0-$36 billion $1.3 billion at 7% Range: $0-$13 billion	$3.9 billion at 3% Range: $0-$36 billion $1.5 billion at 7% Range: $0-$13 billion
	Environmental Benefits	The environmental benefit of carbon sequestration is reduced greenhouse gas emissions. The environmental benefit of DOE's carbon sequestration program depends on what technologies would be implemented if IGCC with sequestration does not become cost-competitive. Given the level of carbon tax in the scenarios evaluated, the least-cost alternatives to carbon sequestration are other zero-emissions technologies. Thus there is no quantifiable environmental benefit of the research program that is separate from the economic benefit of the reduced costs for very-low emissions generation. Under other assumptions, emissions would be reduced.	
	Security Benefits	The security benefit of carbon sequestration is the ability to continue to build electric generation plants that use coal, a domestic resource, minimizing our dependence on imported fuel resources. Given the level of carbon tax in the scenarios evaluated, however, the least-cost alternatives to carbon sequestration are a combination of nuclear generation and renewables; thus there are no quantifiable security benefits associated with the research program separate from the economic benefits. It is possible, however, that absent carbon sequestration, natural gas will be used instead. In such a case U.S. energy security would be decreased.	

[a]The panel judged that carbon sequestration technologies would not be implemented in global scenarios without a carbon constraint. They did not evaluate the program under the Reference Case or the High Oil and Gas Prices scenarios but evaluated it instead under the Carbon Constrained scenario and a fourth scenario defined by the panel to have a higher carbon tax than the Carbon Constrained scenario.

[b]Net economic benefits are calculated as the expected net present value (at 3% and 7% annual discount rates) of the reduction in the cost of electricity for zero- or very-low emissions generation over a 20-year plant life for all IGCC plants with carbon sequestration built between 2006 and 2025.

FIGURE 2-2 Findings for DOE's carbon sequestration program.

Comments and observations. The DOE program has the goal of developing, by 2012, carbon capture and storage technologies that will increase the cost of new energy services by no more than 10%.

Technical risks. The main technological uncertainty considered was the increase in the cost of electricity (COE) associated with the capture and storage of carbon emissions from coal-fired power plants, specifically from advanced IGCC plants.

Market risks. Consideration of uncertainty focused on whether the public (and regulators) would allow large-scale underground storage of carbon. Without such acceptance, carbon capture and storage technologies would not be deployed. The panel's assessments of this uncertainty are summarized in Figure G-4. The average of the probabilities estimated by the panel that the large-scale sequestration would be allowed is .66 without DOE's research support and .77 with DOE's support.

Benefits. The economic benefit of carbon sequestration improvements is the product of the amount of electricity produced by IGCC with sequestration multiplied by the difference between the costs of the IGCC with sequestration and the costs of the cheapest alternative technology. To estimate the quantity of IGCC with carbon sequestration that would be built in each year, a simple cost comparison was made to determine what technology would be least costly for a utility making a decision about what to build. Whichever technology was least expensive was assumed to capture all of the possible low-emissions capacity added in that year. Costs are the net present value of the expected total costs over a 20-year life, discounted at 14%.[a] The panel considered COE in three time periods (2012, 2017, 2022) and for four different levels of cost increases in each period. On a net basis, there is no increase in environmental benefits (reduction in CO_2 emitted) since the alternative technologies that would supply the generating capacity also capture and store carbon. As for security, the least-cost alternatives to the technology under development in the DOE program would be a combination of nuclear generation and renewable electricity that also would reduce natural gas consumption. Compared with these alternatives, the DOE-sponsored technologies would offer no net reduction in consumption of that resource.

Program observations. While individual members had different judgments about the likelihood of achieving the R&D goals and the extent of market penetration of the resulting technology, there was general agreement on these conclusions:

1. Carbon sequestration technology will not be implemented commercially without carbon emissions constraints.
2. A carbon tax of $100 per ton is sufficient to make carbon sequestration competitive with IGCC plants that vent their carbon.
3. The panel judges that DOE's R&D program will speed the attainment of the carbon sequestration R&D goals by about 3 years because there is so much private sector interest and R&D in these technologies.
4. DOE should encourage private sector R&D in conducting its program. The DOE R&D results should be made available quickly to the private sector.
5. If the technology is demonstrated to be reliable and cost-effective, IGCC with carbon sequestration could be widely deployed following the implementation of carbon emissions constraints.
6. The expected benefit of the DOE program is large, roughly four times the R&D costs incurred by the federal government.
7. DOE's carbon sequestration R&D can make a great contribution to society even if it accelerates attainment of the national goals by only a few years.

---

[a]This value was selected to represent what might be used by a utility or merchant generator.

FIGURE 2-2 Continued

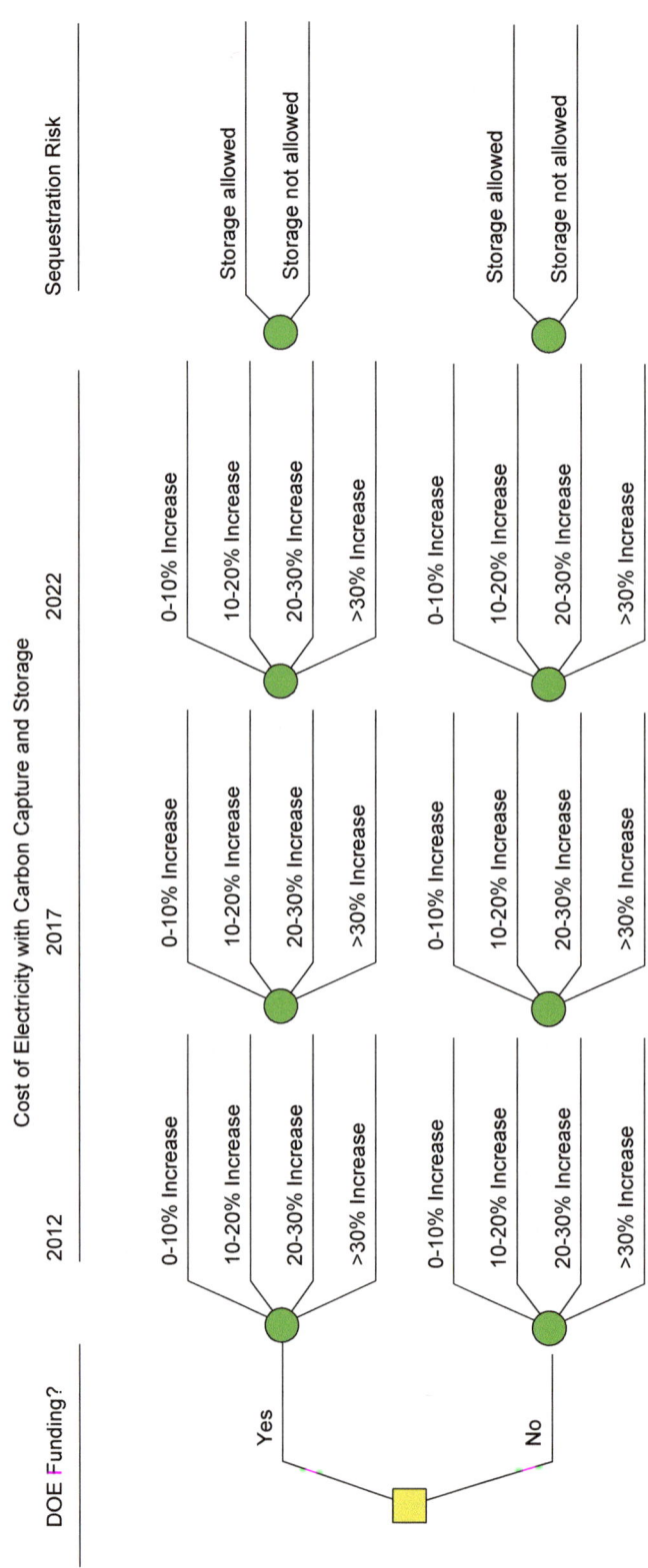

FIGURE 2-2 Continued

Program Name: Natural gas technologies program, exploration and production
Program Goals: Reduce the cost of drilling for unconventional and other gas by 5%, resulting in a 13 Tcf increase in economically recoverable resources by 2015
Program Costs: Approximately $140 million through 2015
Current (FY05) Funding: existing fields program, $1.6 million; drilling, completion, and stimulation, $7.3 million; advanced diagnostics and imaging, $3.8 million; Deep Trek, $1.5 million
Proposed Year (FY06) Funding: Program closeout

		Global Scenario		
		AEO Reference Case	High Oil and Gas Prices	Carbon Constrained
Program Risks	Technical Risks	Technical risks were evaluated at the subprogram level and were defined as the likelihood of achieving specific increases in economically recoverable resources (ERR) by 2015 as a result of DOE's R&D program. Aggregated at the program level, the expected increase in ERR made technically possible by the DOE program in each scenario is shown below. Numbers in parentheses represent the 10th and the 90th percentiles of the uncertainty and assume no market risks.		
		17 (9, 25) Tcf	20 (11, 28) Tcf	20 (11, 29) Tcf
	Market Risks	Market risks were identified for two of the four subprograms evaluated and were defined as the fraction of the total potential addition to ERR that would be realized. The effect of including market risks is to decrease the estimated net increase in ERR in 2015. The expected value and the 10th and 90th percentiles of ERR increase given both technical and market risks are shown.		
		13 (6.8, 19) Tcf	17 (9.5, 25) Tcf	17 (9.5, 25) Tcf
Program Benefits	Expected Economic Benefits[a]	$220 million at 3% ($100-$280 million) $145 million at 7% ($68-$180 million)	$590 million at 3% ($275-$675 million) $380 million at 7% ($180-$440 million)	$300 million at 3% ($150-$370 million) $200 million at 7% ($100-$250 million)
	Environmental Benefits	Several elements of the program have the potential to lead to generally smaller footprints for natural gas production, although this benefit is at least partially offset by the increase in producibility from previously marginal regions. The net impact on the environment in terms of disturbed area has not been estimated. The other potential environmental benefit would be a reduction in carbon emissions if natural gas substitutes for other fossil fuels. At the relatively modest level of increased production envisioned for this program alone, environmental benefits are expected to be modest.		
	Security Benefits	Increased domestic natural gas production partially offsets gas imports. Based on DOE's estimates, about 50% of the increased production results directly in reductions in imports under Reference Case prices (25% in a High Gas Prices scenario). Natural gas imports avoided (2005 to 2025).		
		1.2 Tcf	0.6 Tcf	1.2 Tcf

[a]Benefits are expected values, with 10th and 90th percentiles shown in parentheses. Economic benefits are present values discounted at both 3% and 7% real. Environmental and security benefits are discounted at 3%. The net economic benefits were calculated based on the assumption that the technology from the DOE program affects no existing gas production and that there will be a decrease in the cost of gas that would have been uneconomical to produce without the DOE program. DOE (2004a) estimates the discounted, cumulative economic savings through 2025 to be $50 billion for the Reference Case, although the committee notes that this includes consumer cost savings largely offset by reduced revenue to producers and, as such, is not indicative of the net economic benefit to the nation.

FIGURE 2-3 Findings for DOE's natural gas exploration and production program.

Technical risks. For incremental improvements in the life of existing wells, technical risks are small compared to exploration and drilling programs. Novel drilling R&D aims at breakthrough improvements to increase the depth that vertical and horizontal wells can be drilled and for ultra-deep-water drilling. Technical risk of any one breakthrough drilling project is high, but probability of success on one or two of such projects is high. Technical risk for improved stimulation and completions systems is relatively low. Developing technology to drill and complete deep gas wells is high risk. Advanced diagnostics and imaging development has been well focused and technically successful.

Market risk. Given that existing wells are operated by small independent operators, market penetration of technology is high risk. In the case of novel drilling technologies, market risk is low because drilling technology is quickly implemented throughout the industry. Improved drilling and stimulation for unconventional gas reservoirs face low market risks because technology developments are used from other industries. Since industry is involved in most diagnostics and imaging R&D projects, market risks should be low.

Benefits. The panel estimates that increased annual production of natural gas will result from the DOE R&D program and estimated benefits from 2006 to 2025. The net economic benefits to the nation are judged by the panel to be significant, especially under the High Oil and Gas Prices scenario, which seems a likely state of the world. Energy security benefits also accrue from reduced natural gas imports because of enhanced domestic gas production. Although it was difficult to evaluate environmental benefits, they would also likely accrue because the use of natural gas is relatively clean compared to other fossil fuels. Many of the benefits would not be realized without federal involvement because the sector is composed of many small independent operators without the resources to conduct long-range R&D; thus the federal investment is critical for such a fragmented sector.

Notable accomplishments/gaps, opportunities, spin-offs. The existing fields program considering marginal or stripper wells is mainly a technology transfer program to operators of marginal wells; program effective, but may not be reaching a significant part of the large universe of marginal producers. The drilling, completion, and stimulation program has the potential to make breakthrough improvements compared to incremental improvements made by the service companies. But technical risk is high limiting likely success to only one or two projects. The advanced diagnostics and imaging program potentially should be effective, given past performance. This is an advanced field in the industry and specific balance in public and private R&D in this area should always be sought. The ultradeep drilling envisioned in the Deep Trek program has high technological dependence especially in the area of deep seismic imaging, and materials capability and R&D are essential, both public and private.

FIGURE 2-3 Continued

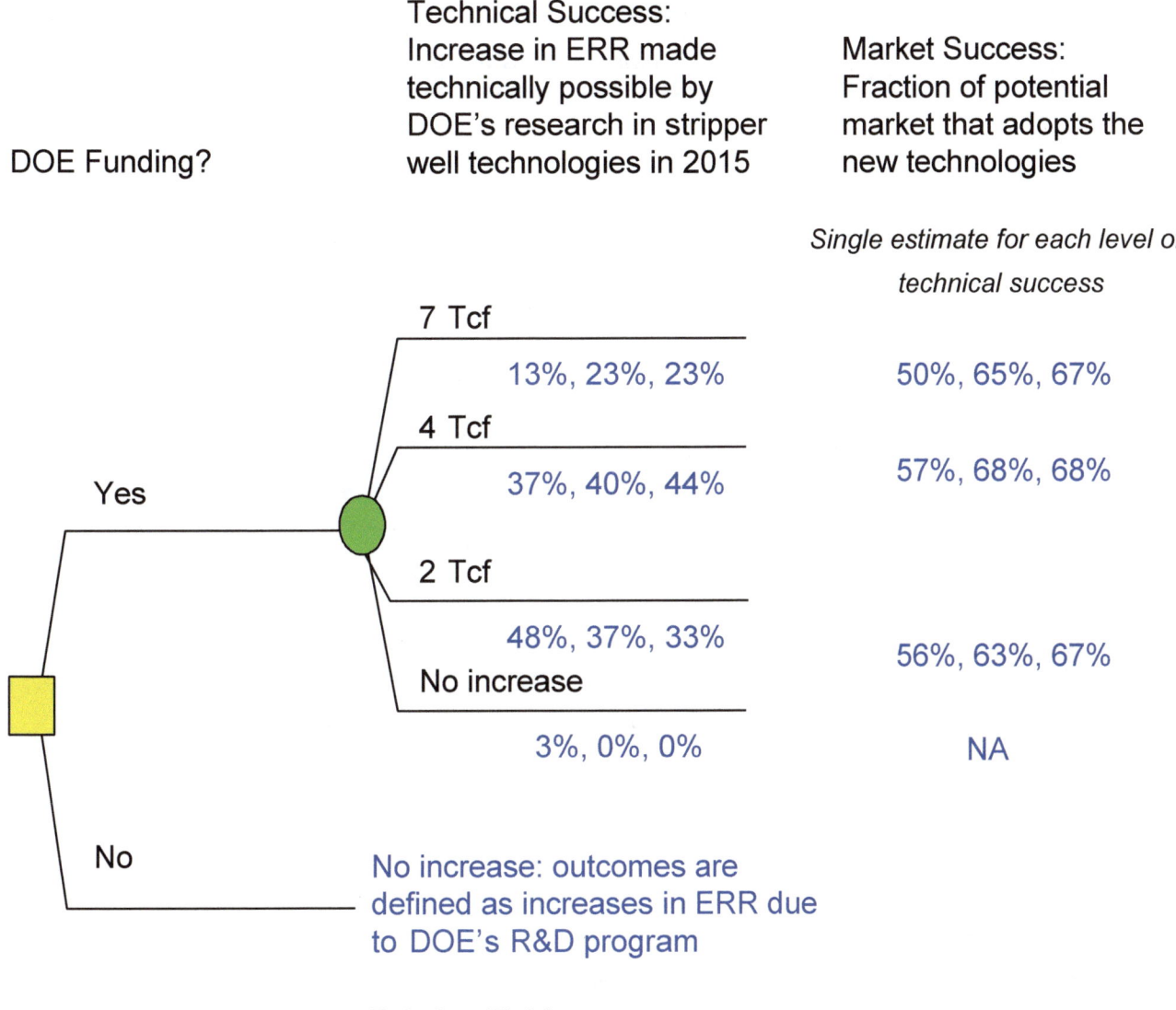

FIGURE 2-3 Continued

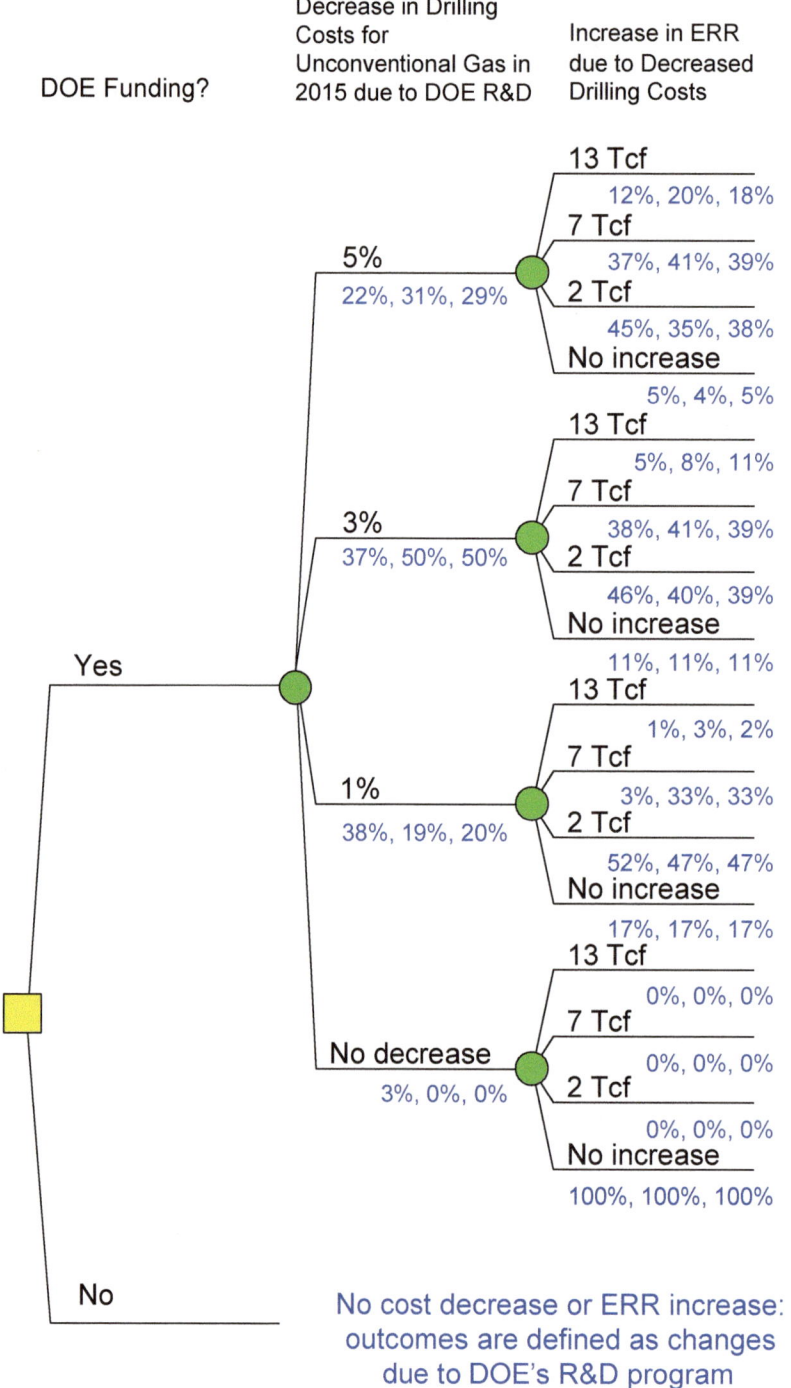

FIGURE 2-3 Continued

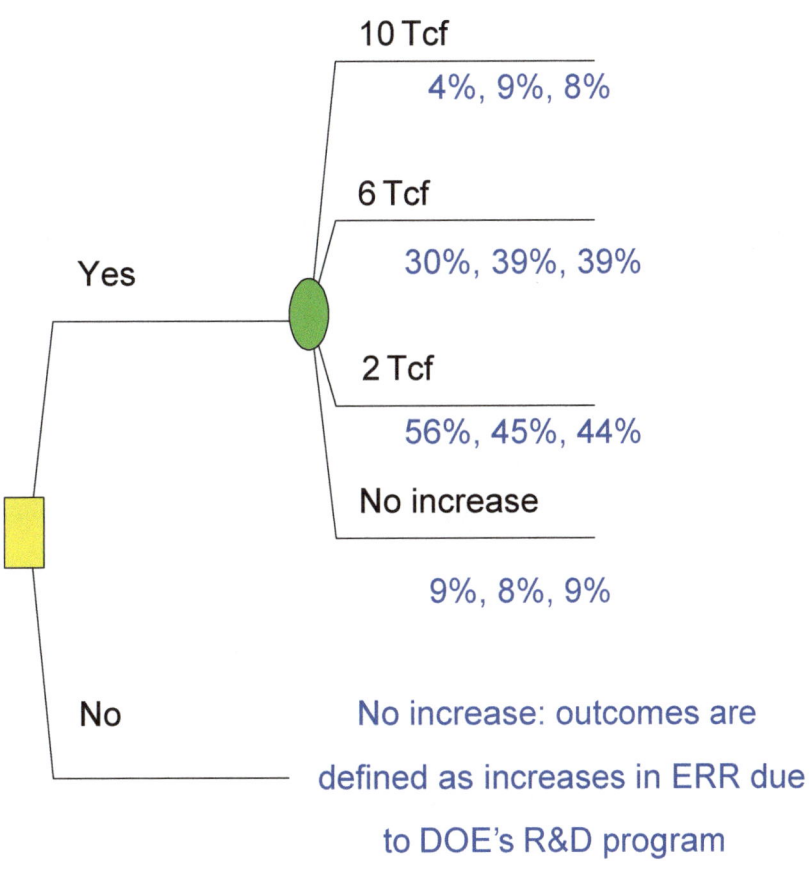

Advanced Diagnostics

FIGURE 2-3 Continued

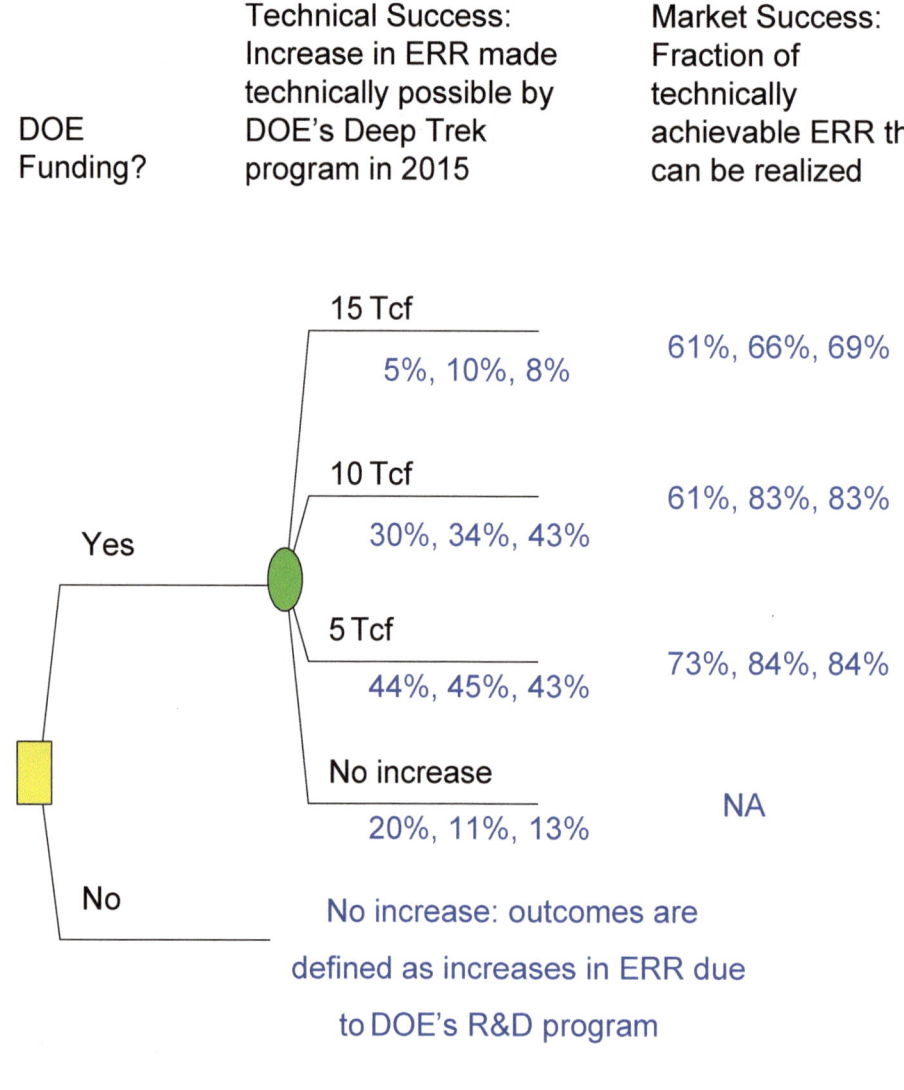

FIGURE 2-3 Continued

Program Name: Distributed energy resources: End Use System Integration and Interface subprogram
Program Goals: Demonstrate four integrated CHP applications each having greater than 70% combined electric and thermal efficiency, which could be manufactured and installed (assuming commercial-scale production) cheaply enough to result in a 4-year payback to customers by 2008
Year Goals Expected to Be Achieved: 2008
Program Costs:
    Current (FY05) Funding: $20.5 million
    Expected Cost to Completion: $205 million through 2015
Industry and Foreign Government Funding: 30% to 50% cost share
Key Complementary/Interdependent DOE Programs: industrial gas turbines, microturbines, advanced reciprocation engines, and related materials and sensors, all within the DER program

		Global Scenario			
		AEO Reference Case	High Oil and Gas Prices	Carbon Constrained	Electricity Constrained[a]
Program Risks	Technical Risks	Technical risks identified as uncertainty in the ability to reach the 4-year-or-less payback target identified by DOE (for a 70+% efficient system). Panel specified two alternative payback periods (5-7 years, 8 years or more) as other possible technical outcomes. Estimated likelihood of achieving a 4-year payback ranged from a low of 20% (without the DOE program under the Reference Case scenario) to a high of 60% (with the DOE program in the Electricity Constrained scenario).			
Program Risks	Market Risks	Market risk characterized by the estimated fraction of the commercial buildings sector that would adopt CHP technologies, given a 4-year or less payback. High market penetration was defined as 50% of new construction and 5% of existing. Moderate market penetration was defined as 25% of new and 5% of existing, and low market penetration was defined as 10% of new and 1% of existing. Estimated likelihood of high market penetration ranged from a low of 15% (without the DOE program in the High Oil and Gas Prices scenario) to a high of 55% (with the DOE program in the Electricity Constrained scenario).			
Program Benefits	Expected Economic Benefits (see note)	Expected net economic benefits are presented as the present value of the annual economic benefit of incremental CHP capacity installed between 2006 and 2025 attributable directly to the DOE program.			
Program Benefits	Expected Economic Benefits (see note)	$57 million at 3%	$46 million at 3%	$64 million at 3%	$83 million at 3%
Program Benefits	Expected Economic Benefits (see note)	$40 million at 7%	$32 million at 7%	$45 million at 7%	$58 million at 7%
Program Benefits	Environmental Benefits	The net environmental impact of CHP is not known. While CHP may result in lower emissions at central generation facilities, the CHP systems themselves are natural gas powered and result in local environmental emissions. At the relatively modest levels of deployment anticipated, the environmental benefits associated with the program are negligible.			
Program Benefits	Security Benefits	Security benefits arise from the displacement of primary energy use by the incremental installed CHP capacity. This benefit is due to dispersed generation being less vulnerable to terrorist and natural disaster events. Reduction in total primary energy use from all incremental CHP capacity installed between 2006 and 2025 attributable directly to the DOE program is			
Program Benefits	Security Benefits	10 trillion Btu	8 trillion Btu	11 trillion Btu	15 trillion Btu

NOTE: Benefits are presented as the difference in the expected value with and without the DOE program. See Figure K-9 for a discussion of the uncertainty surrounding benefits. Economic benefits are shown as present values discounted at both 3% and 7% real; environmental and security benefits are discounted at 3%.

[a]Fourth scenario defined by the panel to represent a future state of the world where there are severe constraints on the electricity system. The scenario assumes that the capacity of the electric distribution system is seriously deficient with respect to meeting the electricity and peak demand needs of its customers in one or more high-demand-load pockets.

FIGURE 2-4 Findings for DOE's distributed energy resources program.

Comments and observations. Of the $56.6 million appropriated for the DER program in FY05, $20.5 million, or 36%, is directed at the activities called End-Use System Integration and Interface—mostly CHP systems—which is the program area subject to this assessment. DOE partners with component equipment and technology manufacturers to develop and demonstrate CHP systems, which provide both electrical power and heating or cooling at customer sites with minimal site-specific engineering. The CHP component of the DER program is designed to demonstrate four integrated CHP applications, each having greater than 70% combined electric and thermal efficiency, which could be manufactured and installed (assuming commercial-scale production) cheaply enough to result in a 4-year payback to customers by 2008.

Key findings of the panel assessment are (1) the 70% efficiency goal is achievable in the stated time frame in the end-use applications targeted by the program, but the payback goal is not, except in areas of the country where electricity is constrained or costs are high relative to natural gas; (2) the program produces a positive net present value economic benefit under the four global scenarios assessed, ranging from $46 million to $83 million; (3) the CHP program as a component of the larger DER program is too small to be modeled accurately using conventional models, and any further application of the committee methodology should focus on larger impact programs.

Technical risks. The panel estimated the highest likelihood of technical success in the Electricity Constrained scenario and the lowest in the High Oil and Gas Prices scenario (as higher natural gas prices will make it more difficult for CHP technologies to achieve a shorter payback). The panel estimated a higher likelihood of achieving a shorter payback with the DOE program than without it and higher likelihoods with locally high electricity prices than with locally low prices.

Market risks. The panel estimated a higher likelihood for strong market adoption with DOE funding than without DOE funding. The panel estimated the highest likelihood of market adoption in the Electricity Constrained scenario.

Benefits. The panel determined that CHP could easily provide a 10% net economic benefit over its costs over the study period, leading to a benefit:cost of 1.1. This translates into a net economic benefit of approximately $230 per kilowatt of installed CHP. This is calculated on an electricity system avoided cost basis and accounts for the initial CHP investment and life-cycle operating costs. Clean and efficient CHP has the potential to reduce overall emissions to the environment since it uses clean-burning natural gas as a fuel and displaces central station generation power and local combustion boilers and furnaces used to generate steam, heating, or hot water. Like all energy-efficiency measures, CHP reduces primary energy use and helps improve the nation's energy security by reducing the likelihood of energy disruptions due to reductions in natural gas imports. Because the modeling work presented to the panel made it difficult to identify specific fuels displace by CHP, the panel preferred to report energy security benefits in energy units, namely, Btu's. However, the unique benefit of CHP is to decentralize power production and locate it at or close to loads, providing the security benefits of safety from terrorist attacks and customer protection from the effects of disruptions to utility electric service.

Program observations. DOE should consider working more closely with utilities to encourage utility ownership of CHP systems. DOE should support replicable pilot applications where there are offsetting capital costs, such as standby generators in hospitals or standby generators and uninterruptible power systems in data centers. DOE should address the problem of institutional barriers and work with regulatory agencies and utilities to eliminate or reduce them.

FIGURE 2-4 Continued

RESULTS OF APPLYING THE METHODOLOGY

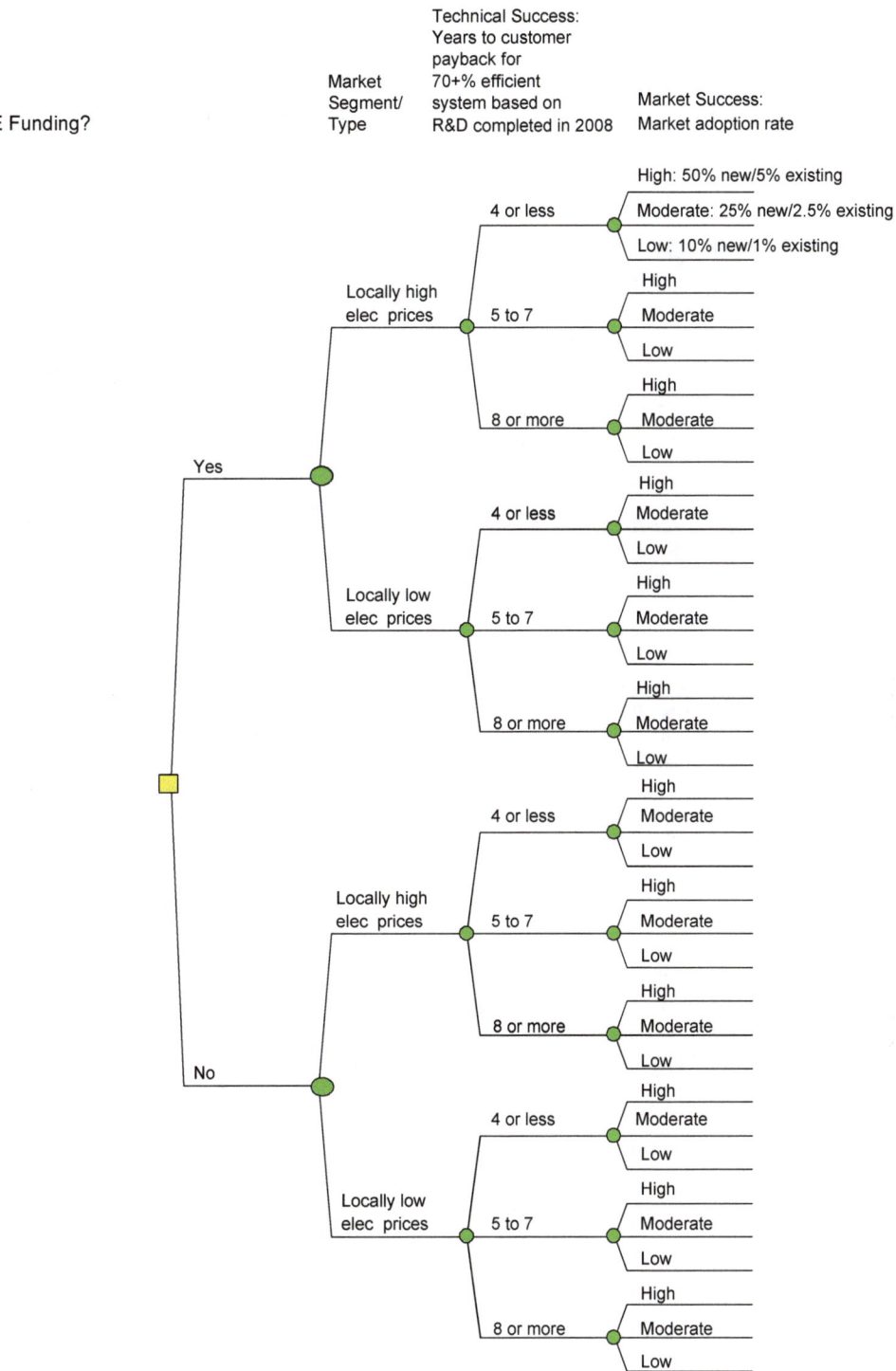

FIGURE 2-4 Continued

Program Name: Light-Duty Vehicle Hybrid Technology R&D Program
Program Goals:
    Storage: By 2010, 300 W/hr storage device capable of 25 kW for 20 s, 15-yr life, and costing $20/kW
    Lightweight materials: By 2012, reduce overall vehicle weight by 50% relative to 1997
    Advanced combustion: By 2010, peak brake thermal efficiency of 45%
Current (FY06) Funding: $81 million
Completion Costs: $567 million through 2012

		Global Scenario		
		AEO Reference Case	High Oil and Gas Prices	Carbon Constrained[a]
Program Risks	Technical risks	Evaluated for four subsets of projects in the program. Technical risks identified as uncertainty in the level of technical advancement, uncertainty in the vehicle cost impact of the new technologies, and, for batteries, uncertainty in the time that the new technologies would be market-ready. See decision trees for the probabilities of technical success in each area.		
	Market risks	The panel addressed two types of market risks: market acceptance of more fuel-efficient conventional vehicles and market acceptance of HEVs. The market acceptance of more fuel-efficient conventional vehicles was assumed to be strictly a function of the trade-off between increased capital costs and decreased lifetime fuel costs. To address the market acceptance of HEVs, two HEV market conditions were defined. In the Low HEV condition, HEV sales increase linearly from 2003 market share to about 12% of new vehicle sales in 2050. In the High HEV condition, HEV sales increase exponentially from 2003 and account for about 40% of new vehicle sales in 2050.		
Program Benefits	Expected economic benefits	Economic benefits are calculated as the reduction in the expected consumer expenditures for vehicles and fuel from 2006 to 2050 attributable to the DOE program.		

		In the Low HEV market condition:
		  $5.9 billion at 3%
		  $3.7 billion at 7%
		In the High HEV market condition:
		  $7.2 billion at 3%
		  $4.2 billion at 7%
			In the Low HEV market condition:
			  $27.5 billion at 3%
			  $15.7 billion at 7%
			In the High HEV market condition:
			  $28.2 billion at 3%
			  $15.9 billion at 7%
				In the Low HEV market condition:
				  $7.3 billion at 3%
				  $4.7 billion at 7%
				In the High HEV market condition:
				  $8.5 billion at 3%
				  $5 billion at 7%

	Environmental benefits	Environmental benefits are calculated as the reduction in total carbon emissions from vehicles from 2006 to 2050 that can be attributed to the DOE program. Difference between Low HEV and High HEV market conditions is less than 2%; only one value is shown.		
		28 million metric tons	51 million metric tons	32 million metric tons
	Security benefits	Security benefits arise from reduced gasoline consumption and associated reduction in oil imports. The estimated reduction in gasoline use by vehicles from 2006 to 2050 that can be attributed to the program for Low HEV and High HEV market conditions are shown.		
		9.2 to 9.4 billion gallons	16.7 to 17 billion gallons	10.4 to 10.6 billion gallons

NOTE: Benefits are presented as expected values. Economic benefits are shown as present values discounted at both 3% and 7% real; environmental and security benefits are discounted at 3%.

[a]For the Carbon Constrained global scenario, the panel assumed Reference Case prices for oil and gas. The differences in expected benefits come from differences in the probability of technical and market risks.

FIGURE 2-5 Findings for DOE's light-duty vehicle hybrid technology R&D.

Comments and Observations. DOE's R&D on technologies for light-duty hybrid vehicles with ICE power trains is conducted under the auspices of the Fuel Cell and Vehicles Technology (FCVT) program. The panel evaluated a portion of the FCVT budget for R&D related to passenger vehicles; specifically: (1) high power energy storage; (2) automotive lightweight materials; and (3) advanced combustion and fuels. These three areas focus on critical technologies for more fuel-efficient light-duty vehicles and have consistently received the highest percentages of the FCVT funding for passenger vehicles.

Technical Risk

High power energy storage. In the panel's judgment, DOE's technical and cost targets are unlikely to be achieved by 2010 because current proven hybrid battery technology (NiMH) is unlikely to achieve the cost targets and 15-year life. Lithium ion batteries still have significant limitations, including safety and low temperature performance, that require further development before volume commercialization.

Lightweight materials. Because of cost and manufacturing issues (such as joining), aluminum has been used primarily for closures in high-volume applications. However, some lower volume vehicles have been aluminum intensive, proving the viability of an all-aluminum body. Carbon-reinforced composites have not seen extensive application in high-volume automotive products owing to high costs and the lack of high-volume fabrication and assembly systems. The panel does not believe that the technical goal of 50 percent reduction in vehicle weight can be achieved by 2012.

Advanced combustion. The goal of achieving 45 percent peak brake efficiency would require significant advancements in many areas, including high-pressure direct-injection fuel injection systems; higher efficiency boosting (turbocharging/supercharging) systems; reduced-friction components; reduction in parasitic and accessory loads; higher strength/lighter weight powertrain materials to accommodate the combustion pressures, which could approach 220 bar or more; advanced fuel injection and combustion process controls to support low-temperature combustion (LTC) processes; and new methods for noise control.

Market Risk Assessment

High power energy storage. Under some driving conditions the limited battery energy storage may be inadequate to maintain adequate performance (e.g., up long hills). Other market risks associated with hybrid vehicle drive trains include unknown durability because of the greater complexity associated with the battery, electric motor, and power electronics. Adequate battery durability (15-year calendar life) is essential.

Lightweight materials. For body and chassis applications of lightweight materials, the most important impediment is the challenge posed by the introduction of new materials into the existing fabrication and assembly processes. The manufacturing footprint is amenable to the introduction of high-strength steels and can accommodate aluminum closures (doors, hoods, deck lids). However, to convert fabrication and assembly systems to accept carbon or glass-reinforced composites would require a major development activity to prove the feasibility of manufacturing composite intensive bodies in high volume.

Advanced combustion. The market risks associated with the combustion engine, emission control, and fuels activity are primarily related to the cost uncertainty of achieving the target performance parameters and/or the ability of the advanced concepts to achieve the durability and reliability requirements of high-volume automotive production.

Benefits estimation. The economic, environmental, and security benefits of DOE's research in this area depend on the degree to which the R&D programs will lead to more fuel-efficient vehicles in the market and on the road, which will result in reduced gasoline consumption. The reduced gasoline consumption leads directly to benefits: economic benefits from reduced consumer expenditures for gasoline, environmental benefits from reduced carbon dioxide and other emissions, and security benefits from reduced demand for oil. To quantify benefits, the panel constructed 145 different possible technical outcomes, each consisting of a unique combination of technical success levels of the three R&D areas it evaluated. These in turn implied different fuel economy and incremental cost differences for conventional and HEVs. The panel considered two different "market success" scenarios for HEVs: one in which the sales of HEVs were estimated to grow relatively quickly ("High HEV") and one where that market growth is significantly slower ("Low HEV"). For each of these cases, the benefits model calculated the total fuel consumption, emissions, and consumer expenditures on vehicles and gasoline by year for each year through 2050.

Program observations. Key findings of the panel assessment are (1) DOE's light-duty hybrid vehicle R&D program is likely to yield important technology advances that could result in fuel economy improvements for light-duty vehicles in the United States; (2) The methods currently used by DOE to assess the potential fuel economy benefits of its light-duty hybrid vehicles R&D tend to be overly optimistic in estimating the impact and timing of technology advances; (3) Important fuel economy benefits could accrue even if DOE's R&D programs on light-duty hybrid vehicles fail to achieve their ambitious cost and performance goals; and (4) DOE's R&D on light-duty hybrid vehicles has benefits over and above the potential for improved fuel economy.

FIGURE 2-5 Continued

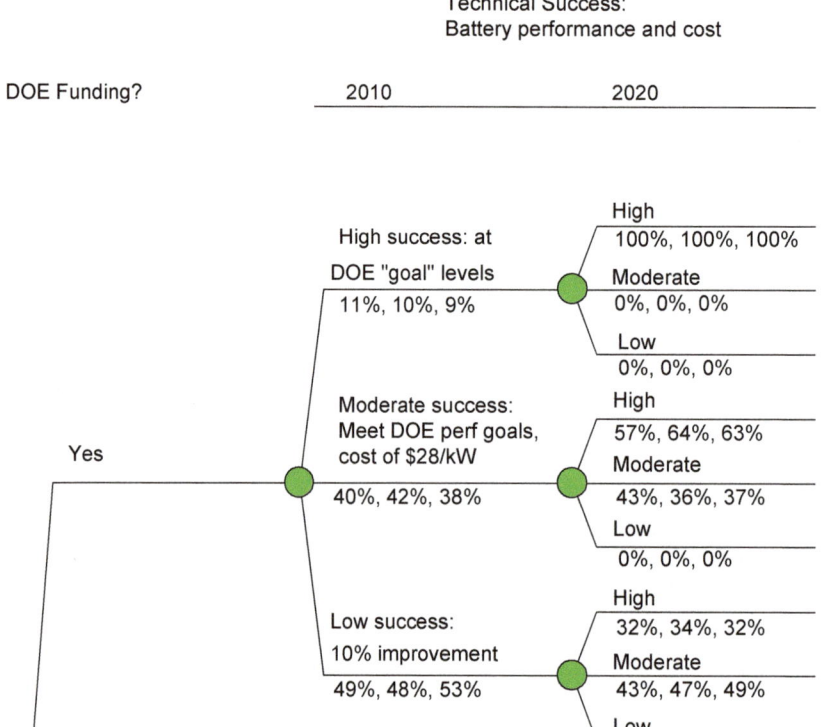

FIGURE 2-5 Continued

Lightweighting

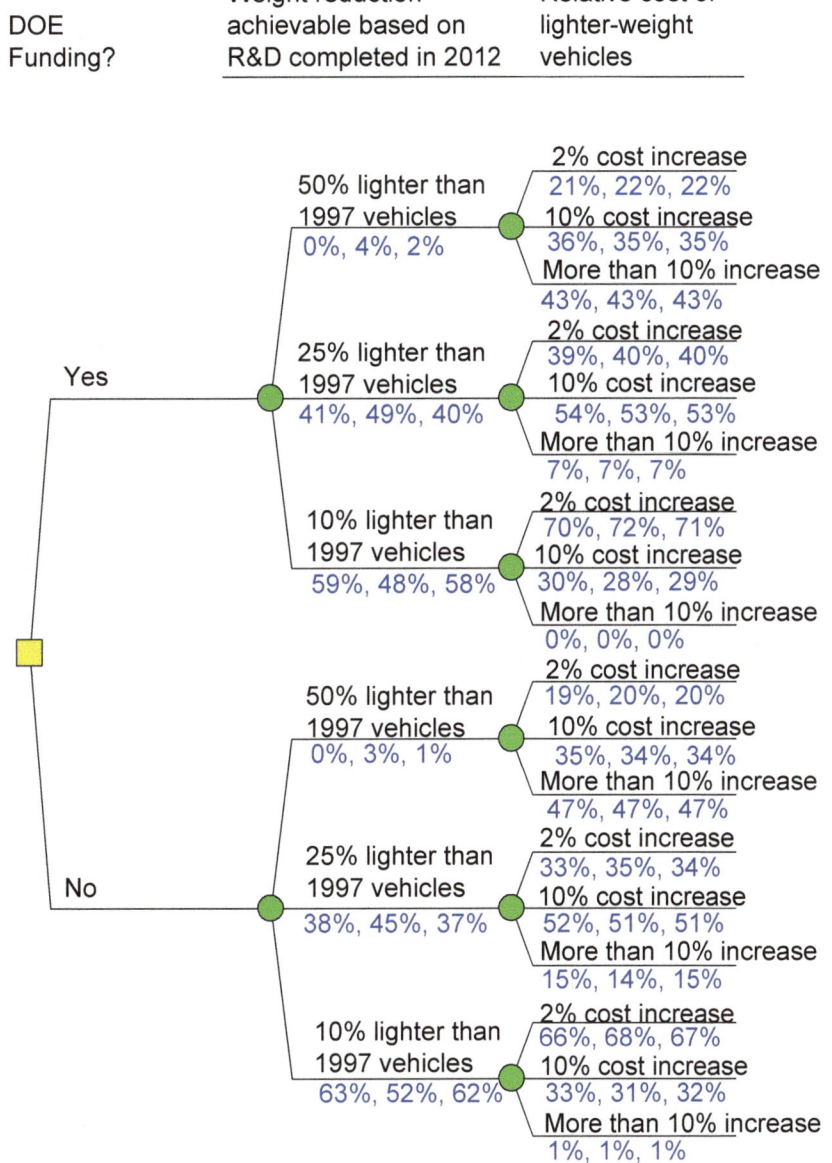

FIGURE 2-5 Continued

Advanced Engines

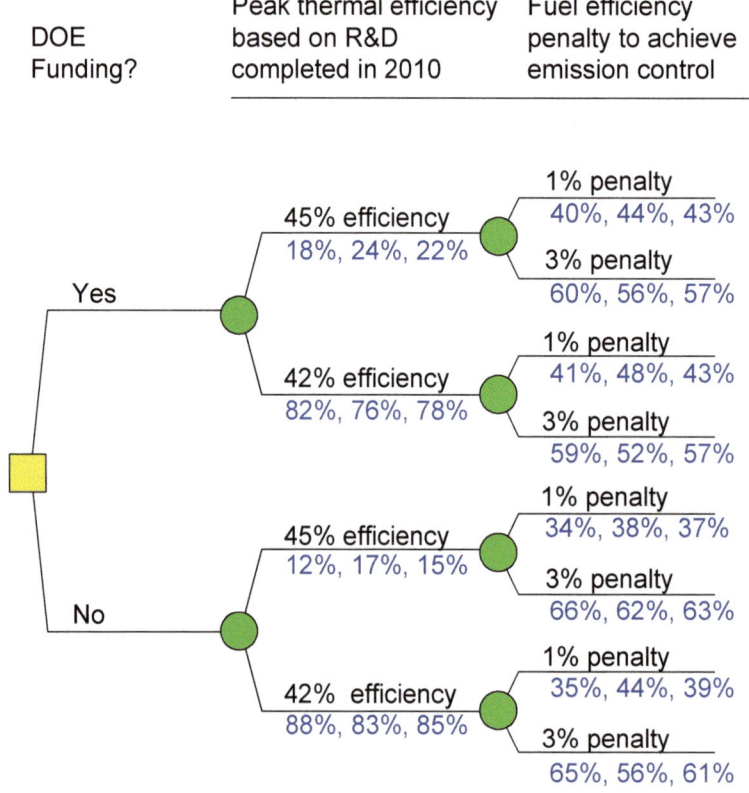

FIGURE 2-5 Continued

Program Name: Industrial Technologies Program–Chemicals
Program Goals: Reduce chemical industry energy consumption by 20% by 2020
Program Costs: Approximately $75 million for complete funding of current portfolio
Proposed FY07 Funding: $6.8 million
Industry Funding: 30%-50% cost share

		Global Scenario[a]		
		AEO Reference Case	High Oil and Gas Prices	Carbon Constrained
Program Risks	Technical Risks	Evaluated for each project in the portfolio. Most projects had technical risks associated with scale-up from laboratory-scale to commercial-scale applications. Estimated probability of technical success for the projects ranges from 5% to 50%.		
	Market Risks	Evaluated for each project in the portfolio. Estimated probability of market success for projects ranges from 10% to 70%.		
Program Benefits	Expected Economic Benefits	Economic benefits are calculated as the expected value of energy savings benefits from 2006 to 2030 resulting from the research portfolio, in 2003 dollars. Ranges shown are for the 10th and 90th percentiles:		
		$534 million at 3% ($0 to 1,550 million) $215 million at 7% ($0 to $640 million)	$950 million at 3% ($0 to $3,000 million) $390 million at 7% ($0 to $1,250 million)	$550 million at 3% ($0 to $1,600 million) $223 million at 7% ($0 to $700 million)
	Environmental Benefits	The same quantity of energy savings, and therefore the same environmental benefits, will be achieved under all three scenarios. Anticipated environmental benefits are estimated emission reductions from 2006 through 2030: CO: 15,100 (0 to 11,800) metric tons[b] CO_2: 1.7 (0-5.4) MMTCE SO_2: 9,200 (−2,000 to 30,000) metric tons[c] NO_x: 14,000 (0 to 42,000) metric tons Particulates: 180 (−50 to 600) metric tons VOCs: 340 (0 to 1,000) metric tons		
	Security Benefits	The same quantity of energy savings, and therefore the same security benefits, will be achieved under all three scenarios. Anticipated security benefits are the reduction in oil and natural gas consumption from 2006 through 2030: Natural gas: 89 (0-330) billion cubic feet Petroleum: 1.3 (0-0.01) million barrels[d]		

NOTE: Benefits are presented as expected values, with 10th and 90th percentile benefits in brackets. Economic benefits are shown as present values discounted at both 3% and 7% real; environmental and security benefits are discounted at 3%.

[a]The panel concluded that the scenarios would produce insignificantly small changes in the volumes of oil and gas saved; therefore, the physical quantities reported for environmental and security benefits are the same for all scenarios. Economic benefits differ because prices differ among the scenarios.

[b]CO reduction benefit derives primarily from a single project. This skews the distribution of benefits such that the expected value is higher than the 90th percentile.

[c]Several projects were estimated by DOE to result in increases of various types of emissions. For SO_2 and particulates, there were a sufficient number of these projects that the 10th percentile of the range of impacts represents an increase in emissions.

[d]Distribution on the reduction in petroleum usage is similarly skewed by a single project accounting for most of the reduction.

FIGURE 2-6 Findings for DOE's chemical industrial technologies program.

Comments and observations. The goal of Chemicals Industrial Technologies Program is to implement a successful DOE research program that helps the chemical industry to use 20 percent less energy in 2020 than in 2001. This translates into a reduction in energy use of the chemical industry of 1.3 quads per year in 2020 and proportional reduction in emissions. To achieve its goal the program has a portfolio of relatively small projects, all of which are competitively awarded and all of which involve 30%-50% cost sharing. Twenty-two projects are currently being funded at a total of $7 million per year. If all projects were funded to completion and all were successful, DOE estimates they would achieve a saving of 0.303 quads per year, or 23 percent of the overall program goal for the Chemical Industrial Technologies Program.

The panel believes that the Chemical Industrial Technologies Program is seizing an important opportunity to produce energy savings in a major industrial segment by supporting early-stage research that industry is unlikely to support. However, to realize its potential benefits, the program must adapt to a seriously constrained budget and a changing domestic industrial environment. Cutting-edge research that can be funded with the restricted budget available will probably continue to produce valuable but relatively small-scale advances. However, the program is also pursuing sweeping changes in process design in the hope that they can yield big energy savings.

The current portfolio of projects has an expected net economic benefit between $215 million and $534 million in the AEO Reference Case. Because the program is composed of early-stage research projects, the range of benefits is from zero to $1.55 billion. Benefits in the High Oil and Gas Prices and Carbon Constrained scenarios are somewhat higher.

Technical and market risk assessment. The panel assessed all of the projects in this program to be high risk and thus consistent with DOE's program strategy. Probability of technical success of individual projects ranged from 5% to 50%, with market success from 10% to 70%. Combining these assessments, the range for success for the entire portfolio is between 1% and 35%, with an average of 7.5%.

Benefits estimation. The economic, environmental, and security benefits of the Chemical Industrial Technologies Program derive directly from the energy savings realizable from the projects. DOE estimates that the research will accelerate the development and implementation of the identified technology by 3 to 5 years. The panel agrees that in almost all cases, DOE support would be a significant accelerating factor. It calculated the expected total benefit of the portfolio by applying its probabilities of technical and market success to DOE's estimates of energy savings for each project in the portfolio and rolled up the individual project estimates into an expected value of gross benefits for the overall program. Net benefits are calculated by assuming that for any project that "succeeds," investments on the order of the net present value of the economic benefits for first 3 years will be required. This assumption reflects the panel's view that the private sector would require a 3-year payback of the investment costs.

Program observations. Based on the experience of its members in managing and conducting research in the chemical industry, the panel wishes to underscore how essential it is for DOE management to focus its limited resources on the most promising opportunities available to it. The danger is more than just the wasting of funds on less promising projects. Equally significant if not more so is the possibility of losing the benefits of high-priority projects because they were not pursued aggressively enough. The panel suggests that DOE frequently review the projects in its existing portfolio to ensure that each project is pursued for as little time and money as it takes to demonstrate feasibility or infeasibility.

FIGURE 2-6 Continued

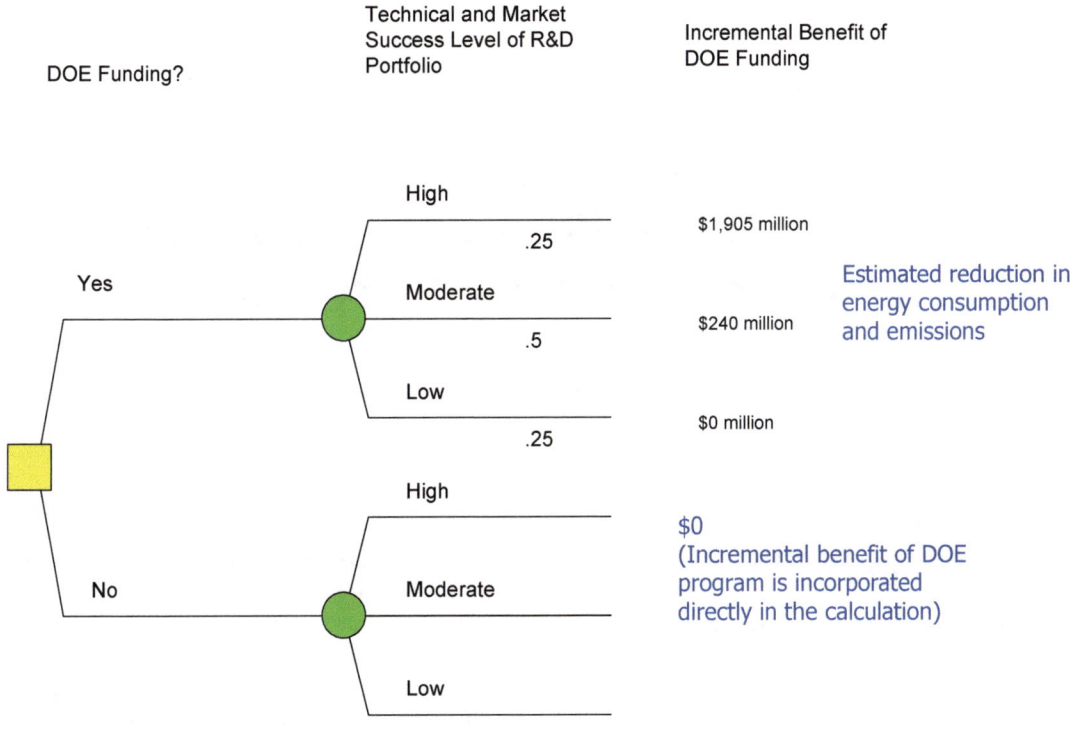

FIGURE 2-6 Continued

3

Methodology for Prospective Evaluation of Department of Energy Programs

The Phase One study of prospective benefits of DOE's applied R&D programs (NRC, 2005a) recommended a specific methodology for benefits calculation based on experience with applying a conceptual methodological approach to DOE's fuel cell, carbon sequestration, and advanced solid-state lighting programs. The present study—the Phase Two study—tests the recommended methodology (see Appendix F for a summary description of the methodology from Phase One) on six programs in a consistent manner. One of its objectives is to refine the methodology and to assess its broader applicability. In addition, the committee examines procedures for estimating the monetary value of environmental and security benefits, a topic that had been deferred from the Phase One study. The results of the committee's analyses of these topics are summarized below, beginning with the valuation of environmental and security benefits.

VALUATION OF ENVIRONMENTAL AND SECURITY BENEFITS

Assessing the prospective benefits of DOE's R&D programs involves challenges common to many public programs but not usually present in private business assessments. Economic benefits—the linchpin of private investments—may be difficult to calculate, and for complex programs critical data are frequently lacking and uncertainty prevails about how successful a technology might turn out to be.[1] Never-theless, there is general agreement among economists about the principles for evaluating economic benefits. Markets that measure the economic value of related technology and (in some cases) futures markets that provide an evaluation of risk can guide an evaluation of economic benefits.

In contrast, public programs often have social benefits that are not valued by markets. Assessing the value of such benefits is inherently difficult: It involves ambiguity and, even as an academic matter, a range of possible answers. For the DOE programs, two broad classes of benefits have this characteristic: the environmental consequences of energy technology and the security implications of energy savings or energy alternatives. These program attributes are in general critical components of the benefits package—indeed, if a program can be justified simply on an economic basis, there may be no rationale for government participation.

This section reviews current practice in evaluating environmental and security benefits and considers ways in which it can be adapted to assessing DOE benefits.

Valuing Air and Water Pollution Emissions

For goods and services purchased in a competitive market (a market in which both buyers and sellers are "price takers" that have no long-term influence on price), the market price represents the economy's best valuation of an additional unit of that good or service. Determining a social value for a good or service is difficult in the absence of a competitive market. Economists have developed a number of approaches to help estimate the social values of goods and services not valued in the market. The advantages and disadvantages of the most relevant approaches for valuing environmental benefits are explored in the following sections. This discussion focuses on the benefit of reducing air pollution, which is both the primary environmental benefit identified in the Phase Two study and the type of benefit for which, as explained below,

---

[1] Economic net benefits are based on changes in the total market value of goods and services that can be produced in the U.S. economy under normal conditions, where "normal" refers to conditions absent energy disruptions or other energy shocks. Economic value can be increased either because a new technology reduces the cost of producing a given output or because the technology allows additional valuable outputs to be produced by the economy. Economic benefits are characterized by changes in the valuations based on market prices. This estimation must be computed on the basis of comparison with the next-best alternative, not some standard or average value (NRC, 2005a, p. 16).

value estimates are most advanced.[2] The committee advocates applying valuations to emissions of air pollutants but not to other types of pollutant emissions.

Damage Function Models

The Phase Two methodology results in an estimate of the quantitative change in the level of air pollutants resulting from the technology being studied. Turning this change in emission levels into an economic value requires two additional steps. The first step in evaluating the benefits of pollution abatement is to estimate the impact of pollutants on things society values, e.g., health, visibility, outdoor recreation, quality and availability of raw materials, ecosystem services, and other measures of environmental quality. Damage functions must be estimated for the most important effects of environmental pollution. In particular, the impact on human health of air pollution includes days lost due to restricted activity, increases in the incidence of asthma and bronchitis, and even premature deaths. The second step is to place a monetary value on the damages (or reductions in damage) to health, visibility, outdoor recreation, materials deterioration, and the natural environment. Both steps present significant problems.

In published studies, the various physical effects are modeled to determine the quantitative effects of changes in ambient levels of pollution. For example, how does the incidence of asthma or bronchitis change as air pollution levels change? Despite decades of damage research, however, both the qualitative and quantitative effects of such pollution are uncertain. One reason for this uncertainty is that our understanding of how the damage is incurred improves over time. For example, the Environmental Protection Agency's (EPA's) benefit-cost analysis of the fine particulate matter ($PM_{2.5}$) and ozone standards for 1970-1990 ascribed all premature mortality associated with air pollution to $PM_{2.5}$, assuming that the contribution of other pollutants was negligible (EPA, 1997a). A further uncertainty surrounds the extent of life shortening associated with a premature death due to $PM_{2.5}$. In addition, estimating damage functions is complicated by interactions with other factors and the changing nature of air pollution across areas and over time. The spatial and temporal variations greatly complicate the task of predicting environmental impacts in a generic manner suitable for application in a prospective planning study. Most of the existing damage functions for water pollution apply to specific sites; there is only a limited ability to produce reliable *generic* damage functions for waterborne pollutants. Hence, where generic regional or national damage functions have been developed, these have typically been for air rather than water pollution.[3] In addition, water availability is a growing issue in many parts of the country. This can often be linked to energy resource recovery, production, and use or to the production of hazardous wastes as part of energy systems' life cycles. Other impacts are more difficult to quantify and include those related to land use, ecosystem impacts, and aesthetics. This complexity and regional specificity hamper efforts to monetize environmental benefits related to water.

Willingness-to-Pay Studies

Once the type and the extent of physical damage due to increased air pollution are known, the next step is to place a monetary value on this damage. However, there is no direct market valuation of a reduced incidence of asthma attacks or of being able to see a landmark from 30 miles away rather than only 10 miles away, because there are no markets for asthma attacks or visibility.

To appraise things like this, economists employ techniques from the field known as nonmarket valuation. Nonmarket valuation seeks to measure in monetary terms the value that people place on items they care about, regardless of whether the item is a conventional marketed commodity (e.g., a loaf of bread, a new car) or something that the person cares about but that cannot be purchased in a market (e.g., a beautiful view at sunset, a pristine wilderness, a historic monument, an excellent public school system, or a healthy body). Conceptually, these nonmarket items are measured in monetary terms by considering the change in income that is equivalent to them, in terms of its impact on the individual's well-being. Thus, while the items themselves are *not* monetary in nature and they cannot be obtained by the individual through the

---

[2]The valuation of pollutants in environmental media other than air is not discussed here. Conceptually, such valuations would be similar to that of air pollutants. In practice, the damage functions apply to specific sites, and data are difficult to elicit.

[3]There are two major contexts in which generic damage functions have been developed, oriented mainly to airborne emissions. One is "environmental costing" by public utility commissions (PUCs). The other is the EPA's retrospective assessment of the effects of the Clean Air Act on the "public health, economy and environment of the United States," mandated by Section 812 of the Clean Air Act Amendments of 1990. EPA's assessment, *The Benefits and Costs of the Clean Air Act, 1970 to 1990*, was published in October 1997. In addition, prior to the adoption of deregulation in the late 1990s, the PUCs in 29 states had adopted or were considering adopting some form of environmental costing for the purpose of comparing electricity generation alternatives in the context of utility planning and regulation. In most cases, this took the form of an approved schedule (or spreadsheet model) of "adders" designed to keep score of the environmental costs associated with different methods of electricity generation under the specific conditions applicable to that state. These adders were not actually used in setting prices nor were they charged to electricity users, but they were used in identifying the least-cost source of electricity generation, based on a consideration of the total social cost, including environmental externalities, rather than just the private cost to the particular utility. The model used by New York State and the issues generally involved in environmental costing are described and discussed in papers presented to a symposium on environmental costing, edited by Shogren and Smulders and published in *Resource and Energy Economics* 18(4) (December 1996).

expenditure of his own funds, their monetary value to the individual is represented by the amount of money that could be exchanged for them while leaving the individual equally well off before and after the exchange. In other words, the economic value of an item can be expressed as the amount of money that an individual would be willing to exchange for that item if an exchange were possible.

Accordingly, economists need to find a trade-off to measure economic value. Either they find a relevant trade-off that occurs naturally (revealed preference) or they create one through a survey or an experiment (stated preference). As an example of the former, economists might examine people's decisions to spend money on various items to protect themselves against the risk of some particular illness: These people are presumably making a trade-off between spending the money and avoiding the illness. Another example of revealed preference is the statistical analysis of how wages vary among occupations and, in particular, of the extent to which higher wages are offered for more risky occupations. The estimated risk differential in wages is used as a measure of how willing people are to trade off more money for a greater risk of death. As an example of stated preference a government that is thinking of introducing a new program (one that will save people's lives, say, or reduce air pollution) surveys people to learn how much they would be willing to pay for the program. Respondents might be told that, if the program is introduced, a household like theirs would have to pay x dollars per year in additional taxes and might then be asked if they would vote for or against it. The tax is varied across different respondents, and the responses are used to trace out a demand curve showing the percent of people who would vote for the program at each different dollar amount. The median of the curve (i.e., the dollar amount that 50 percent of the population would be willing to pay) can be used as a measure of the value placed on the item by the population surveyed.

Both approaches—stated and revealed preference—to the estimation of economic value can have problems in practice. The specific choice being used to exemplify the trade-off in a revealed preference study might not adequately discern what the researcher has in mind, or it might be difficult in practice to disentangle the trade-off of interest from other factors involved in the subjects' decision making. Or, respondents might not find the trade-off credible, or they might think they will not actually have to pay the higher taxes.[4] All of these are challenges that researchers have to deal with.

While uncertainties surround both the damage functions and the economic valuations used to monetize the consequences of energy technologies, the general experience has been that the uncertainties associated with the damage functions are even larger than those associated with the economic valuations.

Just as the physical damage caused by a particular discharge can vary greatly depending on the location and timing of the discharge, with the variability being much greater for waterborne than airborne emissions, so, too, can the economic value that people assign to that particular physical consequence vary greatly depending on the location and timing of the damage. In part, this is due to the spatial (and sometimes temporal) variation in the availability of substitutes for the good or service suffering damage: Other things being equal, the more substitutes that are available, the smaller the associated economic impact. In addition, people's preferences may vary spatially or temporally as a function of differences in norms and expectations. There also tends to be a difference on the economic valuation side of impact assessment with respect to air pollution versus water pollution. Typically, the main impacts of air pollution are on human health, whether morbidity or mortality; impacts on amenities and recreation, materials, and ecosystems tend to play a smaller role.[5] With water pollution, by contrast, the major consequences tend to be recreation and ecosystem impacts, with human health playing a lesser role. The economic values associated with recreation and ecosystem impacts tend to vary spatially and temporally and generally do not lend themselves well to generic assessment. By contrast, the economic values associated with human health impacts lend themselves better to generic assessment. Moreover, for its assessments of the human health consequences of pollution, EPA has developed over the years generalized or consensus economic valuation of a "statistical life," and of morbidity impacts.[6] Such estimates are routinely used by the EPA in a generic manner.[7]

Because of the relative paucity of generic damage functions and generic economic valuations associated with water pollution, as compared to air pollution, the committee recommends that the assessment of environmental impacts

---

[4]Whether or not people respond truthfully to the hypothetical trade-off in a stated preference survey is a matter of controversy. It depends in part on the skill with which the survey has been constructed.

[5]For example, in the retrospective assessment of the Clean Air Act, the EPA determined that human health impacts (specifically, the avoidance of premature mortality) accounted for more than 90 percent of the calculated economic benefit (EPA, 1997a and 1997b).

[6]The value of a statistical life is the value associated with a change of one in the expected total number of annual deaths from some cause; it is a statistical probability—the particular person whose life is affected is not known and will never be individually identified. For example, a statistical life could be the one more random traffic death in addition to the 42,000 killed on the highways each year.

[7]The EPA's Retrospective Analysis of the Clean Air Act used 26 individual willingness-to-pay studies as the basis of its distribution on health effects from premature mortality (EPA, 1997b). Of these 26 studies, five were based on contingent valuation methods that sought the willingness-to-pay of individuals to avoid the risk of premature death. The remaining studies were economic analyses that estimated the additional wages paid to workers for accepting increases in risk of premature death. The EPA used the median estimate of these studies, $4.8 million per life, in its benefits calculations.

associated with DOE's applied R&D programs focus on air pollution consequences and human health impacts.

A Statistical Look at Existing Damage Valuation Studies

To illustrate the effects of the foregoing uncertainties, Table 3-1 provides a sample of the existing economic valuation literature on the damage associated with air pollution. It shows estimates of the damage cost for conventional pollutant and greenhouse gas emissions in constant 1992 dollars per ton ($1992/t) for these releases. Note that the existing body of literature presents a wide range of estimates of the damage resulting from an additional ton of pollution. Uncertainty is evident: The maximum estimates are 6 to 1,000 times greater than the minimum estimates. (The above paragraph is adapted from Matthews and Lave (2000).)

It is important to note that these estimates of the damage costs are not the same thing as the cost of reducing emissions of the pollutant in question—that is, the abatement cost. The marginal abatement cost of an emission reduction can be estimated with much greater confidence than damage costs, especially for pollutants for which a market in abatement allowances exists.

For example, the 1990 Clean Air Act Amendments sets up a plan for buying and selling SO_2 and NO_x allowances. Each allowance allows the owner to put one ton of that pollutant into the air. The Act sharply reduced the allowable emissions of these two pollutants over time. Plants that are able to reduce more than the required amount are allowed to sell their excess allowances. Plants that are not able to meet the required reductions are forced to buy allowances. The prices of these allowances reflect the marginal cost of abatement rather than the social value put on the pollution abated. Such prices depend on the stringency of abatement, the current technology for abatement, and the ability to shift the allowance to the future.

The price of an SO_2 allowance has been as low as $70 per ton (/t) in 1996 and is currently around $850/t (EPA, 2006). EPA projects that allowance prices will rise to $1,200/t by 2020. The NO_x abatement cost is somewhat more complex because it depends on location—areas like Los Angeles have stringent NO_x curtailment, for example, and other areas have much less stringent abatement. Los Angeles NO_x allowance prices have been as low as $200/t in 1998 and are currently selling at a high of around $7,700/t.

Ideally, public policy would constrain the emissions of pollutants until the marginal cost of abatement was equal to the marginal benefit (i.e., the decrease in total damage cost). If so, the allowance price, an estimate of the marginal cost of abatement, should approximate the social value of abating these pollutants. To test this ideal against the real world, Table 3-2 compares data on damage costs and abatement costs. It includes the median damage costs from Table 3-1 and an additional damage cost estimate from a study of European markets (Banzhaf et al., 2002). Although the damage costs differ considerably across studies, all are within the range that Matthews and Lave (2000) identified from previous studies. Of interest is the disparity between allowance prices and damage costs. The allowance price for both NO_x and CO_2 seems, respectively, to be near or above the maximum

TABLE 3-1 Estimates of the Social Damage Costs of Air Emissions

Species	No. of Studies	Cost of Environmental Externality (1992 $/t of air emissions)			
		Minimum	Median	Mean	Maximum
Carbon monoxide (CO)	2	1	520	520	1,050
Nitrogen oxides (NO_x)	9	220	1,060	2,800	9,500
Sulfur dioxide (SO_2)	10	770	1,800	2,000	4,700
Particulate matter (PM_{10})	12	950	2,800	4,300	16,200
Volatile organic compounds (VOCs)	5	160	1,400	1,600	4,400
Global warming potential (in CO_2 equivalents)	4	2	14	13	23

NOTE: CO_2 equivalent is the amount of CO_2 that would cause the same amount of radiative forcing as a given amount of another greenhouse gas. When used with concentrations, it refers to the instantaneous radiative forcing caused by the greenhouse gas or the equivalent amount of CO_2. When used with emissions, it refers to the time-integrated radiative forcing over a specified time horizon caused by the change in concentration produced by the emissions.
SOURCE: Cifuentes and Lave (1993), CEC (1993), Desvousges et al. (1994), Fankhauser (1994), Koomey (1990), Ottinger (1992), OTA (1994), and Zuckerman et al. (1995).

TABLE 3-2 Estimates of Pollution Abatement Costs

Species	Median Social Damage Cost, Taken from Table 3-1	Price of an Allowance in 2006	Estimated Abatement Cost from Banzhaf et al.
Carbon monoxide (CO)	520		
Nitrogen oxides (NO_x)	1,060	7,700	1,160
Sulfur dioxide (SO_2)	1,800	850	3,850
Particulate matter (PM_{10})	2,800		2,060
Volatile organic compounds (VOCs)	1,400		
Global warming potential (in CO_2 equivalents)	14	30	8

social cost estimates. In contrast, the allowance price for SO_2 is closer to the minimum abatement cost.

Finding: Although there are a host of land, water, and perhaps public health impacts to consider, the benefits, in monetary units, of reducing criteria air pollutants is both the primary environmental benefit considered in the present study and the type of benefit for which valuation methods are most advanced.

Recommendation 1: Panels should apply valuations in monetary units to criteria air pollutant emissions in the results matrix, but not to other types of pollutant emissions. The valuations used should be the allowance price forecasts for the future period.

Valuing Energy Security Benefits

Security Benefits Related to Electricity Supply

Security considerations affect the electricity supply in two ways. One security concern relates to the type of fuel used to produce the electricity. The United States relied on imported petroleum for much of its electricity generation in 1973 at the time of the boycott by the Organization of Petroleum Exporting Countries (OPEC), but less than 2 percent of electricity is generated from petroleum at present. Now, almost 20 percent of electricity is generated with natural gas. Increasing the importation of natural gas in the form of liquefied natural gas (LNG) from nations in the Middle East can lead to a concomitant increase in the risk of terrorists or nations trying to influence our policies. The next section includes a discussion of the valuation issues related to oil and gas dependence.

The second security concern is that the electricity generation system at present provides a tempting target for terrorists. The resulting blackout could affect tens of millions of people and lead to billions of dollars of costs. For example, on August 14, 2003, 50 million people, from Cleveland to Toronto to New York City, lost their electrical service (U.S.-Canada Power Systems Outage Task Force, 2004). The immediate effect was gridlock in the cities, since the traffic signals stopped working, and danger and inconvenience as elevators and trains (including subways) stopped operating. The disruptions are estimated to have cost $4 billion to $10 billion (U.S.-Canada Power Systems Outage Task Force, 2004) and would have been much more expensive except that many of the customers who would have suffered the highest cost from the power failure had backup generators.[8]

One way to reduce blackouts is by having distributed generation, particularly where the generators are small and located at the customer's site. Distributed generators can be internal combustion engines (diesel- or natural-gas-fueled), microturbines run on natural gas, or renewable energy systems using biomass, wind, or solar energy. Zerriffi et al. (2002) estimate that having distributed generation (DG) in the system could improve reliability 10-fold, at little additional cost. In some applications, distributed generators can also have economic advantages. They can provide power for peak demand, saving considerable costs. Alternatively, they can be used for combined heat and power (CHP). If rejected heat from the power cycle is utilized as process heat, the energy utilization factor of the CHP system could be close to 100 percent, compared with 30 to 50 percent for a central station generator. In some cases, however, these generators would replace large central station generators powered principally by coal or nuclear fuel. If so, increased reliability and one type of security would be gained by use of imported fuels for DG machines, but another type of security would be decreased.

Thus, there is a complicated relationship between security and the electricity supply system. Large blackouts are relatively frequent and costly.[9] The system is an easy target for terrorists. If terrorists could mount a compound attack, they could paralyze a city by taking out the electricity supply and then attack high-value targets knowing that police, fire, medical services, and security services would find it hard to respond. Adding DG to the current electricity system could increase reliability and security, removing the electricity system as a terrorist target, and could also increase the efficiency with which fuels are used. However, an increase in oil- or natural-gas-powered DG would require more imports of oil and natural gas, increasing the other kind of energy security threat.

Finding: While the complex relationship between the electricity supply and security is becoming clearer, analysts are a long way from having methods for valuing reductions in security threats contributed by technologies such as distributed generation.

Recommendation 2: Panels conducting prospective benefits assessments should describe reductions in threats to energy security related to electricity supply as physical quantities of oil and gas.

Security Benefits Related to Oil and Gas Consumption

In addition to environmental externalities, increases in

---

[8]Rose et al. (2005) investigated the extent to which the initial losses can be made up later, such as by rescheduling production. They estimate that close to 80 percent of the initial losses could be recovered by rescheduling production, shopping another time, or other substitute actions. Thus, the true economic cost of a disruption is likely to be different from the cost estimated without considering these mitigating factors.

[9]Analysis of North American Electric Reliability (NERC) Distribution Analysis Working Group (DAWG) database indicates an average of three blackouts per year that were greater than 1,000 MW and one or two blackouts per year that affected 1 million or more people. See Hines et al. (2006) for an analysis of large blackouts in the United States.

U.S. oil and gas consumption and imports may impose incremental costs on the United States that are not reflected fully in the market price. Therefore, individual households and businesses may choose to consume more oil or gas than is optimal from the U.S. perspective, and reductions in consumption could increase national welfare. Such incremental costs might include (1) increases in the price of imported oil; (2) macroeconomic disruption externalities associated with price volatility; (3) adverse consequences for U.S. foreign relations; and (4) increased military expenditures to secure energy supplies. These costs have been estimated for oil consumption and are reported below. In principle, similar estimates can be made for natural gas, but the committee is unaware of such research having been done.

Collectively, these cost components are sometimes called the "oil premium." A comprehensive review by Leiby et al. (1997) suggested a preferred range for the oil premium at $0-$5/bbl; however, under broader assumptions their range extends to $10/bbl. Another recent NRC committee assumed an oil premium of $5/bbl in its examples, with a range of $1-$10/bbl (NRC, 2002a). This somewhat broader range mainly reflects the updating of Leiby et al. (1997) for higher baseline oil prices. Parry and Darmstadter (2003) put their best assessment of the oil premium at about $5 per barrel and cite a range in the literature of $0-$14/bbl. However, those studies focused on the first two cost components listed above. And, they were completed when there was significantly more excess capacity and lower prices in the oil markets than is the case now. But these two components of the oil premium can be expected to depend heavily on world oil market conditions and to vary sharply over time as these conditions change.

Each of the factors has been discussed qualitatively by the recent report of an independent task force sponsored by the Council on Foreign Relations, *National Security Consequences of United States Oil Dependency* (2006). The following paragraphs discuss each of these components of the oil premium in turn.[10]

Impacts of U.S. Demand on World Petroleum Prices. Reductions in U.S. petroleum consumption may have the effect of lowering world prices, with U.S. consumers paying less to oil producers, and increases in U.S. consumption may increase world prices. From the point of view of U.S. economic welfare, any reduction in payments to foreign oil producers can be considered a terms-of-trade benefit.[11] Previous studies have valued this benefit at $1-$5/bbl (Leiby et al., 1997;

Parry and Darmstadter, 2003), with the range depending primarily on assumptions about the import supply elasticity and the price of oil. Such estimates depend crucially on the elasticity of supply and demand in the world oil market. When there is very little extra capacity, as can be expected over the foreseeable future, shifts in the worldwide supply and demand for oil can have significant impacts on the world oil price. Because world oil markets clear, supply changes and demand changes must be equal, with prices adjusting to preserve the equality.

Assume that in the short run, the elasticity of demand[12] for oil on the world market is between −0.1 and −0.2. Thus, when supply cannot adjust—that is, the elasticity of supply is zero—a 1 percent increase in world oil consumption (0.86 million barrels per day (mmbpd)) can be expected to increase world oil price by between 5 percent and 10 percent ($3.50-$7.00/bbl, at current prices.) A decrease in world oil consumption can be expected to decrease world oil price correspondingly. Assume that the United States is importing 11 mmbpd, or 4.015 billion barrels per year, and paying $70/bbl, the annual cost would be $281 billion per year. If the United States increased its imports to 11.86 mmbpd (annual imports of 4.329 billion barrels) and the oil price increased to $73.50/bbl, the annual cost would increase to $318 billion, up by $37 billion. The total annual cost to the United States would increase by $118 per additional barrel of oil imported during the year, an amount far higher than the market price of $70. The difference between the additional cost to the United States and the market price—$48/bbl—is a term-of-trade cost to the United States and thus, from its perspective, a financial externality. If the oil price were to increase by $7/bbl, the annual cost would increase to $333 billion, up by $52 billion. The total annual cost to the United States would increase by $167 per additional barrel of oil imported during the year; the financial externality would be $97/bbl.

When there is much unused oil production capacity and the OPEC nations are operating so as to keep oil price at a target level, increases in oil demand will have virtually no effect on world oil price; equivalently, the elasticity of supply of oil is very high. In that case this financial externality is near zero. To illustrate, assume that in a time of significant unused oil production capacity, the elasticity of oil supply is 10—that is, an increase in oil consumption of 1 percent (0.86 mmbpd) increases oil price by 0.1 percent. Then such an increase in oil consumption would increase price to $70.07/bbl and import costs would increase to $303 billion, a total increase of $22 billion. The total annual cost to the United States would increase by $71 per additional barrel of

---

[10]Greene and Ahmad (2005) treat the economic costs "as arising from the use of monopoly power in the world oil market," and distinguish their approach from those treating the oil premium as an externality, such as the approach of Parry and Darmstadter (2003).

[11]Terms-of-trade issues exist in markets other than the oil market. However, for most nonagricultural goods, nations have agreed through the World Trade Organization (WTO) and previous trade agreements to operate competitively. But in oil markets, the OPEC has exercised its ability to keep prices higher than competitive levels and such agreements do not operate.

[12]The elasticity of demand measures the ratio of percentage change in demand to percentage change in price, all else equal. Thus for a 10 percent increase in price, an elasticity of demand of −0.1 implies that demand would decrease by 1 percent and an elasticity of demand of −0.2 implies that demand would decrease by 2 percent.

oil imported during the year; the financial externality would be only $1/bbl.

Thus, for those times during which the elasticity of supply of oil is zero, the financial externality could be between $48/bbl and $97/bbl, while for those times during which the elasticity of supply of oil is as high as 10, the financial externality would be as low as $1/bbl.

Oil Supply Disruptions.[13] Wages and prices for many goods are "sticky" and do not adjust seamlessly to changes in input prices or macroeconomic variables. In particular, spikes in oil prices increase input costs and reduce the marginal productivity of labor. In a labor market with sticky wages, such price increases could lead to increased unemployment and reductions in gross domestic product (GDP). Oil price volatility also adversely affects owners of vehicles and other fixed capital that operates on oil, who are limited in their ability to adjust oil consumption when prices rise or fall. Volatility imposes an extra cost on these consumers. These costs are reduced if the economy is less dependent on oil. Estimates of the value of these disruption externalities range from $0 to $8 per barrel (Leiby et al., 1997; Parry and Darmstadter, 2003). As a point estimate, the National Highway Traffic Safety Administration (NHTSA) (2006) uses $2/bbl. See Jones and Leiby (1996) and Jones et al. (2004) for reviews of the impact of oil price volatility on the macroeconomy.

These estimates can be expected to be considerably higher when there is very little unused capacity in world oil markets. When there is a large amount of unused capacity, disruptions in oil supply from one part of the world can be met be compensating increases in supply from other regions—that is, the short-run elasticity of supply can be expected to be very high. In those circumstances disruptions in supply from one region lead to relatively small increases in world oil price and few macroeconomic dislocations. However, when there is very little unused capacity, disruptions in oil supply cannot be met by compensatory increases in supply from other regions; the short-run elasticity of supply can be expected to be very small. In those circumstances disruptions in supply lead to large increases in world oil price and very costly macroeconomic dislocations.

Increased Oil Imports and U.S. Foreign Relations. In the current tight world oil market situation, increases in oil consumption and thus of oil imports limit foreign policy options (for example, in our relationships with Saudi Arabia) and may be putting the United States in a position of competing with China and other oil-importing countries. Additional imports increase the revenues flowing to oil-exporting nations whose political agendas conflict with those of the United States (for example, Venezuela) and give those nations increased ability to use their oil revenues to press their strategic advantage over the United States. Additional U.S. imports tighten the world oil market; a tight world oil market discourages other importing nations from taking strong stands against the actions of major oil exporters (for example, Iran's nuclear program).

Although this factor has been discussed qualitatively in the Council on Foreign Relations' independent task force report *National Security Consequences of United States Oil Dependency* (2006), the report did not provide quantitative estimates of these externalities. The committee is not aware of any study that attempts to quantify the impact of increased oil imports on U.S. foreign relations.

Military Expenditures Related to Securing Energy Supplies. These costs of U.S. oil consumption are not reflected in the price of oil. However, it is not clear that these costs would vary significantly with changes in U.S. oil consumption. For example, military activities in world regions that are vital suppliers of oil undoubtedly serve a wide range of security and foreign policy objectives so that most analysts estimating the benefits of marginal changes in oil consumption and imports have not included these costs in their estimates, either because they believe that the military expenditures would not change significantly with changes in oil imports or because they have no method of rationally quantifying the externality.

Summary. Reasonable estimates of some of the externalities related to oil consumption exist and could be used to assign an economic value to reducing oil consumption. However, the estimates will be highly dependent on time and on expected future oil market conditions. Moreover, the committee is at this time not aware of any good quantification of some of the externalities, so they must be discussed qualitatively unless and until reasonable quantifications are developed.

Finding: Increases in U.S. oil and gas consumption and imports may impose incremental costs that are not fully reflected in the market price. Some of the cost components of this oil premium have been estimated in various studies. In principle, similar estimates could be made for natural gas, but the committee is unaware of such research having been done.

Recommendation 3: Panels should describe energy security benefits related to reduced oil and natural gas consumption quantitatively in the benefits matrix as physical quantities of oil and gas. The time pattern of the oil consumption impacts

---

[13]Macroeconomic externalities cannot be measured by simply looking at the change in prices, costs, or quantities of energy. The committee includes them in a separate category from the economic benefits in part to reflect the difficulty of measurement and the substantial uncertainty that surrounds their occurence, severity, and assessment. Of course the macroeconomic externalities—like the other security and environmental benefits—have real implications for the economic health of the country.

should be made explicit, along with an assessment of the probable state of the oil market during those future times.

METHODOLOGICAL CONSIDERATIONS IN MEASURING BENEFITS

The Phase One prospective benefits study discussed and demonstrated principles for calculating the economic benefits of DOE programs and proposed a decision tree methodology for benefits estimation based on these principles. (At the start of their work, the six panels all received the methodology as it stood at the conclusion of Phase One. See Appendix F.) As noted at the outset of this chapter, one objective of this Phase Two study was to refine the methodology and assess its utility.

As detailed in Appendix F, the general structure of prospective benefit assessment methodology involves (1) characterizing relevant states of the world according to prices, regulations, and other constraints that influence the benefits of the technology (i.e., the scenarios); (2) considering plausible technological outcomes of the program and assigning probabilities to each outcome; and (3) evaluating the extent to which the technology is deployed under a given scenario as well as probable competition from alternative technologies that may have also been developed over the project period. Each plausible technological outcome, together with market and policy conditions, yields economic savings and environmental and security benefits. Estimating the net benefit of DOE's programs requires two additional steps. The first is to work out the difference due to the DOE program—that is, to compare the outcome with the DOE program with the outcome had there been no such program The second step is to estimate the likelihood of each potential technological and market outcome and then weight the benefit of each outcome by this likelihood. This process is summarized in the decision tree analysis associated with each of the panel studies.

In the following sections the committee draws conclusions about the methodology based on the experience of applying it to six DOE applied research programs. Where appropriate, it recommends refinements to the methodology and offers guidance for its implementation in the future. The committee's conclusions and recommendations are based on its review of the experiences of the six panels, which are summarized in the panel reports. The conclusions and recommendations of the committee fall in five areas:

- *Decision tree analysis.* The committee endorses the decision tree framework for use in estimating the benefits of DOE's applied research programs. However, panels must take care, and they will require guidance from a decision analyst (the consultant) to understand how to assign probabilities and isolate the government impact.
- *Global scenarios.* The scenarios developed by the committee for all panels proved to be a valuable tool for characterizing and quantifying the benefits of the DOE R&D programs. However, the Phase Two experience shows that panels sometimes needed to have the scenarios clarified for them to be able to address issues that were important to the specific R&D program.
- *The National Energy Modeling System (NEMS).* This model is important for providing baseline energy prices and demands, but using it to estimate the prices of and demands of all different program outcomes is unlikely to yield refinements to the estimated benefits.
- *Modeling alternative technologies.* The success or failure of competing or complementary technologies can significantly affect the value of a DOE applied research project. Although Phase Two clarified this issue and provided some methodological guidance for dealing with it, more work is required to describe a method for estimating the benefits in DOE's overall portfolio.
- *Implementation issues.* The panels were generally successful in implementing the committee's methodology, but the Phase Two experience turned up three process issues that should be considered in future studies.

Decision Tree Analysis

The primary aims of DOE's programs are these: (1) to reduce technical risk, (2) to reduce market risk, and (3) to accelerate the introduction of the technology into the marketplace. To understand how a government program can yield net benefits and to estimate benefits under different scenarios, the committee involved in Phase One of this project developed a decision tree methodology that integrates the assessments of experts and yields estimates of market risks, technical risks, and project outcomes under the three global scenarios.

Constructing a decision tree that captures a minimal yet representative set of outcomes for a DOE research program is the most important part of the assessment exercise, as well as a very challenging task. It requires knowledge of markets, the technology status, and program alternatives. The benefits estimated using the recommended methodology thus depend crucially on a panel's ability to (1) construct a decision tree that captures the key technical and market risks of an applied research program, (2) make reasonable estimates of those risks, (3) assess the timing of technology development,[14] and (4) assess the differential success of the government research program relative to that of non-DOE efforts in achieving the stated goals of the program. The panels all succeeded in developing an appropriate decision tree for their DOE

---

[14]The 5-year rule had been recommended by the NRC Committee on Benefits of DOE R&D on Energy Efficiency and Fossil Energy R&D (see NRC, 2001) for use in the absence of better information. Because the experience with the panels convinced the committee that the 5-year rule is often inadequate for prospective evaluation, it recommends a more elaborate methodology in this report.

programs. For this reason, the committee continues to recommend the use of the decision tree framework. However, the construction of the tree is a difficult process, so the committee also continues to recommend, as detailed in Chapter 4, the use of a trained decision analysis consultant to facilitate this process.

The panels were also able to specify the probabilities required to complete the tree, but again the task was challenging. The committee is particularly concerned about the experts' ability to estimate very low probabilities. For example, experts may not be able to distinguish between .1 and .01 when assigning probabilities to technical and market success, even though a tenfold difference can greatly influence the calculated values. Alternatively, it is possible that a very small probability—say .1 or .05—overestimate the still lower probability an expert believes in but finds difficult to justify or defend. These issues also point to the need for a skilled facilitator who can help panelists appreciate the impact of these differences and/or perhaps decompose the overall assessment into a series of smaller and easier assessments. For example, rather than directly assessing a very small probability of success, the assessment can perhaps be decomposed into assessments of success for several technical hurdles required for success; none of these hurdles may on its own have a very low probability, yet the overall probability of success (given by the product of these probabilities) may be quite low. In cases with very low probabilities of very large benefits, it is important for the discussion of the risks in the matrix to make this clear.

Another challenge that arose in some panels with respect to decision trees was isolating the effect of government-funded research, particularly when the government program is small in relation to private sector activity aimed at the same goal. The decision tree methodology proposed in Phase One and Phase Two by the committee calls for estimating this benefit as the difference between benefit of the R&D with government support and the benefit without public support. In this case, both benefits estimates are very large and uncertain numbers, and the difference is relatively small and difficult to estimate with much confidence. Consideration should therefore be given to modifying the methodology or to using alternative approaches. In addition, the committee encourages panels to describe explicitly the role of the government effort and its expected benefits when there is already a relatively large amount of non-DOE research.

Global Scenarios and Their Implementation

As did its predecessor, the Phase Two committee defined three global scenarios for each panel to use: a Reference Case scenario, a High Oil and Gas Prices scenario, and a Carbon Constrained scenario:

• *Reference Case.* The Reference Case is consistent with the Energy Information Administration's (EIA's) reference case in its *Annual Energy Outlook 2005* (EIA, 2005b). World oil prices increase from $24.10/bbl in 2003 to about $30/bbl in 2025. Natural gas consumption increases significantly—i.e., from 22 trillion cubic feet (Tcf) in 2003 to 31 Tcf in 2025—with wellhead prices decreasing from $4.98 in 2003 to $3.64 per thousand cubic feet (Mcf) in 2010, then increasing to $4.79 per Mcf in 2025.[15] There is assumed to be an increase in primary energy consumption from 98.22 quadrillion British thermal units (quads) in 2003 to 133.8 quads in 2025. GDP is expected to grow 3.1 percent per year through 2025. U.S. carbon dioxide emissions from energy consumption are assumed to grow, from 5,788 million metric tons in 2003 to 8,062 million metric tons in 2025.

• *High Oil and Gas Prices.* The High Oil and Gas Prices scenario assumes that oil prices will remain very high throughout the period and that constraints on natural gas supply lead to higher natural gas prices and higher electricity prices. The committee doubled the prices forecast in the EIA (2005b) High Price A scenario to arrive at its own set of prices it felt to be more likely given the recent upsurge in prices. For example, the oil price in 2010 in this scenario is $67.98 (including the committee's doubling) versus $25 in the Reference Case, and the natural gas price in 2010 is $7.34 per Mcf versus $3.64 per Mcf in the Reference Case. In the period after 2025, the real price remains constant.

• *Carbon Constrained.* The Carbon Constrained scenario is consistent with that developed by DOE and assumes that U.S. emissions of carbon are constrained in response to environmental concerns. Specifically, this scenario assumes that the Global Climate Change Initiative goal of an 18 percent reduction in national greenhouse gas intensity (below the 2002 level) is achieved by 2012 (White House, 2002). This effort is implemented as a tax of $100 per ton of carbon (/t C) emissions beginning in 2012, increasing at 3 percent in real terms thereafter, and assumes that annual emissions are held constant at that level thereafter.

Each panel was encouraged to define additional scenarios if it thought that such scenarios would more accurately portray the program's benefits. Only one panel needed an entirely new scenario. The Panel on DOE's Distributed Energy Resources Program found that DG technologies, and CHP in particular, are of much greater value to host facilities located in large urban areas of the country. As defined by the panel, the scenario assumes that sufficient power is not available for some extended time (weeks or months) in one or more high-demand load pockets in the affected regions of the country. The critical power shortage can be ameliorated by taking several actions, including reducing voltage, imposing selected rolling blackouts; terminating supply to selected high demand customers with interruptible electricity service contracts; increasing real time electricity prices; implementing special energy efficiency programs; and installing CHP

---

[15] All prices are stated in 2003 dollars and thus do not reflect inflation.

systems in load-constrained areas. This energy-constrained scenario recognizes the value of bringing new electricity to an urban area and increases the expected net benefits of CHP by over 40 percent.

Two other panels found that one or more of the global scenarios required some clarification to address issues encountered by the DOE program in question:

• The Panel on DOE's Carbon Sequestration Program found that a good deal of private sector R&D was taking place and thought that even more R&D would take place if it were widely known that carbon taxes or constraints were imminent. To address the timing of new carbon taxes or constraints, the panel assumed that the carbon-constrained scenario would include an announcement of the carbon tax 5 years before the tax is levied. The announcement would expedite private sector R&D and reduce the impact of DOE funding. This assumption about the timing of the announcement of constraints had a significant effect on the estimated benefit.

• The Panel on DOE's Integrated Gasification Combined Cycle (IGCC) Technology Program found the scenario approach to be valuable but adjusted the scenarios to account for the dependence of outcomes on the success of other R&D programs and the dissemination of competing technologies. Thus, the panel estimated one benefit level if IGCC displaces pulverized coal and a higher level if it displaces natural gas combined cycle (NGCC) power plants. This difference is particularly noticeable for the High Oil and Gas Prices scenario.

• Finally, the Panel on DOE's Light-Duty Hybrid Vehicle Technology Program found the scenarios useful but found it difficult to translate DOE's program goals into expected benefits in the three global scenarios because different elements of the program reacted differently to different scenarios. For example, with the R&D on improved battery performance, the Reference Case led to moderate benefits for the program; the High Oil and Gas Prices led to higher probabilities of achieving the goal and higher benefits; and the Carbon Constrained scenario led to an intermediate result. The benefits and probabilities for the vehicle light-weighting R&D varied by scenario in a different way, and the advanced combustion engine R&D was not sensitive to the choice of scenario.

The National Energy Modeling System

NEMS is the basic tool used by DOE to estimate energy prices and consumption. The committee's recommended methodology proposes to use NEMS to develop prices and quantities for the global scenarios, but also suggested that in most cases a simpler spreadsheet analysis would be sufficient to account for the changes in outcomes caused by introducing a new technology in a given scenario. This approach was taken by the panels in the Phase Two study.

NEMS is a critical tool for deriving price and quantity projections of programs that could have the potential for economy-wide consequences. Its value lies in identifying and characterizing equilibrium and feedback effects. For example, carbon constraints will dramatically increase the cost of electricity generated from coal in the absence of sequestration technology. This will have a number of economic consequences, each of which is relevant to evaluating the economic benefits of the DOE sequestration program. Absent the technology, demand will rise for natural gas (as will its price) as an alternative to coal for generating electricity. The cost increase is large enough to affect both demand and supply: Given the increased costs, electricity demand itself will diminish without sequestration technology. Estimates depend on assumptions and projections about economic growth, industrial use, base load versus peak load deployment opportunities, and other technological developments and cannot be easily calculated without NEMS.

NEMS also yields costs of alternative technologies, thereby providing additional guidelines for the choice of scenarios. Gasification, like sequestration, has the potential to bring about major shifts in the electricity generation mix, electricity prices, and fuel input prices. The cost projections for alternative technologies indicate the range of IGCC cost and performance characteristics over which IGCC will satisfy a large share of baseload power needs, only a moderate share, or no share at all (if it is unable to compete with other technologies).

Although NEMS is invaluable for these purposes, a key consideration in using it is whether small changes in parameters—for example, a 10 percent increase or decrease in the price of the technology—are likely to yield estimates significantly different from, or, possibly, even more accurate than estimates yielded by modifying or interpolating from a few NEMS runs. The model itself, as well as cost estimates for the technologies and the assigned probabilities, is subject to uncertainty. Furthermore, considerable time and resources are needed to use the NEMS model.

The work of the panels in Phase Two suggests that the committee's recommended approach of using simple spreadsheet models calibrated to NEMS results is feasible and workable. In particular, each panel was able to estimate the change in benefits associated with reasonably small variations in cost by interpolating from several NEMS runs rather than requiring separate estimates for each of the alternatives considered in the analysis. In some panels, NEMS was not needed at all. For example, the chemical plant energy efficiency projects considered by the Panel on DOE's Chemical Industrial Technologies Program involved niche markets, so that market prices for energy inputs and overall demand for other goods are unlikely to be significantly affected by the outcome of the program.

Although the spreadsheet approach to calculating benefits was successfully applied in each case, some committee members felt that the panels did not devote enough time and attention to the benefits calculation model. The benefit modeling process, like decision tree modeling and probability assessment, is challenging. It requires considerable expertise and judgment and should not be considered a cleanup task to be taken care of by the consultant and one or two panelists. Rather, the approach and logic underlying these calculations should be presented to all the panel members so they can understand and debate its key assumptions.

Modeling Alternative Technologies

The success or failure of alternative technologies can significantly affect the value of a DOE applied research program. These other technologies may be competing programs, each of which is directed to the same goal, or complementary programs that must also be successful if the program being analyzed is to produce a benefit. Technologies that have a major impact on project benefits should be included in the analysis, but this is methodologically challenging. The Phase Two experience helped to define the problem but did not fully resolve it.

When dealing with competing technologies, the benefits analysis must consider the entire set of competing projects, so benefits will not be double counted. In these cases, investing in competing technologies increases the probability of success but not the benefits associated with success.

A similar situation exists, but with different net program benefits, when two distinct competing technologies are invested in by different parties (e.g., IGCC at DOE and pulverized coal technology by private companies). While the likelihood of a technology's success and its associated benefit increases in this situations, the net benefit of the government program decreases with the likelihood that the private technology will succeed. When the alternative technology programs are conducted outside DOE, the committee's methodology can estimate the benefits satisfactorily by including outcomes for these competing technologies in the decision tree. For example, the success of pulverized coal technology with carbon capture affects the likely market penetration of IGCC technology. That relationship can be reflected in the decision tree directly by assigning probabilities to the relevant outcomes.

Benefits can be more difficult to assess with complementary programs, although the principles remain the same. For example, the main components of the DOE R&D electricity programs in fossil energy and energy efficiency are complementary, most obviously, coal gasification and sequestration. Perhaps less obvious are the complementarities of electricity programs generally with the fuel cell and DG programs. These last two technologies use natural gas and reduce the share of peak power plants in the generation mix. Overall, according to the NEMS results, the program increases both demand for IGCC and its attractiveness relative to NGCC.

While more work is required to define the portfolio analysis in the presence of complementarities, the committee notes that it will be important to consider the connections among DOE programs when grouping programs for a benefits assessment. For example, a review of the full DER effort may be more desirable than a review of the CHP element alone. Aggregating along these lines can reduce the complications of complementarities by subsuming them in a more general decision tree analysis.

The Phase Two studies provide relatively little experience in modeling dependence between DOE projects and technologies. The Panel on DOE's Chemical Industrial Technologies Program examined a portfolio of small research projects but treated them as mutually independent; that is, all could succeed or fail without significantly affecting the benefits of any of the others. The panel showed that in this limited case, it is possible to assign overall success probabilities to each project and to create a cumulative probability distribution of expected benefits using a Monte Carlo simulation. However, further work would be required to analyze jointly a portfolio of projects where the success of one could affect the benefits of others.

In a full implementation of this program assessment process, the committee recommends that in selecting activities for review the interdependencies of program elements be considered. For example, the IGCC, gas turbine, carbon sequestration, and stationary fuel cell activities are all programs within the Office of Fossil Energy that address a carbon-constrained scenario. In addition to these activities is the FutureGen coal demonstration project. If the review of these components is carefully coordinated, decision makers would gain a more integrated view of the benefits of the overall program.

Implementation Issues

The panels were generally successful in implementing the committee's methodology, but the Phase Two experience uncovered three issues that should be considered in future studies:

1. The natural gas and hybrid vehicles panels were not able to calculate benefits for certain scenarios. To a large extent this was due to the inadequate support provided by DOE program staff. To a lesser extent, the strong technical competency of the panel members led them to focus more on technical risks than market risks. The committee recommends that the next assessment engage an additional consultant with benefits modeling expertise to ensure that the benefits are calculated in a consistent manner. The committee determined that it would be helpful as well to have these calculations performed between the first and second panel meetings and made available to the panelists, on a preliminary basis, at the start of the second panel meeting.

2. If the DOE programs account for only a small share of the overall international and/or industrial effort in the area, it is very important for the panel to be knowledgeable about that overall effort so that it is able to reliably assess

the potential impact of the DOE program. Both the Natural Gas Exploration and Production Program and the Chemical Industrial Technologies Program are in this category, and it is these panels that had difficulty assessing the benefits of the two DOE programs.

3. Applying the prospective benefits methodology to DOE's light-duty vehicle hybrid technology R&D program required the panel to specify key items that were not always apparent from the documents and information provided by DOE. In particular, some of the program goals were not described explicitly and completely. For example, setting a cost target of $28/kW for a battery by the year 2010 does not describe the objective adequately for assessment purposes. Does the cost target mean a customer could actually buy a battery at that cost? Does it mean that the technology exists that in principle would allow a commercial firm to make such a product? Does it mean the 500,000th production unit or the first? All these conditions must be specified for the assessment method to succeed, and both reviewers and proponents must state their goals quite explicitly.

4

Expert Panel Process

INTRODUCTION

The program assessment process centers on establishing an expert panel to review selected DOE research and development (R&D) programs or projects. The expert panel conducts a technical assessment based on a brief description by DOE of the program under review and its component projects. The panel assesses the conditional benefits of the program, assuming the program meets its stated goals. The panel also considers other perhaps more appropriate or more likely to be attained goals of the program. The panel estimates benefits using simplified models whenever possible, following the methodology described in Appendix F, "Guidance on Evaluation of Program Benefits," but it can also obtain benefit information from a general equilibrium model like the National Energy Modeling System (NEMS) to provide internally consistent information on price and quantity and their impacts. In this instance, benefits estimates are provided by DOE to the panel as illustrated in Appendix G, "Information to be Requested of the Department of Energy." The panel members' expertise and the decision trees assessment tool described in Appendix F are used to develop the probabilities for technical and market risk for the program as a whole. The results of the probability analysis are used to estimate the expected program benefits. The panel is supported by two consultants: one, perhaps in decision analysis, who has a working knowledge of the benefits assessment methodology being proposed by the committee and the other in modeling, to calculate the benefits of the program using inputs from DOE and the probabilities arrived at by the panel. The panel reports its results, including its decision trees and its comments on the program risks, in the format defined in Appendix F, Figure F-4. In addition, the panels must consider a number of issues and adhere to the quality assurance guidelines described in the last two sections of the present chapter. The committee supports the panels by overseeing and reviewing their work to ensure consistency in the way the panels apply the methodology.

This chapter describes a specific process for the work of the expert panels. It also contains suggestions for establishing an overall quality assurance function.

EXPERT PANEL COMPOSITION

Six to eight experts are empanelled to evaluate a given DOE program and to apply the process, including the risk and benefits assessment methodologies. The size of the panel is dependent on the breadth of the program to be reviewed. Each panel will require a balance of skills and a wide range of expertise and experience to ensure that all relevant issues are identified, fully discussed, and factored into the assessment. The membership of the panel might include the following:

- A manufacturer knowledgeable about both the conventional technology and the new technology being advanced by the program, market issues associated with its commercial adoption, and research, development, and commercial application of the technology domestically and internationally;
- An end user of the technology, possibly a builder or utility representative who can provide a user's perspective;
- A chief technology officer or an R&D manager (or equivalent) from industry familiar with how to take a technology from the laboratory to the market;
- A public policy analyst or decision maker with expertise in energy, environmental, and economic analysis;
- An expert in the technology being reviewed, who could come from industry, academia, or a national laboratory;
- An economist who can review and provide insights on the economic information provided by DOE and who will assure that the evaluation methodology is properly used by the panel; and
- Other related expertise as deemed necessary.

The panel is led by a chair with broad expertise and experience in analyzing energy and environmental issues and technologies. The panel chair should have considerable skills in managing and facilitating meetings and be familiar with benefit assessment of R&D programs. The panel chair should be recognized as having technical and/or assessment knowledge in the program area that the panel will be evaluating. It would also be helpful if the chair is familiar with how DOE conducts R&D activities and with DOE's operating procedures. The chair is identified before the panel is nominated.

To keep the size of the panel to six to eight people, some of the members might possess more than one kind of expertise. Members may come from institutions currently engaged in activities with DOE, but they should not be directly involved with DOE in the program being reviewed. The panels will be chosen by an entity independent of the programs being reviewed.

To assure independence, freedom from conflict of interest, and balanced composition, a process similar to that used by the National Research Council (NRC) to form committees is adopted, whether the review is being performed under the auspices of the NRC, of a DOE-appointed Federal Advisory Committee Act (FACA) committee, or of another organization. While the panel is clearly not being charged with performing a traditional program review or evaluation, it must understand the programmatic issues in order to independently establish probabilities and expected benefits using the committee's methodology.

Panel members are chosen based on their expertise in the specific technology being reviewed, in business development, or in related policy areas. They are not necessarily expert in or even familiar with the methods used by DOE to administer, implement, analyze, or evaluate programs, including the Energy Information Administration's (EIA's) NEMS, decision-tree analysis, or the benefits methodology proposed by this committee. Accordingly, the committee recommends that one or two consultants (depending on the number of programs that are being reviewed) provide support to the panels. The consultant(s) should have expertise in (1) decision analysis and methodology and (2) modeling and economics. They work with the panel members and, if it is an NRC review, with NRC staff. The one or two consultants each work with all the panels and assure consistency among the panels in their use of the methodology and calculation of benefits.

ROLE OF THE PANEL CHAIR, THE CONSULTANT(S), AND THE OVERSIGHT COMMITTEE

Chair

The panel chair schedules, organizes, and facilitates meetings of the panel and is responsible for report completion. In addition, the panel chair

- Recommends potential panel members.
- Meets with other panel chairs before the first meeting of the panel to coordinate and ensure consistency of activities.
- Meets with the consultant and DOE program staff to review the panel's need for data, as illustrated in the program assessment summary (PAS) forms (Appendix G).
- In preparation for the first meeting, draws up, with the consultant(s), an initial request to DOE for information, consistent with the discussion in the section "Interactions with DOE and Information Request" in this report which calls for the chair and consultant(s) to meet with the DOE program management prior to the first panel meeting.
- Discusses with the consultant(s) any questions about the use of the methodology as well as any emphasis (or deemphasis) to be applied to portions of the methodology to make it relevant to the needs of the program being reviewed.
- At the completion of each panel meeting, meets with the oversight committee chair to assess crosscutting and portfolio issues as well as lessons learned.

The panel chair needs to spend a fair amount of time outside the actual panel meetings working with DOE program managers, DOE management—perhaps at the level of assistant secretary—and the independent consultant(s). If the study is conducted by the NRC, the panel chair's primary point of contact is NRC staff.

At the first panel meeting, the chair ensures that all panelists are familiar with the procedures outlined in this chapter and that they know what is expected of them in the study. With support from the consultants, the chair opens the first meeting with a briefing to panelists on the following topics:

- History of the prospective benefits study being undertaken by the panel, brief review of past studies, and review of other panel studies under this phase of work.
- Description of the methodology to be used.
- Emphasis that this is not a review of a past program but is instead an assessment of the current program.
- Role of the consultant(s).

In addition, at the outset of the first panel meeting, the chair outlines the schedule for the study and the tasks that will be undertaken each day, the use of the study results, and panel study focus. The discussion of these topics by the chair, who will impart a firm understanding of the approach to be taken, focuses the panel on the task being undertaken.

Independent Consultant(s)

The primary responsibility of the consultant(s) is to maintain consistency across the panels in applying the committee's methodology. The consultant(s) might also suggest and implement modifications to the process to address the needs of any specific program being evaluated while maintaining consistency with the committee's approach across other panels. Responsibilities of the independent consultant(s) are summarized below, with additional description provided in the next section, "Panel Activities and Process."

• Work closely with the panel chair to plan activities and clarify roles, responsibilities, and expectations for the panels' work.
• Participate in initial meeting(s) with DOE program management, the panel chairs, and NRC staff to review the information needs of each expert panel.
• Review the committee's methodology and recommend modifications for the panel's consideration, as necessary.
• Attend all expert panel meetings. Work with panel members, individually and collectively, to structure and work through the necessary analyses. This will include structuring the decision tree or trees, facilitating the assignment of probabilities to technical and market risk outcomes, and guiding and assisting in the modeling of benefits. Review each report's output to ensure that its analyses and recommendations are consistent with internal panel discussions and modeling and with reports of the other panels, explaining any needed modifications.

Depending on the number of programs being evaluated and the panel's schedule, it may be necessary to use more than one consultant. In that case, it is recommended that one consultant should focus on the decision-tree development and probability assessments and the second should focus on the benefits modeling. Because the technical and market risk assessment and the benefits calculations are tightly linked, if there are two consultants, they would work closely together. To help ensure panel-to-panel consistency, the consultant(s) work with all panels assessing the programs.

Oversight Committee

An oversight committee, similar in responsibilities to the current Committee on Prospective Benefits of DOE's Energy Efficiency and Fossil Energy R&D Programs (Phase Two), will ensure consistency across the various programs being addressed by the expert panels. There are several options for the form of such an oversight committee: (1) a standing committee of the National Research Council; (2) a DOE-appointed FACA committee; (3) a committee of panel chairs; (4) an independent contractor; and (5) an internal DOE committee. The committee concluded that either an NRC committee or a DOE-appointed FACA committee would be most appropriate because, independently, both organizations have institutionalized mechanisms for preventing bias, and both have access to a broad, high-quality pool of potential committee members.

A key role of the oversight committee is to ensure that the panels are performing their assessments in a consistent manner that allows their results to inform decision making.

Several specific functions of the committee are as follows:

1. Review the composition of the panels to determine where additions to the panels and/or information presented to the panels could aid in the evaluation.
2. Evaluate the consultant(s) and ensure that they are knowledgeable about the process/methodology and can work with the panels to develop the decision trees and benefits calculations in a consistent and timely manner.
3. Meet with the panel chairs prior to the first panel meeting and instruct the chairs on the process and the role the consultant(s) will play in the evaluation.
4. Evaluate the panels' progress, working with information from the panel chairs and the consultant(s) between panel meetings. The oversight committee might at this point recommend modification to the panels to ensure consistent assessment. It will evaluate the information being provided to the panels by the DOE program under review. If the DOE information is not sufficient for the panels to perform their assessment, the committee will take action to ensure that DOE is being responsive.
5. Hold a debriefing meeting with the panel chairs to review the process used and to determine if there were inconsistencies in the evaluations.
6. Review the panel reports. If it is necessary to modify the panel assessment, the committee will provide comments to the panel chair and offer suggestions for addressing the suggested modifications.
7. Be responsible for briefing the funding agency and other stakeholders on the results of the benefits assessments.

PANEL ACTIVITIES AND PROCESS

The panel convenes at least two meetings lasting 2 full days each, with the possibility of a third meeting or conference calls, as necessary.

Premeeting Work

The panel chair and the panel consultant(s) can expect to expend a significant effort prior to the first panel meeting. Together they determine the specifics of the information to be requested from the DOE following the general guidelines in the section "DOE Interactions and Information Request." They then meet with DOE staff to review the request and identify any DOE concerns with the information request or the methodology and review the methods by which DOE calculates program benefits. The consultant reviews and develops proposed modifications to the methods appropriate to the program under review, including any modifications to the decision tree implementation and suggestions for how the economic and other benefits should be calculated.

Panel members receive a package of information from the staff supporting the panel at least 2 weeks before the first meeting. This package includes the statement of task and the process/methodology description. The panel members also receive the program and project summaries described in the section "DOE Interactions and Information Request" and a list of panel members and short biographies. Prior to the first panel meeting, panel chairs meet with the consultant(s) to coordinate activities and ensure consistency among panel approaches. The results of this meeting are communicated to the panels by the panel chair on the first day of the panel meeting. It would also be helpful to the work of the panel if a teleconference could be held among panel members before the first meeting to determine whether any additional expertise or members should be added to the panel. This teleconference should be held well in advance of the first meeting to give time for additional appointments to the panel. If this turns out not to be possible, consideration of such matters will be deferred to first meeting.

First Meeting, First Day

In closed session, the panel members are introduced to one another and they assess panel balance and ask for any additional expertise they require. The panel also discusses the process/methodology, schedule, role of the consultant(s), and deliverables. The panel members are advised about their roles and responsibilities for the effort. It is important for this initial discussion and introduction to the process/methodology to take place before the full panel meets with DOE. This focuses the panel on carrying out the specific form of probability analysis and benefits assessment exactly as the committee recommends. This activity is expected to take up the first morning. If time permits, the panel chair develops a list of panel member questions to be shared with DOE prior to DOE's program presentations.

In the afternoon open session, the panel hears presentations by the DOE program manager, who elaborates on information provided to the panel and answers questions from the panel. The panel also hears a presentation on the models, scenarios, assumptions, and other techniques DOE uses to calculate benefits. It emphasizes to DOE that it is not conducting a traditional program review but is assessing the prospective benefits of the program.

Throughout these presentations, in an end-of-day review and during the second day, the panel identifies additional information that it requires from DOE. The panel might choose to meet in closed session at the end of the day.

First Meeting, Second Day

The second day continues in closed session. The panel will have read the project descriptions before the meeting. It will have heard the DOE presentations and had a chance to raise questions. Members are ready to begin discussion of the program goals, timing, budget, and benefits estimates. They decide whether to carry out the review at the program level or the project level.[1] Beginning the review at the project level forces the panel to look at the details systematically and makes for a more informed panel. Eventually, however, the probability of a successful outcome has to be assessed at the program level, and it is unlikely that this assessment can be done by mathematically combining the probabilities assigned to the individual projects. Rather, it would be based on the panel's judgment using knowledge gleaned from the project assessments.

Once the panel finishes this work, the members familiarize themselves with the committee's results matrix (see Appendix F) and discuss what work they need to do to provide inputs to the matrix.

With the help of the consultant(s), a decision tree is constructed and a benefits modeling approach specified. The panel identifies the technical and market risks that need to be looked at in the decision tree and, if time permits, begins discussion of the probabilities of reaching the specified levels of technical and market success. If the panel believes the program goals cannot be completed in the time allotted, it uses a different time frame for assigning probabilities. Similarly, the panel identifies other parameters to consider in the decision tree analysis, such as dependencies with other programs, milestones to be met in order to meet program goals, and reasonableness of the program's budget to meet its goals.

By the end of the second day, the panel agrees to a decision-tree framework, including branches to account for both with and without DOE funding. The panel should also have a good understanding of the manner in which probabilities are to be assigned for each branch of the tree.

Between the First and Second Meetings

The consultant(s) prepares and distributes a questionnaire soliciting probability estimates from each panel member for each major variable in the decision tree. The questionnaire is sent to panel members and, when completed, returned to the consultant(s) before the second meeting. The consultant(s) compiles the data for the panel's review. The questionnaire can be at the project level or at the program level.

The panel also reviews the chair's draft outline for the panel report and receives writing assignments for initial report drafting, to be completed prior to the second meeting.

During this time the consultant(s) uses the panel members' first-cut probability assessments using the decision tree as the basis for an initial estimate of the program benefits. The con-

---

[1] In this chapter, the term "program" refers to the highest level of organization associated with the collection of R&D activities under review. Thus, the panel should consider such terms as technical objective and mission to be synonymous with "program objective." Likewise, the term "project" can refer to a single R&D activity such as a grant or contract or to a collection of activities addressing a single technical task, or to a quantitative technical target that must be met at the system level.

sultant, working with the panel chair and/or individual panel members, develops a prototype benefits model that can be used in conjunction with the decision tree and preliminary probabilities to quantify the estimated benefits of the program.

Second Meeting, First Day

The panel members' compiled responses to the questionnaire are discussed. The panel reviews the decision tree and the preliminary probabilities assigned by individual panelists, identifies and discusses any areas where significant difference of opinion on technical and market risks exists among panel members, and determines whether additional information is necessary to resolve those disagreements or refine individual inputs. Using the decision-tree process, each member of the panel reassesses the probabilities for each node and each relevant outcome in the decision tree. The panel decides to estimate probabilities individually or to develop consensus estimates. DOE is invited back to answer the questions generated at the first meeting and additional questions from the panelists. Using the decision-tree process, each member of the panel once again reassesses the probabilities for each node and each relevant outcome in the decision tree. The panel discusses more fully the next-best alternative (competing technology) to the DOE program, non-DOE technology funding, and other issues related to the decision tree, assignment of probabilities by panel members, and the consultants' estimation of benefits.

Second Meeting, Second Day

Several tasks will be completed on this day:

- Finalize the decision tree by agreeing to the probabilities for each branch of the tree.
- Complete a review of the benefits calculations conducted by the consultant(s) and provide guidance to the consultant(s) to enable them to finalize the calculations and the decision matrix.
- Review the report drafts of panel members and make writing assignments for the final report.

If possible, the panel develops the full matrix for the program, including the explanatory material that accompanies the matrix, consistent with the template provided at the end of Appendix F, "Expected Benefits Results and Report Guidance." This is the critical deliverable of the panel. The panel may also choose to provide expanded commentary on its approach to the evaluation and its use of the methodology. The complete panel report, including the completed version of the two-page template, should be as brief as possible. The panel also determines if it needs any more information from DOE and if a third meeting is necessary. The panel has the option of holding discussions with DOE to clarify any aspects of the program in order to complete these tasks.

Third Meeting or Teleconference Sessions

The panel probably needs follow-up discussions to review new DOE information and to review its draft report. After the report has been completed and reviewed and is ready for publication, the panel chair prepares a summary briefing for DOE management, the Office of Management and Budget, and Congress. These briefings are coordinated and managed by the oversight committee.

Meeting to Ensure Consistency Among Panels

Following the panel meetings, and before finalizing their reports, the panel chairs and consultant(s) meet again with the chair of the oversight committee to discuss panel activities, analyses conducted, issues raised and addressed or left outstanding, and critical assumptions upon which the analysis rests. The group then assesses the consistency of panel approaches and use of the process/methodology, and determines whether any panel report should be modified to ensure consistency with the process/methodology. Any changes to the panel reports are the responsibility of the panel chair. Issues that cannot be resolved by the panels will be brought to the oversight committee for resolution. (See discussion in other sections of Chapter 4 for more details.)

INTERACTIONS WITH DOE AND INFORMATION REQUEST

Information required for the panel's deliberation is provided by DOE at least two weeks prior to the panel's first meeting to give panel members sufficient time to get acquainted with the materials. To help ensure that the information provided is most relevant to the panel and least burdensome for DOE, the panel chair and NRC staff meet with the DOE program manager to discuss the panel's needs and the form in which the information is to be provided. The chair also discusses with DOE the expectations for DOE's presentation of the information to the panel, including presentation template, program information, and data content, results of modeling analysis, and program and project highlights. The request to DOE includes all information and data that the panel believes it needs to complete its task. To the extent possible, but with exceptions as defined by the chair owing to the unique nature of the DOE program, information and data are standardized across programs and projects.

DOE provides the panel with the program's goals, budget, and schedule for achieving its goals as well as program plans and roadmaps. It also provides the panel with individual project goals if such projects are a significant part of the larger program and if the goals represent milestones that need to be met if the larger program is to succeed. The conditional relationship is such that the goals of the larger program depend on the individual projects succeeding in their own right. The panel also needs DOE's estimates of the expected net economic, environmental, and security benefits of the program once goals are met. In addition, the primary assumptions associated with DOE's benefits analyses should be provided. The net benefits analysis requires that benefits are reported as being over and above those of the next-best alternative to the R&D technology or program under review. Information provided by DOE should comply with the following requirements:

- Data should be consistent with DOE's reporting under the Program Assessment Review Tool (PART) and/or the Government Performance Review Act (GPRA) and be the most current available.
- Data should be reported consistently across individual projects in the program to support project data aggregation at the program level.
- Net economic benefits data should be reported in nominal as well as real dollars using the same discount rate across projects and programs and should reasonably account for known life-cycle benefits and costs.
- Net environmental and security benefits should be quantified to the extent possible and qualified as necessary.
- NEMS and MARKAL modeling results and key assumptions should be consistently reported over a like time period for benefits calculations and simply reported numerically and graphically for ease of understanding. DOE should also explain the specific commercialization process and assumptions used in the benefits calculations.
- Technology goals must be clearly stated, and the extent of market adoption of the technology once relevant goals are met and the technology is commercially available must be reported along with the underlying assumptions reflected in the arithmetic market adoption function.
- Information should be provided on external (to DOE) RD&D funding and activities by governments, institutions, and industry to develop and deploy the technologies being evaluated.

The information request and supporting documentation take the form of a brief program assessment summary (PAS), discussed in Appendix G. Individual assessment summaries are prepared for each program under review and each project in the program if projects are also going to be subject to the panel review.

DURATION AND FREQUENCY OF THE EXPERT PANEL REVIEWS

Expert panel assessments of the benefits of each major DOE program occur at least once every 3 years. Programs in which significant changes have taken place are assessed by the expert panels soon after the changes. Between expert panel reviews, DOE comments on and updates the program status annually. Individual expert panels, once convened, aim to finish their work within 3 months of the first meeting, because the reviews and recommendations should tie into and be relevant to the administration and congressional budget processes.

ASSESSMENT OF ACTIVITIES BY NON-DOE ENTITIES

DOE's expenditure of public funds should be employed to "make the difference" in areas where other public entities, other national governments, and the private sector are not succeeding at spurring innovation and advancing critical technology. Therefore, an assessment of DOE's R&D investment needs to also examine the effectiveness and potential for success of the non-DOE programs.

To establish the character of the non-DOE R&D activities the review panel must, the goals, objectives, funding, and milestones of those activities that are relevant to the particular DOE R&D program that is being assessed. Details such as technical and marketing risks of the research sponsored by entities other than DOE must be ascertained.

DOE staff should have some, and in some cases considerable, information on RD&D activities taking place at the state level, or being carried out by foreign governments or by industry. This information is shared with the panels early in the review. In addition, the panel selection process will ensure the selection of knowledgeable professionals involved in other related R&D activities. This should add considerable value to the assessment. There may, however, be occasions where NRC and DOE conflict-of-interest requirements make it impossible to have external experts on the panel who are involved in all of these related activities. On these occasions the panel chair, in consultation with DOE managers and NRC staff, selects external experts to brief the panel during the first 2 days of its deliberation so that members better understand the status of related non-DOE R&D activities.

GENERAL ISSUES

As it assesses a program and reviews DOE activities, a review panel needs to take into account several issues (if the members need additional expertise or information, they may ask for it):

- Showstoppers
    —Identify projects whose success is absolutely critical for program success and determine whether they are receiving sufficient attention and resources from DOE.
    —Identify other projects and programs that are enabling for or complementary to the program under review. If attainment of the program goal is dependent on parallel DOE programs, the panel requires sufficient information to assess this interdependence.
    —Determine if DOE has a termination strategy in the event that a project is not successful and determine the likely effect of that termination on the program under review.

- Program disconnects
 —Determine if success of the projects that constitute the program will translate easily into achievement of the program goal. Identify gaps in the program that would keep that goal from being achieved.
 —Determine, based on the technical and economic expertise of its members, if the program goal is realistic in light of current and/or expected future funding levels.
 —Evaluate a project's funding and determine if it is sufficient to permit the project to proceed to a go/no go decision. Determine if this decision point is clearly defined by DOE.
- Assessment of non-DOE activities
 —Assess industry programs—to the extent they are known to DOE or to the panel itself—that may reach the DOE program goal or its equivalent before or at the same time as the DOE-funded activities.
 —Assess international R&D activities that might support or compete with the DOE program and evaluate their impact on expected benefits.
 —Assess industry projects that DOE is supporting or working on jointly with private industry. If the panel needs proprietary information to do this, it will need to enter into a nondisclosure agreement to access the information.
- Next-best technology
 —Review the next-best technology that would be competing with the new concepts being developed under the DOE program and track the status of competing concepts throughout the conduct of the DOE program.

QUALITY ASSURANCE

Oversight Committee

An oversight committee, which would be similar to the Committee on Prospective Benefits of DOE's Energy Efficiency and Fossil Energy R&D Programs (Phase Two), will have as its primary role assuring consistency over time across the various assessments of DOE programs being conducted by the expert panels. This and other roles of the oversight committee were discussed earlier in this chapter.

Assurance of Panel-to-Panel Consistency

To be of value to the decision makers at DOE, OMB, and the Congress, there must be consistency among the panels in the conduct of their activities, the use of the methodology, and the products that are delivered. Examples of panel activities that are needed to ensure a consistent approach and quality assurance are the following:

- *Panel selection.* The panel members are selected using a procedure similar to NRC's composition and balance procedure. An effort will be made to balance the different biases of the panel and to ensure that no one panel member has a strong positive or negative bias toward the program and technology under consideration.
- *Decision tree and benefit assessment.* A consultant and/or consultants are employed to work with each panel to apply the process/methodology and to work with the panels in their assessments. The consultant(s) perform this function for each panel and assure that the assumptions and data for the assessments are consistent for each panel.
- *Information from DOE.* The information flow to the panels from DOE follows a template developed by the committee (and described in Appendix G). Alternatively, DOE provides all the information requested in the template at a level that is consistent from program to program. DOE is asked to run its own benefit calculation using an internal computer program, NEMS. The consultant(s) assess the results of DOE benefits analyses to ensure that the inputs to the panels are as consistent as possible.
- *Role of the panel chairs.* Before the first panel meeting, the panel chair meets with the oversight committee chair and is instructed on the application of the process/methodology in program assessment. The chair is instructed to guide its panel's assessment following the process outlined by the committee. After the first meeting of the panel, its chair communicates with the oversight committee chair and describes how the process is working for his or her panel and any lessons learned that will assist the other panels in applying the process/methodology. After the panel has completed its evaluation, the chair will again meet with the oversight committee chair, discuss the panel's results, and ensure that the process was applied by each panel in a consistent manner.

Periodically—approximately every 4 months—the entire oversight committee should review for consistency all the panel assessments that were conducted during that time. The consultants and the panel chair should prepare briefings for the committee, highlighting any inconsistencies.

FULL-SCALE IMPLEMENTATION OF THE METHODOLOGY

At the conclusion of the Phase Two assessment, the process/methodology will be ready for full-scale implementation.

The annual R&D budget for all applied energy programs (energy efficiency, renewable energy, fossil energy, nuclear energy, and electricity delivery and reliability) is about $2 billion. Because this assessment approach is most appropriate for efforts that spend at least $10 million per year, efforts would be reviewed at the program level or the major activity level. The estimation of benefits for programs funded at less than $10 million per year is difficult, particularly when using the NEMS or MARKAL models to calculate benefits. In some cases, however, efforts funded at less than $10 million might warrant assessments, particularly energy efficiency R&D, where specific and narrow expertise might be needed to flesh out the program and its expected benefits. The Industries of the Future Program might be an example where separate panels for chemicals, glass, steel, and so on might have budgets less than $10 million each.

About 40 panels would be needed. Since the assessment should occur every 3 years, there would be 12 or 13 assessments and panels each year. It is suggested that three be initiated every quarter. The oversight committee would meet three times a year to review panel reports as they are completed for consistency and provide feedback to the panels.

There are several options for full-scale implementation. First, the DOE needs to decide whether or not it wants to (1) conduct these reviews in-house using an internal committee or DOE-appointed committee, or (2) use an external third-party institution such as the NRC, which does these types of reviews for the National Institute of Standards and Technology and the Army Research Laboratories, or a contractor in the model of JASON.[2]

---

[2]JASON is a third-party review of the DOD weapons program, established in the 1950s, managed by MITRE Corporation and funded by the Department of Defense Research and Engineering. JASON was formed by academic scientists to give advice to the U.S. government. A recently published book gives a very good description of the origin of JASON and how it has evolved during the past 50 years (Finkbeiner, 2006).

5

Conclusions and Recommendations

INTRODUCTION

The Phase Two study continued the work of the Phase One study by refining the methodology and applying it on a consistent basis to the energy efficiency (EE) and fossil energy (FE) programs. The methodology is meant to provide consistent information that will enable decision makers to better allocate the funds available for R&D and to identify programs where funding should be continued, expanded, scaled back, or eliminated. However, the methodology is only one piece of what is needed for an allocation decision, so it would not be appropriate to force-fit it to yield a particular decision, e.g., to increase or decrease marginal funds. The key to the methodology is the use of a panel of experts with a balance of skills and a wide range of expertise and experience to ensure that all relevant issues are identified, fully discussed, and factored into the assessment. The panel was charged not with performing a traditional program review or evaluation but with understanding programmatic issues to independently establish probabilities and expected benefits using the committee's methodology.

Applying the methodology to the six case studies,[1] the committee and expert panels developed recommendations on obtaining the results, using the results, methodological issues, and the continuity of the evaluation activity.[2]

PRIORITIES IDENTIFIED FOR PHASE THREE OF THIS PROJECT

The scope of programs subject to the NRC studies in this series has been limited to fossil energy R&D and the energy conservation portion of the R&D efforts by the Office of Energy Efficiency and Renewable Energy (EERE) R&D. Prior to FY06, funding for this group of programs fell under the jurisdiction of the Interior and Related Agencies appropriation subcommittees owing to the programs' origins in the Department of Interior; the balance of EERE's R&D funding, on energy efficiency, was included on a separate appropriations bill. However, in FY06, the House Committee on Appropriations restructured its subcommittees' jurisdictions and reduced them in number from 13 to 10. Funding for all FE and EERE R&D programs was consolidated into one account subject to the jurisdiction of the House Appropriations Subcommittee on Energy and Water Development and Related Agencies (CRS, 2005).[3] Four of the program funding line items that made up the pre-FY06 accounts were merged into two. In another instance—that of distributed energy—a program was moved out of EERE and into the appropriation account corresponding to the DOE Office of Electric Transmission and Distribution.

Finding: Phase Two showed that the basic structure of the methodology—scenarios, decision trees, technical and market risk assessments, and economic, environment, and security benefits—could be implemented by six panels of experts on a consistent basis. The panels were able to obtain the quantitative and qualitative information they needed

---

[1] The activities selected for review included three within EE—the Chemicals subprogram of the Industrial Technologies Program, the Distributed Energy Resources (DER) program, and Light-Duty Vehicle Hybrid Technology activities within the Vehicle Technologies Program—and three within FE—the Integrated Gasification Combined Cycle (IGCC) subprogram, the Carbon Sequestration program, and the Natural Gas Exploration and Production R&D program.

[2] This chapter contains 14 recommendations, 11 of which have been given numbers and carried forward to the summary (the numbering reflects the order there). Three recommendations are not numbered.

[3] The name of the combined account is Energy Supply and Conservation. Page 97 of House Report 109-275 explains it thus: "Energy Conservation programs previously funded by the Interior and Related Agencies Appropriations Act are now funded by the Energy Supply and Conservation appropriation, and are combined with energy efficiency activities in the Energy Efficiency and Renewable Energy account." See also OMB, 2005, p. 391.

to assess individual programs in fossil energy and energy efficiency.

Recommendation: The committee should undertake the following activities in Phase Three to further demonstrate the robustness of the methodology and maximize its value for decision makers:

- Expand the case studies to include at least one program in renewable energy and one in nuclear energy.
- Determine how the benefits methodology can be applied for portfolio analysis and evaluate a portfolio. Portfolios were not evaluated in Phase Two, but the review of the light-duty vehicle hybrid technologies encompasses three separate program elements and provides an opportunity to aggregate several activities. IGCC and sequestration represent two major components of the FutureGen program.
- Continue to evaluate and refine the quality control process.
- Continue to communicate and have informal conversations with stakeholders throughout the process. These would include discussions with the committee and panel chairs and with some members of the committee about the case studies and process enhancements or modifications.
- Make recommendations and provide for transition to full-scale implementation by either DOE or the NRC.

THE PROCESS FOR OBTAINING RESULTS

The components necessary for completing the assessment include the methodology, the panel of experts, input from DOE, and a quality control process.

Panel of Experts

Panel Composition

Finding: The commitment and the technology background of the panel members determine the quality of the assessment of the program.

Recommendation 4: Panel composition and level of expertise must be critically considered during the selection process. If a panel concludes that certain skills are not possessed by its members, it should consider expanding its membership or using an outside expert to brief it.

Panel Chair

Finding: The leadership role of the panel chair cannot be overemphasized. For the panel to succeed, the chair has to take a lead role in interacting with DOE to ensure that the best possible information is available to the panel before it meets for the first time. Panels where the chair devoted significant time to ensuring that all panel members were fully familiar with the process and methodology produced the best assessments.

Recommendation 5: The panel chair should spend a fair amount of time outside the actual panel meetings working with DOE program managers, DOE management, and the independent consultant(s).

Independent Consultant(s)

Finding: The main responsibility of a consultant is to maintain consistency across the panels in applying the methodology and to facilitate the analysis. This includes structuring the decision trees, facilitating the assignment of probabilities to technical and market outcomes, and assisting in the modeling of benefits. Phase Two made use of a consultant for all the panels, which worked very well.

Recommendation 6: Depending on the number of programs being evaluated and the panels' schedules, it might be necessary to have more than one consultant. One consultant should focus on the decision tree development and probability assessment and the other on the modeling of benefits.

Input from DOE

Finding: The level and timeliness of information provided by DOE to the various panels play a critical role in facilitating the deliberations and conclusions on the panel. Completion of panel evaluations is contingent on the panel's receiving synoptic information and inputs for benefits calculations. The timeliness and quality of this information impacts the quality and utility of the panel evaluation.

Recommendation 7: Since the usefulness of the benefits estimates depends on the quality and timeliness of information available to the panels, DOE management should give its full support for providing the necessary information. DOE at all levels should buy into this process because it is useful for managing and assessing its programs. If this commitment is not clear, the committee should explore all avenues for gaining DOE support.

Quality Control

Finding: Quality control continues to be important in ensuring the consistency, and therefore the utility, of panel evaluations.

Recommendation 8: An oversight committee should apply the quality control process to several elements of the study process, including ensuring appropriate panel membership and composition, orienting the panel chair and consultant, monitoring the panel's progress, monitoring information received from DOE for adequacy and consistency, and reviewing and revising the process itself.

USING THE RESULTS

Impact of Policy Measures Unrelated to Research on Realization of Program Benefits

Finding: Policy measures unrelated to research have an effect on when and whether the benefits of some programs will be realized. For example, the benefits of carbon capture and storage depend on the size and timing of a carbon tax (or an equivalent policy intervention in the market). The scenarios, which should include some of these factors, are a valuable tool for characterizing and quantifying the benefits of the DOE R&D program.

Recommendation 9: Decision makers should consider the impact of other policy measures—that is, policies not related to research—in all domains of action (federal, state, and international, say) when considering the results of prospective benefits evaluations. Having a common set of scenarios is useful in general, although additional scenarios may be called for in some cases. While defining the scenarios more completely would be helpful for interpreting the outcomes of the analysis, at the same time it is essential to preserve flexibility by keeping the scenarios as broad as possible.

Guidance for Budget Formulation

Finding: Phase Two showed that the basic structure, using decision trees, worked well and could be implemented with the panels. The panel evaluations permit calculation of a benefit-to-cost ratio, which is not, however, the correct metric to use when allocating resources among the programs in a portfolio.

Recommendation 10: To allocate resources, DOE should know the marginal benefit of a budget increase on a program-by-program basis. To calculate the marginal benefit, the decision tree should be examined to identify the outcomes that would be most sensitive to changes in budget levels. In the Phase One study, for example, the lighting program proved to be highly budget-dependent. When such sensitivities exist, the decision tree can be re-estimated for a different budget level, using the committee's methodology. The marginal benefit associated with the change in budget level is the difference between the net benefits of the two calculations.

Consideration of Alternative Futures

Finding: The methodology presents benefits for each of three scenarios that describe future states of the world, but does not attempt to combine the three sets of benefits into a single set.

Recommendation: DOE should weigh the alternative scenarios in arriving at judgments about the benefits of the overall research portfolio. The portfolio should contain a balance of projects that will produce acceptable results across the range of scenarios.

Benefits Not Captured by the Methodology

Finding: The methodology estimates public benefits in three areas—economic, environmental, and energy security. While these three types of benefits reflect DOE's strategic goals (DOE, 2005a), the committee recognizes that other kinds of benefits may be important in evaluating some projects. For example, market forces demand that automobile manufacturers produce cars that not only meet the fuel economy standard criterion of importance to DOE but that also have several other attributes. A technology that achieves DOE's objectives must also provide these additional attributes. As another example, DOE's research might have employment impacts.

Recommendation 11: If benefits in areas other than economic, environmental, or energy security are found to occur, they should be noted in the text accompanying the results matrix. However, the matrix should stay focused on the three main types of benefits to facilitate comparisons across programs.

ESTIMATING NATIONAL SECURITY AND ENVIRONMENTAL BENEFITS

Environmental Benefits

Finding: Valuation of benefits in monetary units related to reducing air pollution is both the primary environmental benefit identified in the present study and the class of benefits for which valuation methods are most advanced.

Recommendation 1: Panels should apply valuations in monetary units to criteria air pollutant emissions in the results matrix, but not to other types of pollutant emissions. The valuations used should be the allowance price forecasts for the future period.

Energy Security Benefits: Electricity

Finding: While the complex relationship between electricity supply and security is becoming clearer, analysts are a long way off from having methods for valuing reductions in security threats contributed by technologies such as distributed generation.

Recommendation 2: Panels conducting prospective benefits assessments should describe reductions in threats to energy security related to electricity supply as physical quantities of oil and gas.

Energy Security Benefits: Oil and Gas

Finding: Increases in U.S. oil and gas consumption and imports may impose incremental costs that are not fully reflected in the market price. The cost components of this oil premium have been estimated in various studies. In principle, similar estimates could be made for natural gas, but the committee is unaware of such research having been done.

Recommendation 3: Panels should describe energy security benefits related to reduced oil and natural gas consumption quantitatively in the benefits matrix as physical quantities of oil and gas. The time pattern of the oil consumption impacts should be made explicit, along with an assessment of the probable state of the oil market during those future times.

CONTINUITY: INSTITUTIONALIZING THE EVALUATION PROCESS

Finding: Prospective benefits evaluations would be most useful if DOE would adopt them. This would allow integration with GPRA, the President's Management Agenda, and other tools and systems related to performance and budgeting.

Recommendation: DOE should create a triennial program evaluation cycle using the methodology of this Phase Two study for all the applied energy programs. If DOE chooses to undertake this internally, it would need to create a set of FACA committees managed by the DOE and reporting to the under secretary or higher. If DOE chooses to use the NRC, the NRC would independently appoint the oversight committee and panels to undertake the prospective benefits evaluations. A third possibility would be for a contractor working in the model of DOD's JASON[4] to perform the assessment.

---

[4]JASON is a third-party review of the DOD weapons program, established in the 1950s, managed by MITRE Corporation and funded by the Department of Defense Research and Engineering (Finkbeiner, 2006).

References

Banzhaf, S., D. Burtraw, and K. Palmer. 2002. Efficient emission fees in the U.S. electricity sector. Discussion Paper 02-45. Washington, D.C.: Resources for the Future.

Brewer, Peter G. 2003. Direct injection of carbon dioxide into the oceans. *The Carbon Dioxide Dilemma: Promising Technologies and Policies.* Washington, D.C.: The National Academies Press.

Cifuentes, L., and Lester B. Lave. 1993. *Annual Review of Energy and the Environment.* (18):319.

Commission of the European Communities. 1993. *Externalities of the Fuel Cycle: Extern Project.* Working Documents 1, 2, 5, and 9, Brussels, Belguim: European Commission.

Congressional Research Service. 2005. *Appropriations Subcommittee Structure: History of Changes from 1920 to 2005.* Order Code RL31572.

Council on Foreign Relations. 2006. *National Security Consequences of United States Oil Dependence.* Independent Task Force Report No. 58. New York: Council on Foreign Relations.

Desvousges, W.H., F.R. Johnson, and H.S. Banzhaf. 1994. *Assessing Environmental Costs for Electricity Generation.* Triangle Economic Research General Working Paper No. G-9402.

DOE (U.S. Department of Energy). 2004a. *Estimating the DOE Office of Fossil Energy's Program Benefits: FY2004 Final Report, Volume 1.* December. Washington, D.C.: Office of Fossil Energy.

DOE. 2004b. *Natural Gas Technologies Program Plan.* April. Washington, D.C.: Office of Fossil Energy.

DOE. 2005a. *Performance and Accountability Report.*

DOE. 2005b. *Projected Benefits of Federal Energy Efficiency and Renewable Energy Programs: FY2006 Budget Request.* March. Golden, Colo.: National Renewable Energy Laboratory.

DOE. 2005c. *FreedomCAR and Vehicles Technology Multi-year Program Plan.* Washington, D.C.: U.S. Department of Energy.

DOE. 2005d. *An Economic Scoping Study for CO_2 Capture Using Aqueous Ammonia: Final Report.* Washington, D.C.: U.S. Department of Energy.

DOE. 2005e. *Chemicals Industry of the Future: Fiscal Year 2004 Annual Report.* Washington D.C.: U.S. Department of Energy.

DOE. 2006a. Available at <http://www.fe.doe.gov/aboutus/budget/07/fy07_budget_request_presentation.pdf>.

DOE. 2006b. *FY2007 Congressional Budget Request: Budget Highlights.* Washington, D.C.: U.S. Department of Energy.

DOE. 2006c. Available at <http://www.ne.doe.gov/admin/FY07Budget Rollout.html>. Accessed March 1, 2006.

EIA (Energy Information Administration). 2003. *The National Energy Modeling System: An Overview.* DOE/EIA-0581(2003). Washington, D.C.: U.S. Department of Energy.

EIA. 2004. *Annual Energy Outlook 2004.* DOE/EIA-0383(2004). Washington, D.C.: U.S. Department of Energy.

EIA. 2005a. *Assumptions to the Annual Energy Outlook 2005.* Washington, D.C.: U.S. Department of Energy.

EIA. 2005b. *Annual Energy Outlook 2005 with Projections to 2025.* Available at <http://www.eia.doe.gov/oiaf/archive/aeo05/electricity.html>.

EIA. 2006. *Annual Energy Outlook* (early release). Available at <www.eia.doe.gov>.

EPA (Environmental Protection Agency). 1997a. *Final Report to Congress on Benefits and Costs of the Clean Air Act, 1970-1990.* EPA 410-R-97-002. Washington, D.C.: U.S. Government Printing Office.

EPA. 1997b. *Final Report to Congress on Benefits and Costs of the Clean Air Act, 1990-2010.* EPA-410-R-99-01. Washington, D.C.: U.S. Government Printing Office.

EPA. 2006. Acid Rain Program Web site. Available at <http://www.epa.gov/airmarkets/progsregs/arp/index.html>.

Fankhauser, S. 1994. "The social costs of greenhouse gas emissions: An expected value approach." *Energy Journal* 15(2).

Finkbeiner, A. 2006. *The Jasons: The Secret History of Science's Postwar Elite.* New York: Viking.

General Accounting Office. 1997. *Measuring Performance: Strengths and Limitations of Research Indicators.* GAO/RCED-97-91. Washington, D.C.: Government Printing Office.

Government Accountability Office. 2005. *Budget Glossary.* GAO-05-734SP. Washington, D.C.: Government Printing Office.

Greene, D.L., and S. Ahmad. 2005. *Costs of U.S. Oil Dependence: 2005 Update.* ORNL/TM-2005/45. Oak Ridge, Tenn.: Oak Ridge National Laboratory.

Herzog, H. 2001. What future for carbon capture and sequestration? *Environmental Science and Technology* 35(7):148-153.

Hill, S.T. 2002. *Science and Engineering Degrees: 1966-2000.* NSF02-327. Arlington, Va.: National Science Foundation, Division of Science Resource Statistics.

Hines, P., J. Apt, H. Liao, and S. Talukdar. 2006. The frequency of large blackouts in the United States electrical transmission system: An empirical study. Presented at the Second Carnegie Mellon Conference in Electric Power Systems: Monitoring, Sensing, Software and Its Valuation for the Changing Electric Power Industry, January 11-12, 2006. Available at <http://www.ece.cmu.edu/~electricityconference/Old06/hines_blackout_frequencies_final.pdf>.

IPCC (Intergovernmental Panel on Climate Change). 2005. *Carbon Dioxide Capture and Storage.* Cambridge, England: Cambridge University Press.

REFERENCES

Jones, Donald W., and Paul N. Leiby. 1996. *The Macroeconomic Impacts of Oil Price Shocks: A Review of Literature and Issues.* Oak Ridge, Tenn.: ORNL.

Jones, Donald W., Paul N. Leiby, and Inja Paik. 2004. Oil price shocks and the macroeconomy: What has been learned since 1996? *The Energy Journal* 25(2):1-32.

Koomey, J. 1990. *Comparative Analysis of Monetary Estimates of External Environmental Costs Associated with Combustion of Fossil Fuels.* Palo Alto, Calif.: Electric Power Research Institute, and Berkeley, Calif.: Lawrence Berkeley National Laboratory.

Leiby, P.N., D.W. Jones, T.R. Curlee, and R. Lee. 1997. *Oil Imports: An Assessment of Benefits and Costs.* ORNL-6851. Oak Ridge, Tenn.: ORNL.

Marburger, J.H., and M.E. Daniels. 2003. *Memorandum for Heads of Executive Departments and Agencies: FY 2005 Interagency Research and Development Priorities.* June 3. Available at <http://www.whitehouse.gov/omb/memoranda/m03-15.pdf>.

Matthews, H.S., and Lester B. Lave. 2000. Applications of environmental valuation for determining externality costs. *Environmental Science Technology* 34(8):1390-1395.

Maurstad, O., H. Herzog, O. Bolland, and J. Ber. 2006. Impact of coal quality and gasifier technology on IGCC performance. Presented at the 8th International Conference on Greenhouse Gas Control Technologies. Trondheim, Norway. June.

National Coal Council. 2004. *Opportunities to Expedite the Construction of New Coal-Based Power Plants.* Available at <http://www.nationalcoalcouncil.org/Documents/ExpediteNov30rpg.pdf>.

NHTSA (National Highway Traffic Safety Administration), Department of Transportation. 2006. *Corporate Average Fuel Economy and CAFE Reform for MY 2008-2010 Light Trucks.* Final Regulatory Impact Analysis.

NRC (National Research Council). 2000. *Review of the Research Program of the Partnership for a New Generation of Vehicles: Sixth Report.* Washington, D.C.: National Academy Press.

NRC. 2001. *Energy Research at DOE: Was It Worth It?* Washington, D.C.: National Academy Press.

NRC. 2002a. *Effectiveness and Impact of Corporate Average Fuel Economy (CAFE) Standards.* Washington, D.C.: National Academy Press.

NRC. 2002b. *Making the Nation Safer: The Role of Science and Technology in Countering Terrorism.* Washington, D.C.: National Academy Press.

NRC. 2004. *Decreasing Energy Intensity in Manufacturing: Assessing the Strategies and Future Directions of the Industrial Technologies Program.* Washington, D.C.: The National Academies Press.

NRC. 2005a. *Prospective Evaluation of Energy Research and Development at DOE (Phase One): A First Look Forward.* Washington, D.C.: The National Academies Press.

NRC. 2005b. *Review of the FreedomCAR and Fuel Partnership: First Report.* Washington, D.C.: The National Academies Press.

OMB (Office of Management and Budget). 2001. *The President's Management Agenda.* Available at <http://www.whitehouse.gov/omb/budget/fy2002/mgmt.pdf>.

OMB. 2003. *Budget of the United States Government: Fiscal Year 2004.* Washington, D.C.: Government Printing Office.

OMB. 2005. *Guidance for Completing the Program Assessment Rating Tool.* Available at <http://www.whitehouse.gov/omb/part/fy2005/2005_guidance.pdf>.

Orr, Franklin M., Jr. 2003. Sequestration via direct injection of carbon dioxide deep into the earth. *The Carbon Dioxide Dilemma: Promising Technologies and Policies.* Washington, D.C.: The National Academies Press.

OTA (Office of Technology Assessment). 1994. *Studies of the Environmental Costs of Electricity.* OTA ETI-134. Washington, D.C.: U.S. Government Printing Office.

Ottinger, R.L. 1992. Social costs of energy. Proceedings of an International Conference held at Racine, Wisc., September 8-11. O. Hohmeyer and R.L. Ottinger, eds. Berlin, Germany: Springer-Verlag.

Palmgren, C.R., M.G. Morgan, W.B. de Bruin, and D.W. Keith. 2004. Initial public perceptions of deep geological and oceanic disposal of carbon dioxide. *Environmental Science and Technology* 38(24):6441-6450.

Parry, I.W.H., and J. Darmstadter. 2003. The costs of U.S. oil dependency. Discussion Paper 03-59. Washington, D.C.: Resources for the Future.

Rao, B., S. Rubin, D. Keith, and G. Morgan. 2005. Evaluation of potential cost reductions from improved amine-based CO_2 capture systems. *Energy Policy* 34(18):3765-3772.

Rose, A., G. Oladosu, and S. Liao. 2005. Regional economic impacts of terrorist attacks on the electric power system of Los Angeles: A computable general disequilibrium analysis. Paper presented at the Second Annual Symposium of the DHS Center for Risk and Economic Analysis of Terrorism Events. Los Angeles, Calif.: University of Southern California.

Sharpe, P., and T. Keelin. 1998. How SmithKline Beecham makes better resource-allocation decisions. *Harvard Business Review.* Cambridge, Mass.: Harvard Business School Publishing. March-April.

Shogren, J.F., and S. Smulders. 1996. *Resource and Energy Economics* 18(4):333-509.

U.S.-Canada Power Systems Outage Task Force. 2004. *Final Report on the August 14, 2003, Blackout in the United States and Canada.* Available at <https://reports.energy.gov/BlackoutFinal-Web.pdf>.

White House. 2002. "Fact sheet: President Bush announces Clear Skies and Global Climate Change Initiatives." Office of the Press Secretary, February 14.

Zerriffi, H., H. Dowlatabadi, and N.D. Strachan. 2002. Electricity and conflict: The robustness of distributed generation. *Electricity Journal* 15(1).

Zuckerman, B., and F. Ackerman. 1995. *The 1994 Update of the Tellus Institute Packaging Study Impact Assessment Method.* SETAC Impact Assessment Working Group Conference, Washington, D.C.

Appendixes

A

PART Assessment Questions

1. PROGRAM PURPOSE AND DESIGN

1.1 Is the program purpose clear?

1.2 Does the program address a specific and existing problem, interest or need?

1.3 Is the program designed so that it is not redundant or duplicative of any other Federal, state, local or private effort?

1.4 Is the program design free of major flaws that would limit the program's effectiveness or efficiency?

1.5 Is the program effectively targeted, so that resources will reach intended beneficiaries and/or otherwise address the program's purpose directly?

2. STRATEGIC PLANNING

2.1 Does the program have a limited number of specific long-term performance measures that focus on outcomes and meaningfully reflect the purpose of the program?

2.2 Does the program have ambitious targets and timeframes for its long-term measures?

2.3 Does the program have a limited number of specific annual performance measures that can demonstrate progress toward achieving the program's long-term goals?

2.4 Does the program have baselines and ambitious targets for its annual measures?

2.5 Do all partners (including grantees, subgrantees, contractors, cost-sharing partners, and other government partners) commit to and work toward the annual and/or long-term goals of the program?

2.6 Are independent evaluations of sufficient scope and quality conducted on a regular basis or as needed to support program improvements and evaluate effectiveness and relevance to the problem, interest, or need?

2.7 Are budget requests explicitly tied to accomplishment of the annual and long-term performance goals, and are the resource needs presented in a complete and transparent manner in the program's budget?

2.8 Has the program taken meaningful steps to correct its strategic planning deficiencies?

2RD1 If applicable, does the program assess and compare the potential benefits of efforts within the program and (if relevant) to other efforts that have similar goals?

2RD2 Does the program use a prioritization process to guide budget requests and funding decisions?

3. PROGRAM MANAGEMENT

3.1 Does the agency regularly collect timely and credible performance information, including information from key program partners, and use it to manage the program and improve performance?

3.2 Are federal managers and program partners (including grantees, subgrantees, contractors, cost-sharing partners, and other government partners) held accountable for cost, schedule and performance results?

3.3 Are funds (federal and partners') obligated in a timely manner and spent for the intended purpose?

NOTE: This appendix is based on *Department of Energy PART Assessments*, available at <http://www.whitehouse.gov/omb/budget/fy2006/pma/energy.pdf>.

3.4 Does the program have procedures (e.g., competitive sourcing/cost comparisons, IT improvements, appropriate incentives) to measure and achieve efficiencies and cost effectiveness in program execution?

3.5 Does the program collaborate and coordinate effectively with related programs?

3.6 Does the program use strong financial management practices?

3.7 Has the program taken meaningful steps to address its management deficiencies?

3RD1 For R&D programs other than competitive grants programs, does the program allocate funds and use management processes that maintain program quality?

4. PROGRAM RESULTS/ACCOUNTABILITY

4.1 Has the program demonstrated adequate progress in achieving its long-term performance goals?

4.2 Does the program (including program partners) achieve its annual performance goals?

4.3 Does the program demonstrate improved efficiencies or cost effectiveness in achieving program goals each year?

4.4 Does the performance of this program compare favorably to other programs, including government, private, etc., with similar purpose and goals?

4.5 Do independent evaluations of sufficient scope and quality indicate that the program is effective and achieving results?

B

Committee Biographies

Maxine L. Savitz (NAE), *Chair*, is retired general manager of Technology Partnerships, Honeywell, Inc. She has managed large R&D programs in the federal government and in the private sector. Some of the positions that she has held include the following: chief, Buildings Conservation Policy Research, Federal Energy Administration; professional manager, Research Applied to National Needs, National Science Foundation; division director, Buildings and Industrial Conservation, Energy Research and Development Administration; deputy assistant secretary for conservation, U.S. Department of Energy; president, Lighting Research Institute, and general manager, Ceramic Components, AlliedSignal Inc. (now Honeywell). Dr. Savitz has extensive technical experience in the areas of materials, fuel cells, batteries and other storage devices, energy efficiency, and R&D management. She is a member of the National Academy of Engineering. She has been, or is serving as, a member of numerous public- and private-sector boards and has served on many energy-related and other NRC committees. She has a Ph.D. in organic chemistry from the Massachusetts Institute of Technology.

Linda R. Cohen is professor of economics, Department of Economics, University of California, Irvine, and professor of social science and law, The Law School, University of Southern California, Los Angeles. She was previously chair, Department of Economics, University of California, Irvine, where she has taught in various capacities with increasing responsibility since 1987. Previously, Dr. Cohen was an economist associate at the Rand Corporation, a research associate for economics with the Brookings Institution, a senior economist with the California Institute of Technology's Environmental Quality Laboratory, and an assistant professor of public policy at Harvard University's Kennedy School of Government. She was the Olin Visiting Professor in Law and Economics at the University of Southern California Law School in 1993 and 1998, a fellow of the California Council for Science and Technology in 1998, and a research fellow at the Brookings Institution in 1977. Dr. Cohen has written many articles and coauthored a book on federal research and technology policy. She is currently a member of the editorial board of *Public Choice* and a member of the California Energy Commission's Advisory Panel for the Public Interest Energy Research Program. She has served on a variety of panels and committees and was a member of the NRC Committee on Benefits of DOE's R&D on Energy Efficiency and Fossil Energy. She has an A.B. degree in mathematics from the University of California at Berkeley and received her Ph.D. in social sciences from the California Institute of Technology.

James Corman is an independent consultant and founder of Energy Alternatives Systems, an engineering consulting company. He retired as general manager of the Advanced Technology Department of General Electric's (GE's) Power Generation Business, where he was responsible for development of the next generation of power systems and technical interactions with GE's international business associates. Dr. Corman was previously manager of the Advanced Projects Laboratory of GE Corporate Research and Development; there he led a diverse R&D program in activities ranging from basic technology to pilot-plant demonstration. Dr. Corman is a fellow of the American Society of Mechanical Engineers (ASME). He was a member of several NRC committees. He is chair of the Industrial Advisory Board for Mechanical Engineering at Pennsylvania State University. He has a Ph.D. in mechanical engineering from Carnegie Mellon University.

Paul A. DeCotis is director of energy analysis at the New York State Energy Research and Development Authority (NYSERDA), where he oversees statewide energy planning and policy analysis, corporate strategic planning, program evaluation, and energy emergency planning and response.

Prior to joining NYSERDA, Mr. DeCotis was chief of policy analysis at the New York State Energy Office. He is the record access officer for the State Energy Planning Board and chair of the Interagency Energy Coordinating Working Group, made up of staffs of the state departments of Public Service, Environmental Conservation, Transportation, and Economic Development, which is charged with preparing New York's energy plan. He is also a member of the New York Independent System Operator (NYISO) Management Committee, the Business Issues Committee, and the Energy Working Group of the Coalition of Northeastern Governors (CONEG). Mr. DeCotis is president of Innovative Management Solutions, a management consulting practice specializing in strategic planning and policy development, mediation, and organizational and executive management training and development. He is an adjunct professor in the M.B.A. program at the Sage Graduate School and in the Public Policy Department at Rochester Institute of Technology (RIT), and was formerly at the School of Industrial and Labor Relations at Cornell University. He is currently on the board of directors of the Association of Energy Service Professionals (AESP), serving as executive vice president and U.S. Department of Energy experts review panel chair for the weatherization study program evaluation. Mr. DeCotis was past peer review panel chair of the U.S. DOE Federal Energy Management Program and was also a member of the Committee on Prospective Benefits of DOE's Energy Efficiency and Fossil Energy R&D Programs. He has a B.S. in international business management from the State University of New York College at Brockport, an M.A. in economics from the State University at Albany, and an M.B.A. in finance and management studies from Russell Sage College.

Ramon L. Espino is currently research professor, University of Virginia, Charlottesville; he has been on the faculty since 1999. Prior to joining the Department of Chemical Engineering, he was with ExxonMobil for 26 years. He held a number of research management positions in petroleum exploration and production, petroleum process and products, alternative fuels and petrochemicals. He has published about 20 technical articles and holds 9 patents. Dr. Espino's research interests focus on fuel cell technology, specifically in the development of processors that convert clean fuels into hydrogen and of fuel cell anodes that are resistant to carbon monoxide poisoning. Another area of interest is the conversion of methane to clean liquid fuels and specifically the development of catalysts for the selective partial oxidation of methane to synthesis gas. He served on the NRC Committee on R&D Opportunities for Advanced Fossil-Fueled Energy Complexes, and is currently a member of the NRC Committee on Review of DOE's Vision 21 R&D Program. He received a B.S. in chemical engineering from Louisiana State University and an M.S. and a doctor of science in Chemical Engineering from the Massachusetts Institute of Technology.

Robert W. Fri is a visiting scholar and senior fellow emeritus at Resources for the Future, where he served as president from 1986 to 1995. From 1996 to 2001 he served as director of the National Museum of Natural History at the Smithsonian Institution. Before joining the Smithsonian, Mr. Fri served in both the public and private sectors, specializing in energy and environmental issues. In 1971 he became the first deputy administrator of the U.S. Environmental Protection Agency. In 1975, President Ford appointed him as the deputy administrator of the Energy Research and Development Administration. He served as acting administrator of both agencies for extended periods. From 1978 to 1986, Mr. Fri headed his own company, Energy Transition Corporation. He began his career with McKinsey and Company, where he was elected a principal. Fri is a senior advisor to private, public, and nonprofit organizations. He is a director of the American Electric Power Company and of the Electric Power Research Institute (EPRI) and a trustee of Science Service, Inc. (publisher of *Science News* and organizer of the Intel Science Talent Search and International Science and Engineering Fair). He is a member of the National Petroleum Council and a member of the Biological and Environmental Research Advisory Committee at the Department of Energy (DOE). In past years, he has been a member of the President's Commission on Environmental Quality, the Secretary of Energy's Advisory Board, and the University of Chicago Board of Governors for Argonne National Laboratory. He has chaired advisory committees of the National Research Council (NRC); the Carnegie Commission on Science, Technology and Government; EPRI; and the Office of Technology Assessment. He served as chair of the NRC Committee on Benefits of DOE R&D on Energy Efficiency and Fossil Energy. From 1978 to 1995 he was a director of Transco Energy Company, where he served as chair of the audit, compensation, and chief executive search committees. He is a member of Phi Beta Kappa and Sigma Xi and a national associate of the National Academies. He received his B.A. in physics from Rice University and his M.B.A. (with distinction) from Harvard University.

W. Michael Hanemann is the Chancellor's Professor, Department of Agricultural and Resource Economics and Goldman School of Public Policy, University of California, Berkeley. His previous positions include teaching fellow, Department of Economics, Harvard University; staff economist/consultant, Urban Systems Research & Engineering, Inc. (Cambridge); assistant professor and associate professor, University of California, Berkeley. He is Director, California Climate Change Center, UC Berkeley; member of the U.S. EPA's Environment Economics Advisory Committee; and a university fellow, Resources for the Future. He has served on several National Academies committees; was chair of the Organizing Committee, Second World Congress of Environmental & Resource Economists (2004); and received an honorary Ph.D. from the Swedish University of Agricultural

Sciences in 2003. His research, expertise, and publications span a wide range of topics in environmental and natural resource economics, evaluation of environmental resources, damage assessment, option value analysis, and econometric studies. He received a B.A. in philosophy, politics, and economics, Oxford University, England; an M.Sc. in development economics, London School of Economics; and M.A. in public finance and decision theory, Harvard University; and a Ph.D. in economics, Harvard University.

Wesley L. Harris (NAE) is the Charles Stark Draper Professor and head of the Department of Aeronautics and Astronautics at the Massachusetts Institute of Technology (MIT). His expertise is in fluid mechanics; aerodynamics; unsteady, nonlinear aerodynamics; acoustics; lean manufacturing processes; and military logistics and sustainment. Dr. Harris's background also includes managing major national and international aeronautical and aviation programs and personnel in the executive branch of the federal government. Prior to coming to MIT, he served as associate administrator for aeronautics at the National Aeronautics and Space Administration and vice president and chief administrative officer of the University of Tennessee Space Institute. Dr. Harris earned a B.S. in aerospace engineering from the University of Virginia and an M.S. and Ph.D. in aerospace and mechanical sciences from Princeton University.

Martha A. Krebs is the director, Energy R&D Division, California Energy Commission. Prior to that she was a consultant with Science Strategies. She was a senior fellow at the Institute for Defense Analysis (IDA), where she led studies in R&D management, planning and budgeting. She has extensive experience on DOE's basic and applied energy programs. Dr. Krebs also served as DOE assistant secretary and director, Office of Science, responsible for the $3 billion basic research programs that underlay the Department's energy, environmental, and national security missions. She also had the statutory responsibility for advising the Secretary on the broad R&D portfolio of the Department and the institutional health of its national laboratories. She has been associate director for planning and development, Lawrence Berkeley National Laboratory, where she was responsible for strategic planning for research and facilities, laboratory technology transfer, and science education and outreach. She also served on the House Committee on Science first as a professional staff member and then as Subcommittee staff director, responsible for authorizing DOE non-nuclear energy technologies and energy science programs. She is a member of Phi Beta Kappa, a fellow of the American Association for the Advancement of Science, a fellow of the Association of Women in Science, and received the Secretary of Energy Gold Medal for Distinguished Service (1999). She is a member of the National Academies Committee on Scientific and Engineering Personnel and the Navy Research Advisory Committee. She is also a member of the Committee on Prospective Benefits of DOE's Energy Efficiency and Fossil Energy R&D Programs Phase Two. She received her bachelor's degree and a Ph.D. in physics from the Catholic University of America.

Lester B. Lave (IOM) is the Harry B. and James H. Higgins Professor of Economics and University Professor; director, Carnegie Mellon Green Design Initiative; and codirector, Carnegie Mellon Electricity Industry Center. His teaching and research interests include applied economics, political economy, quantitative risk assessment, safety standards, modeling the effects of global climate change, public policy concerning greenhouse gas emissions, and understanding the issues surrounding the electric transmission and distribution system. He is a member of the National Academies' Institute of Medicine and a recipient of the Distinguished Achievement Award of the Society for Risk Analysis. He has a B.S. in economics, Reed College, and a Ph.D. in economics, Harvard University.

Richard G. Newell is a senior fellow at Resources for the Future (RFF). His previous positions include researcher and teaching fellow, Harvard University; and senior associate, ICF Incorporated. On RFF's research staff since 1997, Dr. Newell is currently focusing on the economic analysis of policy design and performance, with a particular interest in technological change and incentive-based policy. His research applications encompass a range of environmental and natural resource issues, including energy efficiency, climate change, air pollution, valuation of costs and benefits over time, and fishery management. He has served as an adviser to state and federal agencies; international, business, and environmental organizations; and private firms. He is a member of the American Economics Association, the Royal Economic Society, the Association of Environmental and Resource Economics, and the European Association of Environmental and Resource Economics. He has a Ph.D. in public policy (environmental/natural resource economics), Harvard University; an M.P.A., Princeton University, Woodrow Wilson School; a B.S. in materials engineering, Rutgers University, and a B.A. in philosophy, Rutgers University.

Jack S. Siegel is a principal with the consulting firm of Energy Resources International, Inc., and president of its Technology and Markets Group. While at the U.S. Department of Energy (DOE), he held various positions of leadership, including deputy assistant secretary for coal technology and acting assistant secretary for fossil energy. Prior to serving at DOE, he was at the U.S. Environmental Protection Agency and led efforts to regulate and enforce the Clean Air Act of 1970. Mr. Siegel has broad and extensive experience on energy and environmental issues and has recently been involved in studies on markets for and barriers to clean coal technologies, conventional and advanced turbines, renewable energy systems, distributed power systems, the impact of electric power restructuring on fuel and technology choices in the energy sector, options for reductions of greenhouse gases, and energy and environmental analysis in support of a number of foreign countries, the World Bank, and the Global Environment Facility. He served as a member of the NRC's Committee on Challenges, Opportunities, and Possibilities for Cooperation in the Energy Futures of China and the United States and was a member of the previous Committee on Benefits of DOE R&D on Energy Efficiency and Fossil Energy. He received the Presidential Award for Superior Achievement (1992) and the Secretary of Energy's Gold Medal for Outstanding Performance (1994). He has a B.S. in chemical engineering from Worcester Polytechnic Institute.

James E. Smith is an associate professor of decision sciences at Duke University. He teaches courses in probability and statistics and decision modeling. Professor Smith's research interests lie primarily in the areas of decision analysis and real options. More specifically, his research focuses on developing methods for formulating and solving dynamic decision problems and valuing risky investments, taking account of the information provided in futures and options markets. His research has been supported by grants from the National Science Foundation and Chevron Corporation. Professor Smith received B.S. and M.S. degrees in electrical engineering from Stanford University (in 1984 and 1986, respectively) and worked as a management consultant prior to earning his Ph.D. in engineering-economic systems at Stanford in 1990. He has been at the Fuqua School of Business since the fall of 1990, and he received the Outstanding Faculty Award from the daytime M.B.A. students in 1993 and 2000; he has been nominated for teaching awards on several other occasions. He spent the 1998-1999 academic year on sabbatical at Stanford and served as associate dean for the daytime M.B.A. program at the Fuqua School of Business from 2000 to 2003.

Terry Surles is program manager for technology integration and policy analysis in the Hawaii Natural Energy Institute at the University of Hawaii at Manoa. Previously, he was director for the Pacific International Center for High Technology Research. Before joining PICHTR, Dr. Surles was vice president at the Electric Power Research Institute (EPRI) and its subsidiary, the Electricity Innovations Institute. He has also served as program manager of the Public Interest Energy Research (PIER) and assistant director for science and technology of the California Energy Commission. Dr. Surles was the associate laboratory director for energy programs at Lawrence Livermore National Laboratory, following his time at the California Environmental Protection Agency as deputy secretary for science and technology. Dr. Surles was at Argonne National Laboratory for a number of years, holding a number of positions in the energy and environmental systems area, with his last position being general manager for Environmental Programs. Dr. Surles holds a B.S. in chemistry from St. Lawrence University and a Ph.D. in chemistry from Michigan State University.

James L. Sweeney is professor and former chair, Department of Engineering-Economic Systems and Operations Research, Stanford University. He has been a consultant, director of the Office of Energy Systems, director of the Office of Quantitative Methods, and director of the Office of Energy Systems Modeling and Forecasting, Federal Energy Administration. At Stanford University, he has been chair, Institute of Energy Studies; director, Center for Economic Policy Research; and director, Energy Modeling Forum. He has served on several NRC committees, including the Committee on the National Energy Modeling System and the Committee on the Human Dimensions of Global Change. He served on the previous Committee on Benefits of DOE's R&D on Energy Efficiency and Fossil Energy, helping to develop the framework and methodology that the committee applied to evaluating benefits. His research and writings address economic and policy issues important for natural resource production and use; energy markets, including oil, natural gas, and electricity; environmental protection; and the use of mathematical models to analyze energy markets. He has a B.S. degree from MIT and a Ph.D. in engineering-economic systems from Stanford University.

Michael L. Telson is the director of National Laboratory Affairs for the University of California in its Washington Office of Federal Governmental Relations. He previously served as chief financial officer (CFO) of DOE from October 1997 (after confirmation by the U.S. Senate) through May 2001. Before working at DOE, he served as a senior analyst on the staff of the Committee on the Budget, U.S. House of Representatives. He was responsible for reviewing energy, science, and space issues in the federal budget, including the programs of DOE, the NSF, and NASA, government-wide R&D policy, and certain user fee programs (including FCC spectrum auction issues). He also served as staff economist to the House ad hoc Committee on Energy created to enact the 1978 National Energy Act. Dr. Telson is a member of Sigma Xi, Tau Beta Pi, and Etta Kappa Nu. He is a fellow of the AAAS, as well as of the APS, and received the Meritorious Service and Superior Performance awards from Energy Secretary Richardson and the Gold Medal for excellence from Energy Secretary Abraham. In 2002, he was named a senior fellow of the U.S. Association for Energy Economics. He holds B.S., M.S., E.E., and Ph.D. degrees in electrical engineering from MIT and an M.S. in management from the MIT Sloan School of Management.

C

Statement of Task

Two general activities will be the focus of Phase 2: refining the methodology developed in Phase 1, and applying it to additional R&D projects. The specific activities that the committee will undertake in Phase 2 depend to some extent on the progress that the Committee on Prospective Benefits of DOE's Energy Efficiency and Fossil Energy R&D Programs-Phase 1 makes on a prospective benefits methodology and its application to different programs in the energy efficiency (EE) and fossil energy (FE) programs. It is proposed that the Phase 2 will include the following.

1. Based on its experience with applying the methodology developed in Phase 1, the committee appointed in Phase 2 may modify the methodology, as appropriate, before it applies it to evaluating the prospective benefits of additional individual programs/projects in EE and FE. It is expected that more attention will be devoted to improving the methodology for estimating environmental benefits (e.g., from reduced emissions), estimating national security benefits (e.g., from reduced oil imports), and consider the extent to which an options evaluation can be used to represent prospective benefits under a variety of representative scenarios. In addition, the committee may examine how project-by-project benefit evaluations can be used for budget decisions. Other issues may be defined during Phase 1 that should be addressed in Phase 2, resources permitting.

2. After the Phase 1 report is available, the Phase 2 committee will review comments from DOE and others on the methodology developed and its application. Based on these inputs, the committee will consider any changes to its methodology, as necessary, for the Phase 2 effort. The committee may prepare a letter report detailing changes to the methodology as a result of these reviews and possible work completed under task 1 above.

3. As in Phase 1, the work of the Phase 2 committee will be supported by several panels that will be separately appointed by the NRC to apply the methodology developed in Phase 1 and evaluate the prospective benefits of individual EE and FE programs/projects. Since a methodology will have been developed in Phase 1, it is expected that a greater number of panels can be formed in Phase 2 and more time and resources can be devoted to evaluating prospective benefits. It is proposed that approximately 6 panels will be appointed by the NRC.

4. The panels and committee will act as a quality control function to review how DOE is evaluating prospective benefits in the various EE and FE programs/projects to ensure that a credible, consistent and transparent approach is being undertaken.

5. The panels will write reports on the benefit evaluations of the programs/projects examined and deliver these reports to the committee. The committee will write a final report that incorporates the panel reports.

It is the intent of the Congress that the NRC will conduct a number of evaluations of prospective benefits on an annual basis with different programs evaluated each year.

D

Letter Report

THE NATIONAL ACADEMIES
Advisers to the Nation on Science, Engineering, and Medicine

Board on Energy and Environmental Systems

500 Fifth Street, NW
Washington, DC 20001
Phone: 202 334 3344
Fax: 202 334 2019

December 14, 2005

Mr. David Garman
Undersecretary for Energy, Science, and Environment
U.S. Department of Energy
1000 Independence Avenue, S.W.
Washington, DC 20585

Dear Mr. Garman:

The National Research Council (NRC) has established the Committee on Prospective Benefits of DOE's Energy Efficiency and Fossil Energy R&D Programs, Phase Two, and the committee has begun work. The committee's purpose is to continue to develop methodology for estimating the economic, environmental, and energy security benefits associated with DOE's Energy Efficiency and Fossil Energy R&D Programs and to apply its proposed methodology to several DOE programs. The committee's statement of task is provided in Attachment A and its members are listed in Attachment B.

To obtain feedback on its proposed methodology and its then-pending selection of DOE programs for further case study, the committee held a workshop on July 14, 2005, in Washington, D.C., attended by stakeholders. In this letter,[1] the committee discusses the principal comments made during the workshop, the case studies it intends to perform in phase two, and the changes to the process and methodology that have occurred since phase one.

---

[1] This report has been reviewed in draft form by individuals chosen for their diverse perspectives and technical expertise, in accordance with procedures approved by the NRC's Report Review Committee. The purpose of this independent review is to provide candid and critical comments that will assist the institution in making its published report as sound as possible and to ensure that the report meets institutional standards for objectivity, evidence, and responsiveness to the study charge. The review comments and draft manuscript remain confidential to protect the integrity of the deliberative process. We wish to thank the following individuals for their review of this report: William Agnew, NAE, General Motors (retired); David Bodde, Clemson University; Charles Lave, University of California, Irvine; John J. Wise, NAE, Mobil Research and Development Corporation (retired); and James Wolf, independent consultant.

Although the reviewers listed above have provided many constructive comments and suggestions, they were not asked to endorse the conclusions or recommendations, nor did they see the final draft of the report before its release. The review of this report was overseen by John Ahearne, NAE, Sigma Xi. Appointed by the NRC, he was responsible for making sure that an independent examination of this report was carried out in accordance with institutional procedures and that all review comments were carefully considered. Responsibility for the final content of this report rests entirely with the authoring committee and the institution.

EVALUATING THE FEDERAL INVESTMENT IN APPLIED ENERGY R&D

From the time the Department of Energy was formed in 1977, successive administrations in Washington, D.C., have looked to technological innovation as a critical tool for ensuring that the nation has a reliable supply of affordable, clean energy. Recognizing the importance of technological innovation, DOE, the Office of Management and Budget (OMB), and congressional committees have given increasing attention to understanding the effectiveness of federal funding for applied energy research and development (R&D).[2] Evaluating government investment in applied energy R&D programs requires assessing their costs and benefits. Doing so is not a trivial matter. First, the analysis of costs and benefits must reflect the full range of public benefits—environmental and energy security impacts as well as economic effects. Second, the analysis must consider how likely the research is to succeed and how valuable the research will be if it is successful. Finally, the analysis must consider what might happen if the government did not support the project: Would some private entity undertake it or an equivalent activity that would produce some or all of the benefits of government involvement?

Congress provided funds for "a continuing annual review by the [National] Academy [of Sciences] of programs . . . to measure the relative benefits expected to be achieved and to inform decision making on what programs should be continued, expanded, scaled-back, or eliminated."[3] The NRC has completed two studies to date. The first study committee, whose report was published in 2001,[4] conducted a retrospective examination of the first 22 years of DOE-funded R&D on energy conservation.[5] A second NRC committee adapted the methodology developed by its predecessor committee for use in prospectively assessing the benefits of the portfolio of ongoing R&D directed at energy conservation. Its report,[6] published in April 2005, culminated phase one of the prospective study.

The methodology suggested by the phase one committee uses expert panels to review the DOE R&D program and estimate the expected economic, environmental, and energy security benefits of the program in three different global economic scenarios, with the results summarized in the matrix shown in Attachment C. The expert panel evaluation process is facilitated by a decision analysis consultant, and the panels construct simple decision trees to describe the main technical and market uncertainties associated with the program and the impact of DOE support on the probability of various technical and market outcomes. The

---

[2] An applied energy R&D program addresses a specific technology with defined performance and cost targets and milestones, whereas a research program has as its objective increased understanding and knowledge.

[3] House Report 107-564, p. 125. July 11, 2002. U.S. Government Printing Office: Washington D.C.

[4] National Research Council. 2001. *Energy Research at DOE: Was It Worth It?* Washington D.C.: National Academy Press. This report was requested by Congress in the conference report of the Consolidated Appropriations Act for fiscal year (FY) 2000 (House Report 106-479, p. 493. November 18, 1999. U.S. Government Printing Office: Washington, D.C.).

[5] These programs include only those that were at the time under the jurisdiction of the U.S. House Appropriations Subcommittee on the Interior and Related Agencies.

[6] National Research Council. 2005. *Prospective Evaluation of Energy Research and Development at DOE (Phase One): A First Look Forward.* Washington, D.C.: The National Academies Press. Interested readers are referred to this report for a complete description of the methodology for prospective evaluation of R&D benefits, subject to the modifications discussed herein.

benefits of each R&D project are then estimated in each of these technical and market scenarios; the phase one report emphasized the potential need to use simple spreadsheet models in conjunction with more sophisticated models (such as the Energy Information Administration's National Energy Modeling System;[7] NEMS) to estimate these benefits. The overall benefit of the DOE R&D program is given as the difference between the expected benefits with DOE support and the expected benefits without DOE support. To ensure consistency across the panels, the process calls for the use of common scenarios and assumptions across evaluations and an oversight committee that provides guidance to the panels reviewing individual activities.

Phase two of the NRC's prospective study calls for testing, refining, and extending the proposed methodology. The committee intends to apply the phase one methodology for prospective evaluation of six applied energy R&D activities residing within DOE's Office of Energy Efficiency and Renewable Energy (EE) and Office of Fossil Energy (FE). In addition, the committee will continue to revise the methodology as further experience with the panels warrants. The goal of this evaluation process is to enhance the value of DOE's R&D programs by helping to establish a basis for increasing the funding of socially valuable programs and for transferring resources from programs that are less socially valuable, as well as justifying total funding.

PRINCIPAL COMMENTS FROM THE WORKSHOP

At the July 14, 2005, workshop, the committee heard presentations from representatives of OMB, Congress, the Office of Science and Technology Policy (OSTP), and DOE. It reviews each of their comments in turn.

- *OMB*. OMB representatives at the July 2005 workshop were quite supportive of the committee's proposed approach, both the analytic methodology and the proposed process. Specifically, OMB supported the use of simple spreadsheet models in conjunction with NEMS, noting that the approach has the "potential to improve resolution, transparency, [and] ease of sensitivity analysis."[8] It also endorsed the committee's decision tree framework as an "appropriate way to model risk for technical outcome, [and] market acceptance." On the process side, OMB supported the use of balanced external review panels and stressed the importance of the oversight committee in ensuring the consistency of assumptions (about macroeconomic factors, next-best technologies, program funding, and so on) across panels.

OMB's representatives indicated that OMB was quite comfortable with the level of complexity of the proposed analysis and emphasized the need to summarize results in a single page for high-level policy analysis, as proposed in the committee's matrix (see Attachment C). Although pointing out that the quantitative benefit estimates provided were

---

[7]NEMS is a computer-based, energy-economy system for modeling U.S. energy markets that projects the production, imports, conversion, consumption, and prices of energy, subject to assumptions about macroeconomic and financial factors, world energy markets, resource availability and costs, behavioral and technological choice criteria, cost and performance characteristics of energy technologies, and demographics.

[8]Leo Sommaripa, "Prospective Benefits Estimation for DOE's Applied R&D—NRC Phase II; OMB Perspective and Interest," presentation to the Committee on Prospective Benefits of DOE's Energy Efficiency and Fossil Energy R&D Programs, Phase Two, July 14, 2005. OMB was also represented at the workshop by Rob Sandoli, program examiner, Energy Branch.

certainly helpful, OMB staff noted that the qualitative issues identified by the review panels in their reports were also very helpful in OMB's reviews of DOE programs. While the NRC's current process for evaluating DOE R&D programs focuses on measuring net benefits for the U.S. economy, OMB indicated that it would be helpful if the NRC's benefits evaluations also distinguished between producer and consumer surpluses so that beneficiaries of DOE's R&D programs could be more readily identified. OMB's representatives also made some suggestions regarding programs to review in phase two that are discussed below.

- *Congress.* A congressional view presented at the July 2005 workshop was also quite positive about the NRC's proposed approach to prospective evaluation of DOE R&D programs.[9] It echoed many of OMB's comments, citing the benefits of the independent external reviews, the more transparent modeling, and the accessible short summaries. It noted that different users in Congress may have different preferences for the quantitative and/or qualitative information provided by panel reports and emphasized the need for both kinds of information. Also expressed was the desire that the panels' analyses more explicitly identify likely beneficiaries of DOE applied energy R&D programs, as well as a reservation regarding whether this kind of analysis was appropriate for the NRC panels.

- *OSTP.* OSTP's representative at the July 2005 workshop gave an overview of the OSTP mission and its role in setting energy R&D policy and talked about the value of a rigorous approach to estimating benefits, such as that proposed by the NRC phase one study, and the potential use of such an approach in portfolio allocation and program management.

- *DOE.* DOE was represented primarily by two staff members—one from the Office of Fossil Energy and one from the DOE Office of Energy Efficiency and Renewable Energy and the Office of Fossil Energy.[10] A number of other DOE representatives and contractors attended the workshop and participated in the discussions throughout the day.

DOE's FE representative emphasized two issues. First, he considered the use of expert panels, questioning whether a single panel can effectively evaluate the broad range of technologies involved in a major system, such as a zero-emission coal plant or the hydrogen fuel program. The second issue was the use of NEMS. The phase one committee report criticized NEMS as being opaque and cumbersome to run and noted that DOE analyses frequently considered consumer savings while neglecting impacts on producers. As indicated above, the phase one committee report proposed the use of simple models in conjunction with NEMS to estimate net benefits in a given scenario. Expressing concerns that these simple models "may take too many shortcuts," he invited Kevin Forbes of Catholic University to describe an approach for calculating net benefits using multiple NEMS runs. DOE's FE representative concluded by calling for more interaction between the DOE and NRC panels during the evaluation process and for better documentation of the evaluation panels' discussions and the logic underlying their risk assessments.

---

[9]Kevin Carroll, House Committee on Science, Subcommittee on Energy, July 14, 2005.

[10] Sam Baldwin, DOE Office of Energy Efficiency and Renewable Energy, and Jay Braitsch, DOE Office of Fossil Energy, "EERE-FE Observations on the NRC Report: Prospective Evaluation of Applied Energy Research and Development at DOE (Phase One): A First Look Forward," presentation to the Committee on Prospective Benefits of DOE's Energy Efficiency and Fossil Energy R&D Programs, Phase Two, July 14, 2005.

DOE's EE representative offered a number of observations on the phase one report that were further documented in an accompanying memorandum.[11] He echoed his colleague's concerns about the use of a single panel of experts for each program and about the use of simplified models in conjunction with NEMS; he also called for more interaction between DOE and the NRC panels during the evaluation process and better documentation. He expressed concerns about the consistency of the process, the lack of clear metrics, the use of single-point estimates, and decision trees not fully capturing the "flexibilities of actual management practices."[12] He went on to describe the activities of a risk team within DOE whose goal is to develop "scaleable risk analysis methods" that can be used by project/program managers, portfolio managers, and political leaders. He described a prototype Monte Carlo simulation-based tool for analyzing wind turbine systems but noted that "many challenges remain to develop/implement these tools."

The committee was pleased to hear about DOE's efforts to improve its ability to calculate the benefits of R&D, both through properly calculating net benefits in NEMS and through developing sophisticated risk analysis models that can be used for program management and evaluation. Whenever available, the results of these analyses should inform panel evaluations of program benefits. The committee agrees that the individual program evaluations would benefit from improved interactions between DOE and the NRC panels and also agrees that DOE and DOE laboratories can contribute meaningfully to the ongoing development of the proposed methodology. Indeed, the modifications to the methodology (described in the final section of this letter) are focused primarily on improving these interactions. The committee also agrees that review panels should discuss the logic underlying their risk assessments.

Although it is sympathetic to DOE's concerns about the use of expert panels and overreliance on simple models, the committee remains optimistic that the proposed process can lead to evaluations that are useful to decision makers. The committee emphasizes that the proposed process is quite similar to processes used routinely to evaluate applied R&D projects in industry. For example, Sharpe and Keelin[13] describe a process used for evaluating R&D projects at SmithKline Beecham that uses simple decision tree models for projects and uses independent review panels to review these assessments. Like applied energy R&D projects, modern pharmaceutical R&D projects are also quite complex and require consideration of both scientific and market issues. Sharpe and Keelin's discussion of the SmithKline Beecham experience emphasizes how the independence and consistency of the evaluation process led to improved communication and credibility: "by tackling the soft issues—such as information quality and trust—SB improved its ability to address the hard

---

[11] "EERE and FE Observations on the NRC Report: Prospective Evaluation of Applied Energy Research and Development at DOE (Phase One): A First Look Forward," background paper delivered to the Committee on Prospective Benefits of DOE's Energy Efficiency and Fossil Energy R&D Programs, Phase Two, July 14, 2005.

[12] Sam Baldwin, DOE Office of Energy Efficiency and Renewable Energy, and Jay Braitsch, DOE Office of Fossil Energy, "EERE-FE Observations on the NRC Report: Prospective Evaluation of Applied Energy Research and Development at DOE (Phase One): A First Look Forward," presentation to the Committee on Prospective Benefits of DOE's Energy Efficiency and Fossil Energy R&D Programs, Phase Two, July 14, 2005.

[13] P. Sharpe and T. Keelin. 1998. "How SmithKline Beecham Makes Better Resource-Allocation Decisions." *Harvard Business Review*. Cambridge, Mass.: Harvard Business School Publishing. March-April.

ones: how much and where to invest."[14] In that case, senior management ultimately concluded that increased R&D funding would be a worthwhile investment. The committee believes that with the cooperation and support of DOE, the proposed process developed in phase one can be similarly successful and can improve communication with stakeholder groups and the credibility of program evaluations.

CASE STUDIES SELECTED FOR PHASE TWO

For phase two, the committee has selected six DOE applied energy R&D activities to be the subject of a prospective assessment of benefits. The selected activities are from the FE and EE programs and are as follows (where applicable the specific subprogram that will be the focus of the assessment is noted in parentheses):

- FE
    - Integrated gasification combined cycle,
    - Sequestration, and
    - Natural gas technologies (exploration and production).
- EE
    - Distributed energy program (end-use system integration and interface),
    - Vehicle technologies program (hybrid and electric propulsion, advanced combustion R&D, and materials technology—excluding projects related to heavy duty vehicles), and
    - Industrial technologies program (chemicals).

Prior to its selection of these six activities for case studies, the committee held discussions with major stakeholders, including congressional committee staff, DOE, and OMB. The discussions with congressional appropriations staff occurred before the July 14, 2005, workshop. At these meetings, congressional and agency staff recommended that since the funds for the prospective benefits studies (phases one, two, and three; phase three will apply the benefits methodology to a new set of case studies) had been appropriated by the Appropriations Subcommittee on the Interior and Related Agencies, the case studies in phase two should be drawn from the energy conservation programs within FE and EE, even though all funds for DOE now fall under the jurisdiction of the newly reorganized Appropriations Subcommittee on Energy and Water Development. At the July 14 workshop, OMB[15] and DOE[16] offered suggestions for the case studies. Letters (see Attachment D) were sent by the

---

[14]P. Sharpe and T. Keelin. 1998. "How SmithKline Beecham Makes Better Resource-Allocation Decisions." *Harvard Business Review*. Cambridge, Mass.: Harvard Business School Publishing. March-April. Page 45.

[15]Leo Sommaripa, "Prospective Benefits Estimation for DOE's Applied R&D—NRC Phase II; OMB Perspective and Interest," presentation to the Committee on Prospective Benefits of DOE's Energy Efficiency and Fossil Energy R&D Programs, Phase Two, July 14, 2005.

[16]Sam Baldwin, DOE Office of Energy Efficiency and Renewable Energy, and Jay Braitsch, DOE Office of Fossil Energy, "EERE-FE Observations on the NRC Report: Prospective Evaluation of Applied Energy Research and Development at DOE (Phase One): A First Look Forward," presentation to the Committee on Prospective Benefits of DOE's Energy Efficiency and Fossil Energy R&D Programs, Phase Two, July 14, 2005.

committee to DOE on July 22, 2005, indicating which activities had been selected for case studies and identifying information the review panels would need for their deliberations. There was final agreement on the case studies after an August 4, 2005, meeting with the undersecretary for energy, science, and environment and DOE staff.

The sequestration R&D program was selected as a case study for phase one and again for phase two. Although the phase one study was not intended to produce accurate quantitative results, the committee thought it would be useful to test in phase two the currently proposed methodology and compare the results with those from phase one. Environmental and energy security benefits are being further defined during phase two. Evaluation of energy security benefits will be relevant for the chemicals subprogram of the industrial technologies program (ITP), distributed energy resources R&D, hybrid vehicle technology R&D, and natural gas technologies R&D. Evaluation of environmental benefits will be relevant for integrated gasification combined cycle R&D and sequestration R&D.

The chemicals subprogram of ITP and the hybrid vehicle technologies program include many separate program elements and provide the opportunity to aggregate several activities. Thus the committee should have an opportunity to comment on how to aggregate programs, the usefulness of spreadsheet models, and methods to account for competing and complementary benefits.

MODIFICATIONS TO METHODOLOGY

The methodology proposed by the phase one committee was developed after its review of the results of three pilot case studies. In phase two, the committee will test the methodology in the new case studies and work to make it extensible to consideration of environmental and energy security benefits. The feedback on methodology received in the July 2005 workshop and other venues has raised many issues that the committee and the review panels will have to bear in mind in conducting the case studies and in further developing the methodology. However, this feedback has not led the committee to propose fundamental changes to the methodology before applying it to these new case studies. The primary change to the proposed process and plan is the decision to work to improve communications and interactions with DOE, to the extent permissible under the Federal Advisory Committee Act, Section 15, and the rules of the National Research Council. Some examples of opportunities for improved communication and interaction are as follows:

1. *Selection of case studies.* Prior to selecting the case studies, the committee obtained input from DOE, which was also asked to suggest experts for the various panel chairs. Per the recommendations in the phase one report (pp. 33-34),[17] the panel chairs are meeting in person or via telephone with DOE program managers to discuss the methodology and the information being requested. DOE was asked to make presentations at the first panel meeting following the template[18] that was developed by the NRC committee in order to ensure that

---

[17]National Research Council. 2005. *Prospective Evaluation of Energy Research and Development at DOE (Phase One): A First Look Forward.* Washington, D.C.: The National Academies Press.

[18]The template is given in Appendix K of National Research Council. 2005. *Prospective Evaluation of Energy Research and Development at DOE (Phase One): A First Look Forward.* Washington, D.C.: The National Academies Press.

the information is provided to all panels in a consistent and complete manner. At the end of the first panel meeting, a decision tree will be constructed that will be sent to DOE to obtain its probability estimates and any suggested modifications of the decision tree. The second panel meeting will include open sessions with DOE participants at which DOE's probability estimates and any suggested modifications will be discussed. The primary change in the process is that DOE has the opportunity to comment on and share its views on each panel's proposed process, before the end of the panel study.

2. *Development and refinement of methodology.* The further development of the evaluation methodology will also involve DOE as well as experts within DOE's laboratories. The July 14, 2005, workshop provided a forum for offering feedback on the phase one methodology and for supplying initial input regarding energy security and environmental benefits. At the September 13, 2005, meeting, FE and EE described their activities related to estimating environmental and energy security benefits. As refinement of the methodology continues, there will be informal conversations with the committee chair, some committee members, and DOE and other stakeholders about the proposed process and enhancements or modifications to it. It is in the interest of the NRC, DOE, and other participants, sponsors, and stakeholders to develop a methodology for evaluating prospective benefits of DOE R&D that is both rigorous and transparent.

The committee looks forward to its work with DOE in the months ahead and welcomes your feedback on its proposed processes.

Sincerely,

Maxine Savitz, *Chair*
Committee on Prospective Benefits of DOE's Energy Efficiency and
    Fossil Energy R&D Programs, Phase Two

Attachment A

Statement of Task

PROSPECTIVE BENEFITS OF DOE'S ENERGY EFFICIENCY AND FOSSIL ENERGY R&D PROGRAMS—PHASE 2

Project Scope:

The Phase 2 activity follows the completion of Phase 1, which resulted in the issuance of two reports on methodology for estimating prospective benefits and evaluating energy R&D programs at DOE. These reports [*Energy Research at DOE: Was It Worth It?*, and *Prospective Evaluation of Applied Energy Research and Development at DOE: A First Look Forward*] are posted in the project record with project identification number BEES-J-03-01-A in the Current Projects System.

At least three issues will require attention as part of the Phase 2 Task. These issues include: (a) further improving the estimation of the value of environmental benefits (e.g., reduced emissions), (b) further improving the estimation of the value of security benefits (e.g., reducing oil imports or ensuring more reliable electricity supplies), and (c) determining how to estimate the overall benefits of the options under a variety of scenarios. The first two issues involve the public good rather than direct economic benefits. The committee will build on the foundation of work from Phase 1 and the body of literature that exists to determine appropriate values for these factors. The committee might commission white papers defining the state of knowledge and suggesting how the methodology could incorporate these estimates. For (c), options evaluation, the committee will consider the extent to which an analytical foundation is appropriate, building on the Phase 1 work and incorporating the full range of benefits for representative scenarios. In addition, the committee will consider mechanisms for quantifying knowledge benefits and include them as appropriate in the overall evaluation. The committee will also provide a peer review of how DOE is evaluating prospective benefits of various Energy Efficiency (EE) and Fossil Energy (FE) programs/projects. As in Phase 1, several panels will be separately appointed to assist the committee in Phase 2.

A workshop will be held early in Phase 2 to discuss the Phase 1 reports and methodology, following which the committee will write a letter report that will set the stage for the work to be accomplished in Phase 2. A final report will be issued at the conclusion of Phase 2, about the end of April 2006. The panels will write panel reports documenting the results of the analyses of the prospective benefits of the various programs/projects in EE and FE chosen by the committee to evaluate. These panel reports may be issued separately or incorporated into the Phase 2 final report.

The project is sponsored by the U.S. Department of Energy.

The approximate starting date for this project is March 15, 2005.

Project Duration: 14 months

Attachment B

Committee Roster

COMMITTEE ON PROSPECTIVE BENEFITS OF DOE'S ENERGY EFFICIENCY AND FOSSIL ENERGY R&D PROGRAMS, PHASE TWO

NAE **Maxine L. Savitz** (*Chair*)
General Manager, Technology Partnerships
Honeywell, Inc. (retired)

Linda Cohen
Professor, Department of Economics
University of California, Irvine

James Corman
President, Energy Alternatives Studies, Inc.

Paul DeCotis
Director, Energy Analysis
New York State Energy Research and
 Development Authority (NYSERDA)

Ramon Espino
Professor, Department of Chemical Engineering
University of Virginia

Robert W. Fri
Visiting Scholar
Resources for the Future

W. Michael Hanemann
Professor, Department of Agricultural and
 Resource Economics
University of California, Berkeley

NAE **Wesley Harris**
Head, Department of Aeronautics
 and Astronautics
Massachusetts Institute of Technology

Martha A. Krebs
Director, Energy R&D Division
California Energy Commission

IOM **Lester B. Lave**
Professor, Tepper School of Business
Carnegie Mellon University

Richard G. Newell
Council of Economic Advisors

Jack Siegel
President, Technology & Markets Group
Energy Resources International, Inc.

James E. Smith
Professor, Fuqua School of Business
Duke University

Terry Surles
Director
Pacific International Center
 for High Technology Research

James L. Sweeney
Professor, Management Science and
 Engineering
Stanford University

Michael Telson
Director of National Laboratory Affairs
University of California

Attachment C

Committee's Template for Presenting Panel Results

PANEL NAME:

Program Name:

Program Goals:

Year Goals Expected to be Achieved:

Program Costs:

Funding to Date: $

Current Funding: $

Proposed Year Funding: $

Expected Cost to Completion: $

Industry and Foreign Government Funding: $

Key Complementary/Interdependent DOE Programs:

All benefits are cumulative through 2050 and are reported in 20XX year dollars.

		Global Scenarios		
		Reference Case	High Oil and Gas Prices	Carbon Sensitive
Program Risks	1. Technical Risks	See decision tree for discussion of probabilities		
	2. Market Risks			
Expected Program Benefits	1. Economic Benefits			
	2. Environmental Benefits			
	3. Security Benefits			

Comments and Observations: One to two paragraphs

> e.g., provide a summary of the panel's completed assessment and estimate of expected benefits of the DOE program.

Technical Risks: 5 to 10 lines

> e.g., describe the program's risks in sufficient detail and clarity, noting program interdependencies, technical and infrastructure innovations and breakthroughs needed, and competitive alternatives, and so on.

Market Risk: 5 to 10 lines

> e.g., note factors that might affect market acceptance, including customer preferences, pricing, competitive domestic and foreign activities; next-best technologies issues, regulatory concerns, and so on.

Benefits: 5 to 10 lines

> e.g., discuss specifically the estimation of benefits, uses and interpretations, caveats, outstanding issues, and so on.

Program Observations: 5 to 10 lines

> e.g., notable accomplishments/gaps, opportunities, spin-offs, and so on.

Attachment D

Letters Sent to DOE by the Committee

Board on Energy and Environmental Systems

500 Fifth Street, NW
Washington, DC 20001
Phone: 202 334 3344
Fax: 202 334 2019

July 22, 2005

Allan Hoffman
EE-3B Forrestal Building
U.S. Department of Energy
1000 Independence Avenue
Washington D.C. 20585

Dear Dr. Hoffman:

At last week's meeting, there was discussion as to which DOE activities would be the subject of benefits assessments during phase two. Accordingly, the committee has selected three EE activities for benefits assessment. The selected activities are included in the following list, with the area that will be the focus of the assessment noted in parentheses:

- Distributed energy program (*end-use system integration and interface*);
- Vehicle technologies program (*hybrid and electric propulsion*—excluding projects related to heavy vehicles; *advanced combustion R&D*—limited to the combustion and emission control R&D activity, only; and *materials technology*—excluding projects related to heavy vehicles and excluding the high temperature materials laboratory activity); and
- Industrial technologies program (*chemicals*).

The committee requests that, for the above activities, DOE provide the necessary program description and model runs (using NEMS, for example) as set forth in Figure K-1 of *Prospective Evaluation of Applied Energy Research and Development at DOE (Phase One): A First Look Forward.* In addition, a brief history of the activity is requested.

It is suggested that EE designate a point of contact for each activity listed above to facilitate requests for information. Please contact Martin Offutt of the NRC at 202-334-2904 or moffutt@nas.edu with the names of these contacts. In addition, it is requested that a meeting take place in the near future between the program managers from DOE and the NRC panel chairs to discuss the information request.

Thank you for your assistance with this request.

Sincerely,

Maxine Savitz
Chair, Committee on Prospective Benefits of DOE's Energy Efficiency
  and Fossil Energy R&D Programs (Phase Two)

Enclosure:
  Appendix K, *Prospective Evaluation of Applied Energy Research
  and Development at DOE (Phase One): A First Look Forward*

cc:
  David Garman, Undersecretary for Energy, Science, and Environment
  Rob Sandoli, Office of Management and Budget
  Terry Tyborowski, Committee on Appropriations, U.S. House of Representatives

APPENDIX D

Board on Energy and Environmental Systems

500 Fifth Street, NW
Washington, DC 20001
Phone: 202 334 3344
Fax: 202 334 2019

July 22, 2005

Jay Braitsch
FE-24 Forrestal Building
U.S. Department of Energy
1000 Independence Avenue
Washington D.C. 20585

Dear Mr. Braitsch:

At last week's meeting, there was discussion as to which DOE activities would be the subject of benefits assessments during phase two. Accordingly, the committee has selected three FE activities for benefits assessment. The selected activities are as follows, and where applicable the specific subprogram that will be the focus of the assessment has been noted in parentheses:

- Integrated Gasification Combined Cycle;
- Sequestration; and
- Natural gas technologies (*exploration and production*).

The committee requests that, for the above activities, DOE provide the necessary program description and model runs (using NEMS, for example) as set forth in Figure K-1 of *Prospective Evaluation of Applied Energy Research and Development at DOE (Phase One): A First Look Forward.* In addition, a brief history of the activity is requested.

It is suggested that FE designate a point of contact for each activity listed above to facilitate requests for information. Please contact Martin Offutt of the NRC at 202-334-2904 or moffutt@nas.edu with the names of these contacts. In addition, it is requested that a meeting take place in the near future between the program managers from DOE and the NRC panel chairs to discuss the information request.

Thank you for your assistance with this request.

Sincerely,

Maxine Savitz
Chair, Committee on Prospective Benefits of DOE's Energy Efficiency
  and Fossil Energy R&D Programs (Phase Two)

Enclosure:
   Appendix K, *Prospective Evaluation of Applied Energy Research
     and Development at DOE (Phase One): A First Look Forward*

cc:
   David Garman, Undersecretary for Energy, Science, and Environment
   Leo Sommaripa, Office of Management and Budget
   Terry Tyborowski, Committee on Appropriations, U.S. House of Representatives

E

Committee and Panel Activities

**COMMITTEE MEETING
WASHINGTON, D.C.
JULY 13-15, 2005**

Summary of Phase One Report
Robert Fri, Phase One Chair

Identification of User Needs
Kevin Carroll, House Committee on Science, Subcommittee on Energy; Rob Sandoli and Leo Sommaripa, Office of Management and Budget; Kevin Hurst, Office of Science and Technology Policy; Sam Baldwin, DOE Office of Energy Efficiency and Renewable Energy; and Jay Braitsch, DOE Office of Fossil Energy

Assessing the Benefits of R&D: A Framework of Analysis and an Application Using NEMS
Kevin Forbes, Catholic University of America

**COMMITTEE MEETING
WASHINGTON, D.C.
SEPTEMBER 15-16, 2005**

Metrics for Energy Security Benefits
Russell Lee, Oak Ridge National Laboratory (ORNL)

Measuring Oil Security Benefits
Dave Greene, ORNL

Energy Security Benefits of Coal R&D
Darren Mollot, DOE Office of Fossil Energy

Energy Security Benefits of Oil and Gas R&D
Rodney Geisbrecht, DOE National Energy Technology Laboratory

Estimating Environmental Benefits
Russell Lee, ORNL

Environmental Benefits of Coal R&D
Darren Mollot, DOE Office of Fossil Energy

Environmental Benefits of Oil and Gas R&D
Bill Hochheiser, DOE Office of Fossil Energy

**MEETING OF PANEL ON DOE'S CARBON
SEQUESTRATION PROGRAM
PITTSBURGH, PENNSYLVANIA
SEPTEMBER 29-30, 2005**

Program Elements and Objectives Most Closely Related to Quantitative Estimates of Prospective Program Benefits
Sean Plasynski, DOE

Relationship of Sequestration Projects to Program Objectives: CO_2 Capture Projects
Jose Figueroa, DOE

Relationship of Sequestration Projects to Program Objectives: CO_2 Storage Projects
Karen Cohen, DOE

Relationship of Sequestration Projects to Program Objectives: Monitoring, Mitigation, and Verification
Karen Cohen, DOE

Relationship of Sequestration Projects to Program Objectives: Regional Partnerships
John Litynski, DOE

NEMS-Based Approaches
Juli Klara, DOE

CarBen Spreadsheet Model
Sarah Forbes, DOE

Economics—Impact of Technology Advances on COE
Jared Ciferno, DOE

DOE Progress to Date on Sequestration Decision Tree and Probability Analysis
Jay Braitsch, DOE

MEETING OF PANEL ON DOE'S LIGHT-DUTY VEHICLE HYBRID TECHNOLOGY R&D PROGRAM
WASHINGTON, D.C.
OCTOBER 3-4, 2005

Prospective Benefits of DOE's Energy Efficiency and Fossil Energy R&D Programs (Phase Two)
Ed Wall, DOE

Estimating Benefits of EERE Light Duty Vehicle R&D
Philip Patterson, DOE/EERE; Frances Wood, OnLocation, Inc.; and Chip Friley, Brookhaven National Laboratory

MEETING OF PANEL ON DOE'S INTEGRATED GASIFICATION COMBINED CYCLE TECHNOLOGY R&D PROGRAM
WASHINGTON, D.C.
OCTOBER 5-6, 2005

Overview of DOE Benefits Analysis
Darren Mollot, DOE Office of Fossil Energy

DOE IGCC Program
Gary Stiegel and Richard Dennis; DOE Office of Fossil Energy

DOE Methodologies for Estimating IGCC Benefits
Julianne Klara, DOE National Energy Technology Laboratory

DOE Progress to Date on Decision Tree Analysis
Jay Braitsch, DOE Office of Fossil Energy

MEETING OF PANEL ON DOE'S CHEMICAL INDUSTRIAL TECHNOLOGIES PROGRAM
WASHINGTON, D.C.
OCTOBER 10-11, 2005

Overview of the Chemicals Subprogram
Dickson Ozokwelu, DOE Lead Technology Manager

Estimating Benefits of EERE Chemical Industrial Technologies R&D
Joan Pellegrino, Energetics

MEETING OF PANEL ON DOE'S NATURAL GAS EXPLORATION AND PRODUCTION PROGRAM
WASHINGTON, D.C.
OCTOBER 12-13, 2005

FE Methodologies for Estimation of Oil and Gas Program Benefits
Rodney Geisbrecht, DOE

Natural Gas Research and Development
Bob Silva, DOE

Assessing the Benefits of R&D: A Framework of Analysis and an Application Using NEMS
John Pyrdol, DOE

MEETING OF PANEL ON DOE'S DISTRIBUTED ENERGY RESOURCES PROGRAM
WASHINGTON, D.C.
OCTOBER 24-25, 2005

DOE Program
Pat Hoffman, DOE Office of Energy Efficiency and Renewable Energy

Estimating Benefits of EERE Combined Heat and Power R&D
Frances Wood, OnLocation, Inc.; Chip Friley, Brookhaven National Laboratory; Chris Marnay and Kristina Hamachi LaCommare, Lawrence Berkeley National Laboratory

COMMITTEE MEETING
WASHINGTON, D.C.
OCTOBER 26-29, 2005

DOE Experience with Decision Trees
Darren Mollot, Office of Fossil Energy

DOE Energy Working Group
John R. Sullivan, Associate Under Secretary for Energy, Science and Environment

MEETING OF PANEL ON DOE'S CHEMICAL INDUSTRIAL TECHNOLOGIES PROGRAM
WASHINGTON, D.C.
NOVEMBER 3-4, 2005

Discussion of DOE's ITP Chemicals Subprogram
Dickson Ozokwelu, DOE

MEETING OF PANEL ON DOE'S LIGHT-DUTY VEHICLE HYBRID TECHNOLOGY R&D PROGRAM
WASHINGTON, D.C.
NOVEMBER 7-8, 2005

Discussions of DOE's Vehicle Technologies Program
Ed Wall, DOE

MEETING OF PANEL ON DOE'S NATURAL GAS EXPLORATION AND PRODUCTION PROGRAM
WASHINGTON, D.C.
NOVEMBER 9-10, 2005

Discussion of DOE's Natural Gas E&P Program
Bob Silva, DOE

**MEETING OF PANEL ON DOE'S
INTEGRATED GASIFICATION COMBINED CYCLE
TECHNOLOGY R&D PROGRAM
WASHINGTON, D.C.
NOVEMBER 29-30, 2005**

Q&A and Discussion of DOE's IGCC Program
Jay Braitsch, DOE

**MEETING OF PANEL ON DOE'S DISTRIBUTED
ENERGY RESOURCES PROGRAM
WASHINGTON, D.C.
DECEMBER 13-14, 2005**

Q&A and Discussion of DOE's Program
Pat Hoffman, DOE

**MEETING OF PANEL ON DOE'S CARBON
SEQUESTRATION PROGRAM
WASHINGTON, D.C.
DECEMBER 15-16, 2005**

CO_2 Capture R&D Pathways for IGCC
Jared Ciferno, DOE

**COMMITTEE MEETING
IRVINE, CALIFORNIA
FEBRUARY 10-11, 2006**

Closed Meeting

F

Guidance on Prospective Benefits Evaluation

OVERVIEW

A methodology is presented for prospectively evaluating the benefits of applied energy research and development (R&D) programs. The types of benefits to be evaluated are based on three fundamental objectives that have guided energy policy since the energy crisis of 1973-1974: economic improvement, environmental protection, and energy security. Two principal sources of uncertainty in the benefits calculation are considered: (1) the risk associated with the technical success of an R&D program and (2) the risk associated with market acceptance of a technology. The methodology calls for the use of a decision tree as a framework for organizing the benefits calculation, which is then presented in a results matrix that uniformly summarizes important data and estimated benefits for all technology programs.

THE RESULTS MATRIX

Prospective evaluation is complicated by uncertainty about how the future will unfold. A standard way to take uncertainties into account in cost-benefit analysis is to consider the "expected benefit," which is the probability-weighted average of the benefits associated with all the possible outcomes of a program.

The benefits framework proposed for prospective evaluation incorporates these characteristics of the possible outcomes and attendant investment risk. This framework is summarized in matrix form in Figure F-1. The bottom three rows represent the same three kinds of benefits—economic, environmental, or security benefits—considered in the retrospective analysis, *Energy Research at DOE: Was It Worth It?* (NRC, 2001).

Economic net benefits are based on changes in the total market value of goods and services that can be produced in the U.S. economy under normal conditions, where "normal" refers to conditions absent energy disruptions or other energy shocks. Economic value can be increased either because a new technology reduces the cost of producing a given output or because the technology allows additional valuable outputs to be produced by the economy. Economic benefits are characterized by changes in the valuations based on market prices. This estimation must be computed on the basis of comparison with the next-best alternative, not some standard or average value.

Environmental net benefits are based on changes in the quality of the environment that have occurred, will occur, or may occur as a result of the technology. These changes could occur because the technology directly reduces the adverse impact on the environment of providing a given amount of energy service, for example by reducing the sulfur dioxide emissions per kilowatt-hour of electric energy generated by a fossil fuel fired power plant, or because the technology indirectly enables the achievement of enhanced environmental standards, for example by introducing the choice of a high-efficiency refrigerator. Environmental net benefits are typically not directly measurable by market prices. They can often be quantified in terms of reductions in net emissions or other physical impacts. In some cases market values can be assigned to the impacts based on emissions trading or other indicators.

Security net benefits are based on changes in the probability or severity of abnormal energy-related events that would adversely impact the overall economy, public health and safety, or the environment. Historically, these benefits arose in terms of "national security" issues, initially the assurance of energy resources required for a military operation or a war effort. Subsequently they focused on dependence on imported oil and vulnerability to interdiction of supply or cartel pricing as a political weapon. More recently, the economic disruptions of rapid international price fluctuations from any cause have come to the fore.

These three classes of benefits have been chosen to reflect the programmatic goals of the U.S. Department of Energy (DOE) offices for which the study was conducted. The three classes are not meant to be comprehensive. For example,

Program Name:

Program Goals:

Year Goals Achieved:

Costs:

Current Funding Cycle:

Expected Cost to Completion:

		Global Scenario		
		Reference Case	High Oil and Gas Prices	Carbon Constrained
Program Benefits	Technical Risk			
	Market Risks			
Expected Program Benefits	Economic Benefits			
	Environmental Benefits			
	Security Benefits			

FIGURE F-1 Results matrix for evaluating benefits and costs prospectively.

they do not include the benefits of fundamental research sponsored by the Office of Science, health benefits, or other quality of life benefits that could be unintended but real consequences of some applied R&D programs.

SCENARIOS

The benefit of a new technology will often depend on developments quite unrelated to the technology itself. For example, the benefit of energy-efficient lighting will depend on the cost of electricity, which in turn depends on the costs of fuels like natural gas and coal used to generate electricity. Similarly, the economic benefits associated with carbon sequestration will depend on carbon emissions being constrained or taxed and the level of constraint. Thus, assumptions about the future of prices and environmental constraints, among other things, can have a significant effect on the prospective benefits of a technology.

The scenarios in Figure F-1 represent three possible future states of the world that are likely to affect the benefits associated with a wide variety of DOE applied R&D programs. It is recommended that the same three scenarios be used to evaluate all the programs. The use of a common set of scenarios across program evaluations will allow reviewers to consider many programs without having to learn definitions for multiple scenarios and will facilitate comparisons across programs. To ensure consistency, it is important for the scenarios used in the review of different programs to be built on precisely the same assumptions (e.g., the same oil and gas price assumptions); it is not sufficient for them to be similar in some high-level or vague sense.

The three global scenarios are as follows:

1. *Reference Case.* This is the scenario developed by the Energy Information Administration and described in the *Annual Energy Outlook* (AEO) (EIA, 2004). The AEO provides detailed forecasts of U.S. energy supply, demand, and prices through 2025. This scenario represents the government's official base-case forecast. The 2004 Reference Case assumed as follows:

—World oil prices decrease from their current levels to about $24 in 2010 and then increase to about $27 per barrel in 2025.

—Natural gas consumption increases significantly— i.e., 23 trillion cubic feet (Tcf) to 26 Tcf in 2010 and 31 Tcf in 2025—with prices decreasing from current levels to $3.49 per thousand cubic feet (Mcf) in 2010, then increasing to $4.47 per Mcf in 2025.[1]

—Primary energy consumption increases from 97.7 quadrillion British thermal units (quads) in 2002 to 136.5 quads in 2025.

—GDP grows 3.0 percent per year to 2025.

—Carbon dioxide emissions from energy consumption grow from 5,729 million metric tons in 2002 to 8,142 million metric tons in 2025.

2. *High Oil and Gas Prices.* This scenario assumes that oil prices will remain very high throughout the period and that constraints on natural gas supply lead to higher natural gas prices and higher electricity prices. For example, the oil price in 2012 in this scenario is $33.41 versus $23.98 in the Reference Case, and the natural gas price in 2012 is $4.53 per Mcf versus $3.92 per Mcf in the Reference Case.

3. *Carbon Constrained.* This scenario, developed by the DOE, assumes that U.S. emissions of carbon are constrained in response to environmental concerns. Specifically, this scenario assumes that the Global Climate Change Initiative goal of an 18 percent reduction in national greenhouse gas intensity (below the 2002 level) is achieved by 2012 (White House, 2002) and that annual emissions are held constant at that level thereafter. Relative to the Reference Case, this leads to increased demand for natural gas and increased prices (for example, $6.79 per Mcf in 2012 versus $5.54 per Mcf in 2012 in the 2004 Reference Case) as well as greater reliance on renewable electricity, unless carbon sequestration technologies are successful.

The three scenarios considered in the prospective benefits matrix (Figure F-1) are not intended to capture everything that could happen in the future. Indeed, there are an uncountable number of different possible futures, including other levels of oil and gas prices and carbon constraints. Therefore, reviewers are discouraged from specifying probabilities for these scenarios in order to calculate a single "expected benefit" that represents a probability-weighted average across the three scenarios. Rather, the three scenarios are intended as a representative set of scenarios that highlight particular policy issues and provide a form of sensitivity analysis for the benefits analysis. Displaying the results across the three scenarios rather than collapsing the scenarios into a single expected value allows reviewers to focus on scenarios that they view as more likely or as representing particular policy objectives or interests. For example, a policy maker contemplating constraints on carbon emissions might look carefully at benefits in the Carbon Constrained scenario to see what DOE is doing to help prepare the United States for this possibility.

Projects and programs may yield benefits in some but not all scenarios. In the study *Prospective Evaluation of Applied Energy Research and Development at DOE (Phase One): A First Look Forward* (NRC, 2005a), the carbon sequestration program was viewed by the expert panel as providing benefits only in the Carbon Constrained scenario. The panel did not believe the sequestration technology would be deployed under the two other scenarios and so would yield zero benefit. But even if the benefit of a program is zero for a given scenario, it is important that all program evaluations consider benefits in all of the scenarios. While carbon constraints or high oil and gas prices might not be particularly relevant to the performance of lighting technologies, the economic benefit associated with efficiency enhancement might vary across scenarios as electricity prices change.[2] Even if the benefit does not change, the scenarios should be considered and expected benefits reported, because at some point reviewers may combine benefits across projects or programs for a particular scenario, and the absence of an estimate for benefit in a given scenario should not be confused with zero benefit.

While these three global scenarios should serve to illustrate the potential benefits of most DOE applied R&D programs, if a particular program is designed to provide benefits under some other set of circumstances, the DOE and/or review panels are invited to consider benefits in additional scenarios. In such a case, the alternative scenario should be described clearly and its benefits for this scenario should be calculated and reported in a manner consistent with the principles outlined elsewhere in this report. It should be considered in addition to the three global scenarios considered in the prospective benefits matrix—that is, it should not be

---

[1] All prices are stated in 2002 dollars and thus do not reflect inflation.

[2] Specifically, the probability of market acceptance of the energy efficient technology may vary from scenario to scenario. Most directly, differences in the price of energy change the attractiveness of energy-efficient technologies to consumers. In addition, high energy prices could drive investments in complementary technologies or cause reassessments of regulatory standards or the relative risks of different technologies.

viewed as a substitute for or a modification of one or more of those scenarios.

THE DECISION TREE FRAMEWORK

Introduction

The estimated benefit of a program is subject to multiple sources of uncertainty, both as a result of what happens in the program itself and what happens to the alternative technologies or in the policy environment. The basic principle of program assessment remains the same as in a retrospective analysis: Find the difference between social benefits with the government program and without it. However, the implications of the various uncertainties must be considered carefully. In this section, the formal mechanism for estimating benefits—the decision tree framework—is described. First, some of the key factors to consider in benefit assessment are discussed.

Key Factors in Benefit Assessment

It is essential first to define the outcomes for which it is worth calculating the benefits. Three factors typically determine the alternative outcomes of an applied R&D program. One (or two) of the factors might not be important in a specific case, but their ubiquity implies that each assessment should look at all three so that analysis teams can determine which of them need to be explicitly incorporated into the benefit calculations. Each is discussed briefly.

Estimating the Net Benefit of Government Support

An estimation of the expected net benefit of a government program involves an explicit or implicit comparison of the possible outcomes with the government program and the possible outcomes absent the program. For example, a government program might lead to a research team making a significant technology advance. The expected benefits of that advance could be estimated. However, to determine the net benefits of the government support, it is necessary to consider the extent to which this or other research teams are likely to have achieved the same technology advance absent the government support and to estimate the benefit and probability of the advance. The expected net benefit of the government program is the difference between the expected benefits with and without government support, not the expected benefit of the technology advance with government support.

Considering the Uncertainty Surrounding the Next-Best Technology

Similarly, estimation of the expected net benefits of a government program requires either explicit or implicit consideration of how the market would evolve without the technology being developed by the DOE research program. Take, for example, the estimation of net benefits of a successful government program for new solid state lighting technology. One could assess the benefits of the new lighting technology advance assuming it is coupled with a government program designed to hasten its market adoption. One would have to compare these benefits with those of a program designed to hasten the adoption of the next-best lighting technology—say, the next generation of compact fluorescent lights. It would not be appropriate to estimate the expected net benefits based on a comparison with the existing generation of compact fluorescents or (and this would be even less appropriate) the existing generation of incandescent lighting technology.

In some cases, there might be considerable uncertainty about the benefits that accrue due to advances in alternative technologies; in some cases the benefits might change radically depending on potential, but uncertain, advances in the next-best technology. Commercial penetration of a moderately successful fuel cell car, for example, might be substantially reduced if the fuel efficiency of hybrid-electric vehicles improves dramatically. In cases such as this, it may be necessary to consider explicitly different levels of success for the next-best technologies and the probabilities that these levels will occur.

Considering Enabling/Complementary Technologies

In many cases, estimating the expected net benefits of a government program requires either explicit or implicit consideration of enabling or complementary technologies. Whenever the market acceptance of a particular technology depends strongly on the existence of other technologies complementing the technology in question or enabling it, it is necessary to assess the probability that enabling or complementary technology will be successful. For example, a successful government program to advance the technology of hydrogen fueling stations for light-duty vehicles would have little benefit unless the various technologies for hydrogen vehicles—particularly fuel cells and onboard storage—were to advance enough that many people would choose to purchase hydrogen fuel cell vehicles; thus it is necessary to assess the probabilities that these complementary technologies will be successful in order to accurately estimate the net benefits of the government program to advance the technology of hydrogen fueling stations.[3]

In some cases, the benefits of a program are greatly enhanced by complementary technologies but the program

---

[3] Note that this treatment of complementary technology allows the evaluation of specific programs (e.g., vehicle fuel cells) without having to conduct a complete evaluation of the entire technology package in which they are embedded (e.g., the hydrogen economy). Without this simplification, benefits evaluations would become unwieldy.

will provide at least some value even in their absence. If the distinction is important, then an assessment of the probability of success for the complementary technology and the expected benefits of the program considering both potential outcomes—the benefits of the program with and without the complementary technology—will be needed.

Decision Tree

Estimating benefits requires the application of a decision tree process that includes consideration of the three key components of government energy R&D program evaluation and clarifies the relationships among them. Figure F-2 illustrates the possible relationships among these three key components using a decision tree, where the first node (the decision node) represents the government action—to pursue the program or not; the second node (first chance node) is the possible outcome of the program; and the third node (second chance node) represents the multiple factors that determine market acceptance, including developments in the next-best technology and the success of enabling and complementary technology programs.

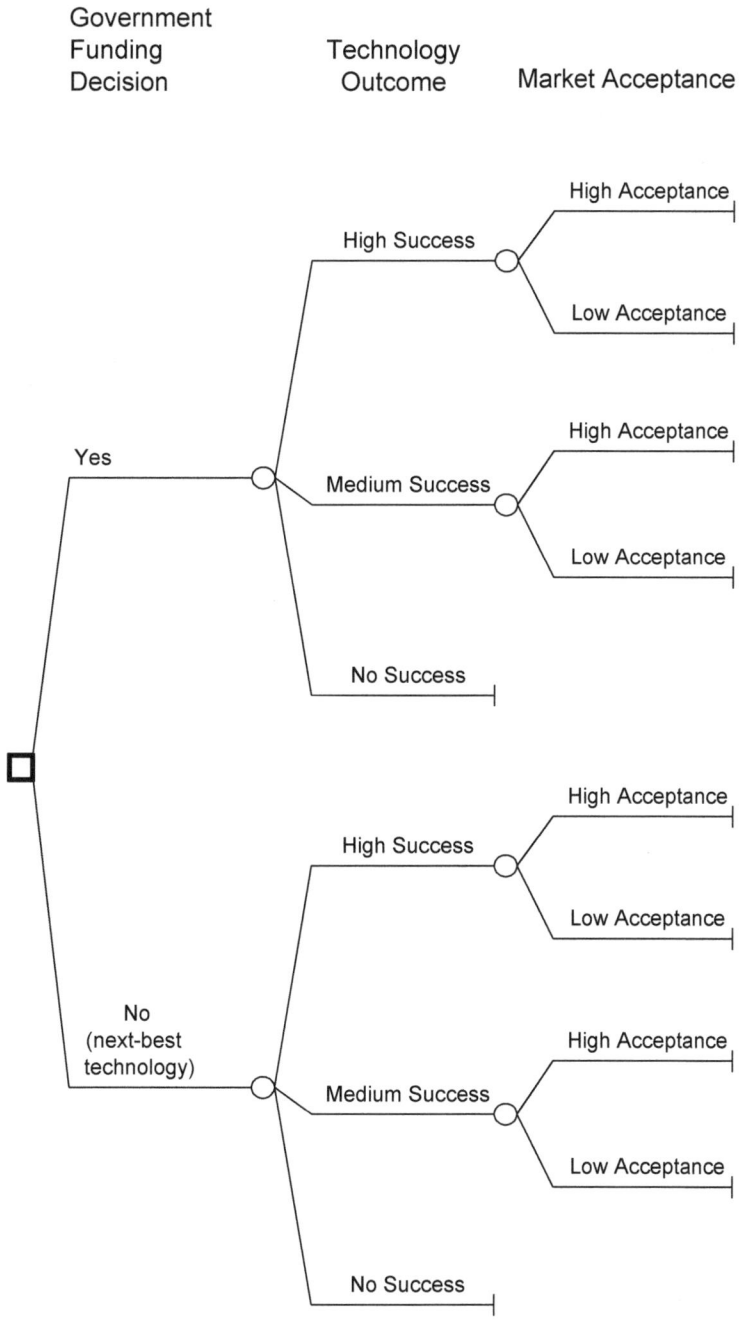

FIGURE F-2 Decision tree.

The decision tree provides a framework for organizing the benefits calculation. The expert panel must specify probabilities for the various uncertainties in the tree. Because the probabilities need not be the same in each global scenario, the expected benefit could differ in each scenario. Then, as discussed in the preceding section, the value of the government program is the difference between the expected benefit of the with-government-support alternative and that of the without-government-support alternative.

Figure F-3 illustrates the decision tree applied to the advanced lighting program, with numerical values included. These numerical values are provided only to show the general structure of such a decision tree and to illustrate the calculations that would be used.

In this decision tree, the government has one basic decision, to invest in the R&D program or not to invest. In either case, three possible levels of lighting efficacy could be achieved by U.S. industry: 150 lumens per watt (lpw), 100 lpw, or no change from the current situation. If DOE invests, then the probability of the greatest advance, 150 lpw, would be increased to 10 percent (from 0 percent in the absence of DOE research); the probability of the medium advance, 100 lpw, would be increased to 50 percent (from 30 percent in the absence of DOE research); and the probability that there would be no advance would be decreased to 40 percent (from 70 percent in the absence of DOE research). Note that these probability assessments do not show that DOE investment will guarantee success—they show that DOE investment increases the probability of successful outcomes.

DOE investment also can change the probability that other

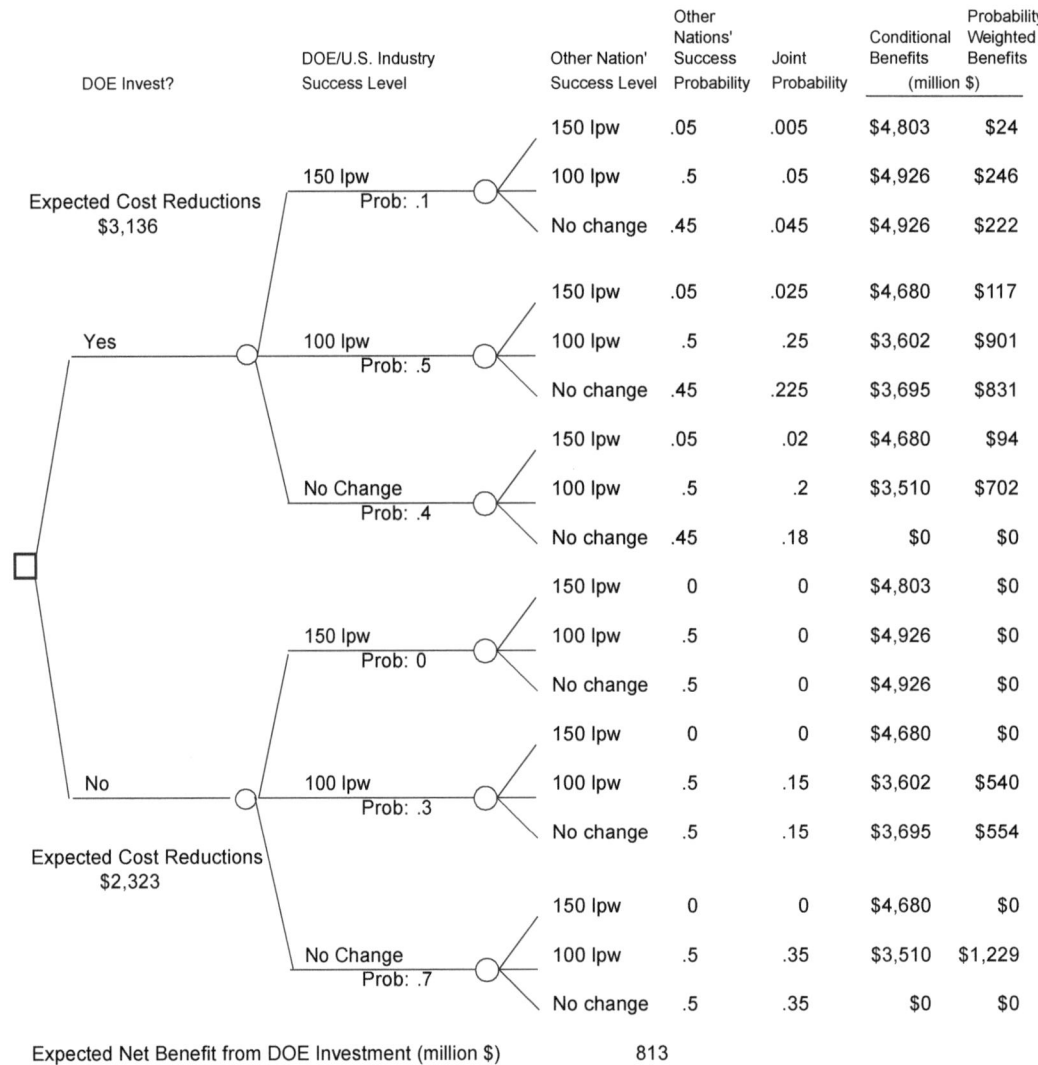

FIGURE F-3 Example of decision tree applied to advanced lighting program. The net expected benefit is the difference between the upper and lower half, or $813 million. The expected benefit of the upper half of the tree (for which DOE invest? = yes) is $3,136 million—the sum of the first nine probability-weighted benefits in the column on the far right. The expected benefit of the lower half of the tree (for which DOE invest? = no) is $2,323 million—the sum of the second set of nine probability-weighted benefits. See also discussion in the text.

nations will have R&D successes. These probabilities are shown in the first column. This decision tree assumes that there will be a 5 percent probability that other nations will achieve 150 lpw if DOE conducts a research project but no probability of that success in the absence of DOE research. It assumes that there will be a 50 percent probability of other nations achieving 100 lumens per watt, with or without DOE investment. With DOE research, there is a 45 percent probability that other nations will not have any incremental improvement; without DOE research, that probability would be 50 percent.

Given those probabilities of U.S. and foreign success, the probabilities of the combinations of outcomes are given in the second column. For example, the probability of U.S. industry achieving 150 lpw and other nations achieving 100 lpw is .05, calculated as the product of .10 and .50.

The third column provides the conditional benefit estimates, which vary according to the combinations of outcomes being considered. Here again, the numbers are presented only for the purpose of illustration. For example, the benefit to the United States, conditional on U.S. industry's achieving 150 lpw and other nations' achieving 100 lpw is $4,926 million.

The contributions to the expected value are given in the third column, as the product of the numbers in the first and second columns. The contribution to expected value of the combination—U.S. industry achieving 150 lpw and other nations achieving 100 lpw—is $246 million, the product of .05 and $4,926 million.

The expected value of all possible combinations of outcomes, given DOE R&D investment, is the sum of the top nine numbers in the column on the right. This sum is $3,136 million. The expected value of all possible combinations of outcomes if DOE makes no such R&D investment is the sum of the bottom nine numbers in the column on the right. This sum is $2,323 million.

Note that as a result of U.S. industry R&D and R&D in other nations, the United States would capture benefits even without DOE investment. But those benefits would be smaller than they would be if DOE did invest in R&D.

The difference between the benefits with DOE investment—$3,136 million—and the benefits absent DOE investment—$2,323 million—is the expected value of the gain resulting from DOE investment. This difference of $813 million is the overall benefit from the DOE investment. The overall benefit is not the benefit with DOE investment—$3,136 million—since a large share of those gains to the United States would occur even without DOE investment.

Expert Evaluation of Probabilities

Evaluating the probabilities associated with the different branches on the tree—the likelihood of technical success for the program and the likelihood it will achieve different levels of market penetration—is a critical part of the work of the expert panels. The following subsections include guidelines for assessing probabilities at each of the nodes in the decision tree.

Government Program Support (Decision Node)

The difference between paths with and without the government program is a measure of the government role. As indicated earlier, evaluation seeks not just to measure change that accompanies a government program, project, or other activity, but also to determine whether the change is attributable to specific government intervention. Evaluation must rule out alternative, competing explanations for an observed or predicted change. For example, a government program may be launched to develop a new energy-conserving technology, and, indeed, it may succeed. But the same outcome, or a portion of it, may have resulted without the government program. The decision tree methodology suggests using a counterfactual approach in the attempt to isolate or demonstrate the effects of the program under evaluation. In a decision tree analysis, one branch describes the path if a specified government program with identified characteristics, including a defined level of funding, is adopted. An alternative branch of the tree describes the path without that government program. Other branches may describe the path of the government program at various funding levels. The probabilities and values for the paths along each branch are based on the assumption that the program is funded at the specified level. By comparing the resulting expected value benefits for alternative branches, the differential expected benefits attributable to the government program can be identified.

A government program may affect societal benefits in multiple ways, including the following:

1. *The program might result in technology development and use that otherwise would not have occurred.* If it is thought that nothing would have happened in the foreseeable future without the government program, this is captured in the benefits stream for the without-government-program branch. The only benefits associated with the without-government-program branch are those attributable to the next-best technology available when the government program would have been completed.

2. *The program might accelerate technology development and use.* If the government program is viewed as causing the development and use of a technology that otherwise would be developed in the same way but at a slower pace, this effect is captured in the timing of the estimated stream of benefits underlying the benefit calculations at the end of the with-government-program branch as compared with the timing of the benefits stream at the end of the without-government-program branch. For example, if the with-government-program branch is expected to result in the technology being developed and adopted into use, say, 3 years sooner

than the without-government-program branch, the stream of benefits will be shifted forward by 3 years and the government's contribution will be credited with the difference in the expected present value of the two benefits streams.

3. *The program might improve a technology and make it more attractive to users.* If the government program provides significant advances over the next-best technology, this can be captured in several ways. The advance may increase the probability of market acceptance. It may improve market penetration. It may generate larger unit benefits in use. Each of these effects can be captured in one of the branches of the decision tree.

4. *The program might increase the probability of technical or market success.* If the government program reduces the risk associated with achieving technical or market success, this is captured in the comparative probabilities assigned to the with- and without-government-program branches of the tree in the technical and/or market risk decision nodes. If the government role is to reduce technical risk, a higher probability of technical success assigned to the with-government-program branch will result in greater expected benefits projection for that branch and a larger expected benefit will be attributed to the government program, calculated as the difference between the expected values of the with- and without-government-program branches.

5. *The program might enlarge the scope of the technology to make it more "enabling."* Enabling technologies are characterized as generating larger-than-average spillover benefits; that is, benefits that accrue broadly in society rather than more narrowly to direct private-sector investors in the research. If the government program enlarges the scope of the technology in ways that create a more enabling technology platform, the effect is captured in the benefits stream of the with-government-program branch of the decision tree, which includes a larger societal benefit than the without-government-program branch.

6. *The program might increase collaborative and multidisciplinary research.* The government program may promote collaboration among researchers by helping to overcome existing barriers to collaboration. Networks of collaborative R&D activity are increasingly seen as playing an important role in innovation. If the government program increases collaboration and if increased collaboration increases the likelihood of technical or market success, this effect is captured by assigning different probabilities of success to the with- and without-government-program branches. If, on the other hand, the change in collaborative effort increases the scope of the research, increasing its enabling characteristics, this effect is captured in the with- and without-government-program branches as differences in the estimates of societal benefits.

7. *The program might produce a combination of effects.* If the government program is expected to have more than one of the effects listed above, these may affect both the comparative probabilities assigned to technical and market success and the comparative projected benefit streams estimated for the with- and without-government-program branches of the decision tree. Multiple effects can easily be accommodated in the context of the decision tree analysis. As in the other cases, it remains necessary to take the difference between the with- and without-government-program branches to find the expected benefits of the government program.

Technical Outcomes (First Chance Node)

In retrospective evaluations, one knows if a specific research program was successful in terms of meeting its technical (and commercial) objectives. In prospective evaluations, it is not known in advance whether a program will be successful. To calculate benefits one must therefore consider the likelihood that the program will be successful. While technical success can be viewed in relation to achievement of a single goal or outcome, a given research program might be judged successful across a range of outcomes, each with its own probability of technical success. As is discussed above, if the possibility of multiple outcomes is ignored, research benefits may be underestimated. Moreover, the use of a single "representative" outcome will usually be inaccurate and will mask assumptions about the relationships between the program outcomes and the alternative technologies.

DOE often identifies single goals for programs, and these goals may be stretch goals in the sense of being at the high end of possible outcome value but having a relatively low probability of technical success. Attaining less optimistic goals with delayed or only partial benefits may still have considerable value as well as a higher probability of technical success. Consideration of stretch goals alone is therefore also likely to lead to an underestimate of benefits. The number and nature of the outcomes to include will vary by project.

Probabilities of technical success (at any outcome level) are likely to depend on the level (and, in some cases, the quality) of resources—financial, technical, and managerial—expected over the life of the program, so that probabilities change with funding/resource level and the calculated expected net benefits are conditional on that level.

Market Risk (Second Chance Node)

"Market risk" refers to the probability of a given technology being moved out of the lab and deployed into use.[4] This risk is expressed as a probability for each case under evaluation and entered into a third node for each relevant branch of the decision tree. It is critical to the estimation of expected

---

[4]"Market risk" may be distinguished from other risk terms commonly used in finance and economics, including "financial risk," where the general category is subdivided into "systematic risk," referring to risk associated with changes in the business cycle, and "unsystematic risk," referring to risks associated with the variation in performance associated with a particular firm. "Economic risk" generally refers to the likelihood that the benefits from an investment may not be sufficient to cover its costs.

benefits, as an applied R&D program yields benefits only if the technology is used.

Probabilities are applied to market penetration estimates. The probability that a technology will be deployed is distinct from the rate at which it is deployed—that is, its market penetration. It is necessary to estimate the rate at which the technology will come into use.

Many factors influence whether, and to which extent, a technology will be deployed into use. These factors do not apply to all programs, and it is recommended that the chief factors that influenced the probability estimates be identified in either a row at the top of the results matrix or in an appendix. These factors include the following:

• *Market demand.* To what extent does the technology meet a current demand? Does it enable users to solve an important problem or exploit a promising opportunity?

• *Competition.* What are the competing technologies that the target technology must overcome in order to be accepted in the market? Note that there are always competing technologies. Are others working on the same technology? Are others working on different technologies that might meet the same need? In the United States? Outside the United States? Who is performing the work?

• *Window of opportunity.* How long will the need for the technology exist? Are there factors that might eliminate or reduce the need for this technology? For example, are advances in a competing technology likely to outstrip those in the target technology, wiping out its advantages? When might this occur?

• *Potential hazards.* Are there potential environmental or safety concerns that might limit the use of the technology? For example, are there any by-products that might create an environmental or biological hazard, incurring mitigation costs or exposure to liability?

• *Ease and cost of implementation.* To what extent does adoption of the technology require changing existing systems or ways of doing things? For example, it would be easy to introduce a new catalyst into an existing reactor. On the other hand, a hydrogen-powered automobile would require a new fuel distribution system if the hydrogen is centrally produced, making implementation more difficult. To what extent will adoption of the technology require large capital investment? To what extent will adoption cause disruptions and downtime in current operations? To what extent will adoption require worker retraining?

• *Resistance by special interests.* To what extent will those adversely affected by the new technology lobby to retard its adoption, and how successful are they likely to be?

• *New regulations.* To what extent are new regulations likely to promote or impede the adoption of the technology? For example, unexpectedly stringent environmental regulations on diesel emissions made obsolete the fuel and engine research programs designed to meet the more modest objectives that had been anticipated. Conversely, adoption of appliance efficiency standards can lead to development of new technologies, as in the case of refrigerators (NRC, 2001, pp. 97-98).

• *Complementary and prerequisite technologies.* Is adoption of the new technology dependent on the availability of other technologies? Are these technologies still in the pipelines of other R&D programs? Will they be available in time to support the technology under evaluation?

Market risk factors are often critical to evaluating the potential of an R&D program. Indeed, for investments in fairly specific technologies, the risks associated with market acceptance may overwhelm those associated with technical success. Alternatively, fundamental programs that yield results applicable to a range of technologies and market conditions may be less susceptible to the market risks discussed here: They may be applicable in a wide range of regulatory regimes and may contribute to multiple technologies, at least some of which are likely to be available and of interest in the relevant time frame. Of course, long-term fundamental programs may be very risky in that the likelihood that all of their goals are achieved is remote.

Long-term, fundamental R&D requires further development and often further research before reaching a commercial outcome. Applied R&D such as the DOE programs considered in this report aims to develop technologies with specific performance and cost criteria that will be commercial in a time frame consistent with the schedule of the applied R&D project. It is this latter type of R&D having more immediate commercial applicability that the methodology has been designed to evaluate. The types of benefits evaluated in the methodology—for example, economic benefits—are consistent with the goals of technology development programs and the expectations of those making the investments. A further difference between fundamental and applied R&D is that the former has knowledge as its primary goal while the latter has knowledge as a by-product. Knowledge benefits may include unanticipated and not closely related technological spin-offs that are made possible by the research programs. Because the methodology proposed here is for applied R&D it has economic, environmental, and security benefits as its primary goal and does not give credit for knowledge generated in the course of a technology's development.

REPORTING OF RESULTS

All panels will report their results consistently using the benefits results template discussed in this Appendix (see Figure F-4). The format requires that each panel present similar information about the program under review and its findings so that users of the results can make informed program and funding decisions. The benefits results template consists of three sections. The first section lists important program information provided by the Department of Energy (DOE). The second section presents the panel's results matrix of

PANEL NAME:

Program Name:
Program Goals:
Year Goals Expected to Be Achieved:
Program Costs:
Funding to Date: $
Current Funding: $
Proposed Year Funding: $
Expected Cost to Completion: $
Industry and Foreign Government Funding: $
Key Complementary/Interdependent DOE Programs:

All benefits are cumulative through 2050 and are reported in 20XX year dollars.

		Global Scenario		
		Reference Case	High Oil and Gas Prices	Carbon Constrained
Program Risks	Technical Risk			
	Market Risks			
Expected Program Benefits	Economic Benefits			
	Environmental Benefits			
	Security Benefits			

FIGURE F-4 Template for presenting panel results.

Comments and Observations: One to two paragraphs

> e.g., provide a summary of the panel's completed assessment and estimate of expected benefits of the DOE program.

Technical Risks: 5 to 10 lines

> e.g., describe the program's risks in sufficient detail and clarity, noting program interdependencies, technical and infrastructure innovations and breakthroughs needed, and competitive alternatives, and so on.

Market Risk: 5 to 10 lines

> e.g., note factors that might affect market acceptance, including customer preferences, pricing, competitive domestic and foreign activities; next-best technologies issues, regulatory concerns, and so on.

Benefits: 5 to 10 lines

> e.g., discuss specifically the estimation of benefits, uses and interpretations, caveats, outstanding issues, and so on.

Program Observations: 5 to 10 lines

> e.g., notable accomplishments/gaps, opportunities, spin-offs, and so on.

FIGURE F-4 Continued

the expected benefits of the program using the committee's methodology applied to DOE's expectation of program outcomes as well as the alternative outcomes anticipated by the panel. The third section of the template provides the panel's findings and conclusions and expert opinions regarding the application of the methodology to the program.

In addition, panels must prepare a brief report—about 10 pages—to include background, summary of the DOE program, technical and market risk assessment, results and discussion, technical and market success, benefits estimation, role of DOE funding, decision tree, benefits calculations, results matrix, summary, and conclusions (including recommendations).

DOE Program Information

The first section of the benefits results template will provide summary information characterizing the DOE program. This information will include an identification of the program goals; funding, including outside funding; and the program's critical interdependencies with other DOE program efforts. This information is provided directly by DOE.

Results Matrix

The results matrix in the template is for presenting the panel's views of the program's technical and market risks and the expected benefits of the R&D program. Market risks reflect the panel's expert opinion of the program's market acceptance assuming that the goals have been met. Assessing the market risk includes consideration of price, infrastructure development and support, ease of use, competition from other technologies or innovations, time, and end-user preferences. In the committee's decision tree framework, there is no single point estimate of technical or market risk, owing to the many possible outcomes. The quantitative estimates of probability are recorded in the decision tree, as discussed in Appendix F, whereas the results matrix should be annotated with a discussion of the key factors that contribute to the technical and market risk.

The benefits estimated by the panel represent an expected value benefit. This is calculated by adjusting the benefits that DOE assumes will be accrued if the program goal is met for the technical and market risks identified by the panel. Figure F-4 provides expected benefits for three scenarios to reflect a bounding of benefits for three possible future states.

Panel Comments

The panel's opinion and quantification and discussion of the risks are provided in the third section of the template in Figure F-4. The information required to be included in this section is listed in the template itself. The panel must complete the matrix using sufficient clarity and transparency to allow readers and potential users of the matrix to make informed and reasoned decisions about future goals, funding levels, and expected benefits. The template should be used by all panels to ensure consistency in reporting, use, and interpretation.

G

Information to Be Requested of the Department of Energy

Each panel applying the methodology to the various Department of Energy (DOE) programs will need to gather information about its particular programs. The information request and supporting documentation should take the form of a brief program assessment summary (PAS) (see Figure G-1). Individual assessment summaries should be prepared for each project in the program portfolio. It is suggested that DOE should provide the following information for each program:

1. Program roadmap and logic;
2. Articulation and quantification of program goals (near term, intermediate, final);
3. Annual program budgets—to date, current, and needed to achieve the program's goals;
4. Co-funding—to date, current, and needed to achieve the program's goals;
5. Identification of complementary or competitive foreign and nonfederal domestic programs;
6. Identification of other programs that comprise enabling and complementary technologies to the program under review;
7. Key accomplishments (milestones met) to date;
8. Barriers to program goal accomplishment and an identification of their importance to the program;
9. Technological or infrastructure innovations or breakthroughs needed to meet the program's goals and identification of competitive technology; and
10. Other information and data as might be requested by the panel.

Similarly, DOE should provide the following information for each project in the program portfolio:

1. Description of how the project aligns with and supports the program's goals;
2. Articulation and quantification of project's goals (near term, intermediate, final);
3. Annual project budget(s)—to date, current, and needed to achieve the program's goals;
4. Project co-funding—to date, current, and needed to achieve the project's goals;
5. Identification of those projects, in addition to those under review, that comprise enabling and complementary technologies;
6. Key accomplishments (milestones met) to date;
7. Barriers to project goal accomplishment and an identification of their importance to the project and program; and
8. Technological or infrastructure innovations or breakthroughs needed to meet the project's goals.

Program Name, Description, and Goals Narrative

[ ]

Program Budgets Narrative

[ ]

Program or Project Interdependencies/Related Foreign and Domestic Programs Narrative

[ ]

Accomplishments to Date and Critical Milestones Narrative

[ ]

Barriers to Overcome and Innovations Required Narrative

[ ]

FIGURE G-1 Three-page program assessment summary (PAS) form, to be completed by DOE.

APPENDIX G

National Energy Modeling System (NEMS) Scenarios Descriptive Narrative

NEMS Calculations of Benefits Narrative

Key NEMS Input Assumptions

FIGURE G-1 Continued

Program Name:

Program Goal:

Goal Quantification											
Target (goal quantified)											
Year Expected to Be Met											
Life Expectancy											
	Cumulative to Date	Current Year	Needed to Meet Goals—Future Years								
			1	2	3	4	5	6	7	8	9	10	Total
Program Funding											
DOE Funding											
Program Co-funding											
Out-of-Program Funding											
Calculation of Benefits			Year								
			% of Mrkt Sales								
Market Penetration (annual market % penetration to full acceptance) from First Year to Year of Maximum Expected Penetration			1	2	3	4	5	6	7	8	9	10
	Five-Year Average	Current Price	Annual Growth Rates (2- to 3-year periods)								
			1	2	3	4	5	6	7	8	9	10	Total
NEMS Modeling Assumptions											
Energy Prices											
Electricity											
Natural Gas											
Oil											
Other—specify											
			Annual or for specified 2- to 3-year periods								
			Year 1	2	3	4	5	6	7	8	9	10	Total
Energy Savings (units)											
Electricity											
Natural Gas											
Oil											
Other—specify											
			Annual or for specified 2- to 3-year periods								
			Year 1	2	3	4	5	6	7	8	9	10	Total
Environmental Benefits											
CO_2											
NO_x											
SO_2											
PM											
Hg											
			Annual or for specified 2- to 3-year periods								
			Year 1	2	3	4	5	6	7	8	9	10	Total
Discount Rate											
Other as Requested by the Panel											

FIGURE G-1 Continued

H

Report of the Panel on DOE's Integrated Gasification Combined Cycle Technology R&D Program

INTRODUCTION AND OBJECTIVE OF THE STUDY

Integrated gasification combined cycle (IGCC) is a technology that can use a variety of feedstocks to produce electricity, synthetic gas, and other by-products while minimizing the environmental impacts of doing so. The underlying coal gasification technology has been in commercial use since the beginning of the 20th century and has been demonstrated throughout the world to produce a variety of valuable end and intermediate products using a variety of configurations, technologies, and feedstocks. IGCC—that is, the coupling of gasification with a combustion turbine for electricity generation—has been in commercial development in the United States since the early 1980s. Various improvements to the gasifier and associated technologies in the IGCC system are under active investigation in programs funded by the private and public sectors.

The Panel on DOE's IGCC Technology R&D Program was created by the NRC to apply the methodology developed in Phase One by NRC's Committee on Prospective Benefits of DOE's Energy Efficiency and Fossil Energy R&D Programs (NRC, 2005a) to assess the potential benefits of DOE's R&D activities that are focused on IGCC.[1] The panel consisted of experts with experience in coal-based electric power generation technologies and markets, combustion and gasification systems, environmental control technologies, and other relevant areas. Biographies of panel members are provided in Attachment A to this appendix.

The panel applied the committee's methodology by (1) assessing the probability of success of meeting DOE's time frame and targets for technology development, (2) considering alternative paths of development with and without DOE funding, (3) reviewing DOE estimates of the economic, environmental, and national security benefits of its program on IGCC technology, and (4) estimating the benefits under the alternative future states of the world (scenarios) specified by the committee.

The panel held two 2-day meetings and a conference call to complete its assessment. It was supported by a consultant and data and program information provided by DOE.

IGCC PROGRAM BASELINE AND GOALS

In IGCC technology, coal is gasified to produce a synthesis gas—principally carbon monoxide (CO) and hydrogen (H_2)—which is cleaned to remove particulates and other contaminants (e.g., sulfur and mercury compounds), then burned in a gas turbine to generate electricity. Heat is recovered from the combustion gases to generate additional electricity. The DOE research activities that are key to the success of IGCC as a technology are the advanced IGCC subprogram and the advanced turbine subprogram. There are DOE coal programs in fuel cells and carbon sequestration that have some relationship to the ultimate implementation of the IGCC technology, but the panel confined itself to analyzing the goals of these two main subelements and their potential benefits. A key goal of the DOE program is to develop IGCC systems capable of separating carbon dioxide (CO_2) for subsequent sequestration. This has a bearing on the performance of various components of the IGCC system other than the gasifier, principally the CO_2 separation system and the gas turbine itself, which must be modified to operate on a hydrogen-rich fuel gas (i.e., after CO_2 separation).

During the meeting with the panel on October 5 and 6, 2005, representatives of DOE made presentations on the performance goals of the advanced IGCC subprogram[2] and

---

[1] For the purposes of this assessment, the panel defined the IGCC program as coal to electric power systems. It did not include IGCC/fuel cell configurations or carbon capture and sequestration technology options since those were being analyzed by other panels.

[2] Gary J. Stiegel, DOE, Gasification Technology Program Manager, "Gasification Program Overview," Presentation to the panel on October 5, 2005.

TABLE H-1 Baseline and 2010 Goals for Total IGCC System as Given in DOE's Advanced IGCC Research

Baseline/Goal	Efficiency (%)	Capital Cost[a] ($/kW)	O&M ($/year)	COE ($/MWh)	Availability (%)
Goal set by DOE advanced gasification program	Increase of 2% to 4%	5% decrease	Decrease of $1 million		Increase of 5%
Baseline assuming entrained gasifier	39.8	1,517		49.3	85
Goal assuming compact gasifier	43.1	1,297		40.2	94
Baseline in DOE systems analysis	37.5	1,300		47.0	75
2010 goal in DOE systems analysis	48.0	1,000		32.0	85
Baseline input to NEMS cases	41.1	1,400			
2010 goal input to NEMS cases	50.0	1,000			

[a]Overnight costs in 2003 dollars.
SOURCE: Gary Stiegel, Gasification Technology Manager, DOE, National Energy Technology Laboratory, "Gasification Overview: Prospective Benefits Study," Presentation to the panel, October 5, 2005; and Julianne Klara, Senior Analyst, DOE, National Energy Technology Laboratory, "NEMS-based benefits of FE gasification R&D," Presentation to the panel, October 5, 2005.

the advanced turbines subprogram[3] and on how DOE uses the Energy Information Administration's (EIA's) National Energy Modeling System (NEMS) to assess the economic benefits of the gasification program.[4] Based on the presentations, there are quantitative goals, generally expressed in terms of the total IGCC system, for six principal criteria:

- Thermal efficiency,
- Capital cost,
- Operation and maintenance (O&M) cost,
- Cost of electricity (COE),
- Availability (the fraction of time during which the plant is generating electricity), and
- Emissions.

The principal research activities in the gasification program that are intended to provide the technology to meet the performance goals are these:

- Warm gas cleanup,
- Instrumentation (e.g., temperature measurement),
- Materials (e.g., refractory),
- Air separation by means of, for example, ion transport membrane (ITM),
- Dry coal feeding (e.g., Stamet pump), and
- Advanced gasifiers (transport gasifier, Rocketdyne gasifier).

In conducting the benefits analysis of the program, it became clear to the panel that there was some inconsistency in the goals and their timing as depicted in the DOE presentations. Furthermore, it was not clear from the information presented how much each of the major research activities listed above was expected to contribute, quantitatively, to reaching the goals. Therefore, the panel asked DOE to fill out a spreadsheet specifically for the gasification element of the program to clarify the program goals and improvements expected if the major program activities are successful.

Table H-1 was prepared by the panel based on information in the DOE presentations. Taken together, the baselines and goals indicate that the DOE program is seeking improvements of 5 to 10 percentage points in thermal efficiency (up to about 50 percent overall), $200 to $500 per kilowatt (/kW) in overnight capital cost (down to $1,000/kW), and 5 to 10 percentage points in availability (up to 90 percent).

Tables H-2, H-3, and H-4 show the data and reference notes DOE provided on the improvements its R&D is expected to make in IGCC system performance. The committee distinguished three categories of improvements:

- *Evolutionary improvements.* These research activities are part of the DOE program but also likely to be developed to some extent by non-DOE efforts. The panel added two activities under the heading "non-DOE or non-gasification program advancements" to help quantify the improvement one might expect absent the DOE program (see Table H-2).
- *Evolutionary improvements—major DOE programs.* These project activities are principally within the DOE program. The panel added a line for the goals of the complementary DOE turbines program (see Table H-3).
- *Revolutionary or long-term improvements.* The panel concluded that these project activities of the DOE program would need to be successful, in addition to the activities that achieve "evolutionary" improvements, to achieve the more aggressive goals of the DOE program (e.g., 48 percent thermal efficiency). The panel added a line to Table H-4 for potential improvements to thermal efficiency resulting from gasifier research being done outside the United States, principally in China and Japan. It noted that advances in gasification-related technologies being developed in the

---

[3]Richard A. Dennis, "FE turbine program: Delivering benefits to future coal based power systems," Presentation to the panel on October 5, 2005, Washington, DC.

[4]Julianne M. Klara, senior analyst, FE, DOE, "NEMS-based benefits of FE gasification R&D," Briefing to the panel on October 5, 2005.

TABLE H-2 Evolutionary Improvements Due to DOE Advanced IGCC Research

Research Activity	Change in Cost and Performance of IGCC System Attributable to R&D					
	Increase in Efficiency (%)	Capital Cost Reduction ($/kW)	O&M Reduction ($/year)	Reduction in COE ($/MWh)	Availability Improvement (%)	Emissions
DOE gasification program[a]						
Warm gas cleanup	1 to 2	70 to 100	Minimal	1.8	0	500 ppb sulfur[b]
Instrumentation (temperature measurements)[c]	0.5 to 1	0	Minimal	Minimal	1 to 2	—
Materials (refractory)[d]	0	0	2 million	0.5	4 to 6	—
Non-DOE or nongasification program						
Heat recovery	3					
Industry learning and evolution	1 to 2					

NOTE: COE, cost of electricity; kW, kilowatt; MWh, megawatt-hour; ppb, parts per billion; ppm, parts per million.
[a]Data courtesy of Gary Stiegel, DOE, National Energy Technology Laboratory.
[b]Reference case is 10 ppm sulfur.
[c]Efficiency gain through high carbon conversion.
[d]Assuming one turnaround per year.

TABLE H-3 Evolutionary Improvements Due to DOE Advanced Gasification Research

Research Activity	Change in Cost and Performance of IGCC System Attributable to R&D				
	Increase in Efficiency (%)	Capital Cost Reduction ($/kW)	O&M Reduction ($/kWh)	Reduction in COE ($/MWh)	Availability Improvement (%)
DOE gasification program					
Ion transport membrane air separation	1	75	Minimal	1.4	0
Stamet pump	0.5	40 to 100	Minimal	1.4 to 1.8	0
Non-DOE or nongasification programs					
DOE turbine program[a]	2 to 3 (for combined cycle power island)	60 to 100			

NOTE: COE, cost of electricity; O&M, operation and maintenance, kW, kilowatt; kWh, kilowatt-hour; and MWh, megawatt-hour.
[a]The DOE turbine program is considered complementary to its advanced IGCC subprogram.
SOURCE: Gary Stiegel, DOE, National Energy Technology Laboratory.

TABLE H-4 Revolutionary or Long-Term Improvements Due to DOE Advanced Gasification Research

Research Activity	Change in Cost and Performance of IGCC System Attributable to R&D				
	Increase in Efficiency (%)	Capital Cost Reduction ($/kW)	O&M Reduction (c/kWh)	Reduction in COE ($/MWh)	Availability Improvement (%)
DOE gasification program[a]					
Transport gasifier	1 to 2				
Rocketdyne gasifier[b]	1 to 2	50	Minimal	1	2 to 5
Chemical looping	2.5	130	0.2	3	0
Non-DOE or nongasification program					
Non-U.S. gasifiers	2 to 3				

NOTE: COE, cost of electricity; O&M, operation and maintenance; c/KWh, cents per kilowatt-hour; kW, kilowatt; kWh, kilowatt-hour; and MWh, megawatt-hour. Improvements in the parameters in this table are deemed necessary in order to achieve 45-48 percent efficiency for the IGCC plant.
[a]Data courtesy of Gary Stiegel, DOE, National Energy Technology Laboratory.
[b]Performance is relative to slurry-fed quench gasifier.

United States and other countries were not likely to be additive because they represent distinctly different technologies, and therefore separated the transport and Rocketdyne gasifier targets. The panel also concluded, based on a follow-up conversation with DOE, that the chemical-looping program, although funded out of the gasification budget, was separate from and further out in time than the core gasification program (see Table H-4).

The panel used this categorization of improvements and the estimated outcomes from each improvement as the basis for evaluating the potential outcomes of DOE's program in terms of thermal efficiency, capital cost, and availability for the IGCC plant and for estimating the probabilities of the various outcomes. For example, if the DOE program were to achieve all its goals for capital cost reduction through the projects in warm gas cleanup, ITM air separation, the Stamet pump, and turbines, and if all of these were additive, the net improvement would be $245-$375/kW.

GENERAL OBSERVATIONS ON DOE IGCC GOALS

In conducting this analysis, the panel identified two issues that affected its ability to quantify the benefits of potential success of the DOE program.

First, DOE considers that its goals will have been achieved when the research has been completed that will lead to the claimed improvements and when the technology has been demonstrated and commercially deployed through operation of the nth plant. While this reflects the reality of the DOE R&D budget, which funds research separately from demonstrations and does not address deployment at all, it does create some possible confusion about the time at which the expected benefits of the R&D program can be realized. In its analysis as discussed below, the panel made an estimate of the time to demonstrate and commercially deploy the advanced IGCC technologies being developed by the DOE research. Clearly, the accuracy of this time-to-commercialization estimate has a significant impact on the discounted value of the research program.

Second, the panel observed that many of the research projects were expected to result in improvements under more than one performance criterion, so that the effects of failure or success of a single project could be amplified in the overall outcome. Also, success of some of the projects would not be additive (i.e., the success of the transport gasifier would not confer an additive benefit to the success of the Rocketdyne gasifier), and some of the technologies were not equally beneficial for all possible coal feeds (bituminous, subbituminous, lignite). Therefore, there was some question about how to combine probabilities of success for different elements of the program, and about their corresponding benefits. The panel attempted to do this through the decision tree analysis.

ASSESSMENT OF DOE IGCC PROGRAMS

Technical Risks

Overview

DOE has focused its R&D program for IGCC on achieving substantial reductions in plant capital cost, plant thermal efficiency, and improvements in operating reliability. Meeting these goals would make this technology competitive with or superior to other forms of coal-fired power generation.

The R&D program was formulated based in large part on the results of a series of workshops with key stakeholders, who identified and prioritized R&D needs. Progress in meeting cost and performance goals is evaluated periodically both in-house, by DOE's National Energy Technology Laboratory (NETL), and by independent contractors. It should be noted that many of these projects deliver both cost and performance improvements. Failure to achieve the goals of any single project means that it may not be possible to achieve either cost or performance improvements. Some projects have synergies with others. As a result, simply adding up the improvements achieved by individual projects might not accurately represent the cumulative value of multiple projects.

Rocketdyne Gasifier

United Technology/Boeing has proposed the development of a compact gasifier that incorporates a number of innovative concepts based in part on rocket engine technology. If successful, this effort would reduce the estimated capital cost of an IGCC plant by about 15 percent and increase its efficiency by about 3 percent. The concept involves utilizing mechanical devices to pressurize the coal (dry feeders), feeding the pressurized coal into a compact reactor through multiple nozzles (rapid mix injectors), and removing heat from the reactor through membrane panels cooled by circulating water to prevent the reactor vessel from overheating (actively cooled wall liner).

A series of individual development activities have been formulated to test the feed system, the injectors, and actively cooled wall systems. If these three programs are successful, it has been proposed that an integrated pilot plant be built, followed by a full-scale demonstration. The technical risks are as follows:

• One or more of the component development activities (feed system, injector system, actively cooled wall) might fail.
• The promised cost and efficiency savings might disappear as the development program identifies unanticipated problems.

Transport Gasifier

The transport gasifier system has been developed with major sponsorship from DOE and technical leadership from the Southern Company. The program has been in operation since 1995 at a nominal 50 tons of coal per day (T/D) scale, demonstrating the concept of both air- and oxygen-blown gasification of a number of coals, including bituminous and subbituminous coals and lignite. The facility has also been used as a test bed for slipstream tests of various DOE-developed components and subsystems.

A 285-megawatt (MW) commercial-scale plant to be located at an Orlando Utilities Commission site in Florida is being designed by the Southern Company and its team under the DOE Clean Coal Power Initiative (CCPI) program. This plant will operate in an air-blown mode with Powder River Basin subbituminous coal. It is scheduled to begin operation in 2010. The technical risks are as follows:

- One or more of the design concepts based on the experience obtained at the Power Systems Development Facility (PSDF) might not scale up as anticipated, and redesign or replacement of equipment might be required.
- The coal feed injector (into the reaction zone) concept might require further development.
- Some of the solids circulation systems might not work as initially designed.

ITM Air Separation

Production of oxygen of 95 percent or greater purity from air by current liquefaction technology requires about 10 percent of the gross power output of the plant and accounts for about 15 percent of an IGCC plant's capital cost. This technology is very mature and has few unexploited areas that could promise further improvement.

DOE has been sponsoring work with Air Products and Chemicals, Inc. (APCI) on the ITM method of separating oxygen from air at high temperature by transporting oxygen ions through a high-temperature ceramic membrane. The source of the air is a bleed stream from the gas turbine compressor. Because the air must be heated to approximately 1000°F for transport through the membrane and the 99.99 percent pure product oxygen must be cooled and compressed prior to feeding to the gasification reactor, integration with the gas turbine cycle must be carefully optimized.

Currently, a 5-T/D pilot plant is in operation at an APCI facility in Sparrows Point, Maryland. Scale-up to 150 T/D has been proposed as the next step. The technical risks are these:

- The production cost goals for ITM modules might not be achieved.
- It might not be possible to scale up the technology successfully with integrated operation of multiple ITM units with one or two large gas turbines.
- Some of the currently available gas turbines that use air for airfoil cooling (typical F series) might not provide sufficient air to the ITM unit; those turbines that use steam for airfoil cooling (typical advanced G and H series) could, however, provide the needed air.

Stamet Dry Coal Pump

Gasification processes that feed coal to the gasifier as a dry, fine powder (as, for example, in the Shell process) or as a dry, crushed solid (as in the KBR Transport reactor) utilize a series of lock hoppers and a transport gas, usually nitrogen, to inject the coal into the reactor. These lock hopper systems are expensive, require extensive maintenance, and require energy to compress the pressurization gas.

DOE has been sponsoring work by Stamet to develop a single-stage mechanical device or pump capable of pressurizing the reduced-size coal to a level that will allow it to then be transported and injected into the gasification reactor. Stamet has already developed and commercialized solids feeders that are successfully used by industry to pressure solids to low differential pressures. Laboratory work at Stamet facilities to date has successfully demonstrated that coal can be injected into a vessel operating at 500 psi. An initial test at the PSDF facility in Wilsonville, Alabama was successful, and further testing is planned. The technical risks are these:

- There could be excessive mechanical wear of the Stamet rotating components.
- It might not be possible to disperse the pressurized coal within the delivery vessel for transport to the gasification reactor.

Warm Gas Cleanup

Cooling product gas prior to sulfur removal is thermodynamically inefficient; the lower the temperature, the greater the inefficiency. Current cleaning technology requires cooling the gasifier effluent gas to between 0°F and 100°F prior to sulfur removal depending on the absorption solvent that is used for sulfur removal. DOE is attempting to increase efficiency by developing technology that is effective at higher temperatures.

Experiments are being conducted in a slipstream unit at the Eastman Chemical coal gasification plant in Kingsport, Tennessee, at a scale equivalent to 0.5 to 1 MW. The system utilizes a finely divided, zinc-based solid sorbent to capture sulfur compounds in the form of hydrogen sulfide (H_2S) and carbon oxysulfide (COS) in the product gas. This work is being carried out in a two-vessel transport reactor system operating at about 600 pounds per square inch gauge (psig), with the absorber operating at 800°F and the regenerator at 1050°F. Recent results have shown that clean gas with a total sulfur content of 100 parts per billion (ppb) can be obtained. The sorbent has also demonstrated an ability to remove arsenic from the gas without any deleterious impact on the sorbent. This attrition resistance of the sorbent has been adequate in limited experiments to date.

Successful development of this technology would eliminate a number of current IGCC plant equipment items, including a number of high-temperature heat exchangers, the COS hydrolysis unit to convert COS to H_2S, the Selexol or other solvent unit to remove H_2S, and the Claus/SCOT units for elemental sulfur production and tail gas cleanup. DOE has proposed scaling up this system to 50 MW, before it qualifies as a commercial offering. The technical challenges or "risks" include these:

- Maintaining sorbent performance at desired levels while achieving sorbent consumption rates (due to activity loss and attrition) at economically attractive levels; and
- Reducing the regenerator temperature to 700°F or lower to maximize efficiency.

Instrumentation (Primarily Temperature Measurement)

Improved instrumentation for use directly in gasifiers and in gasifier supporting systems has the potential to improve overall system efficiency and minimize operation and maintenance (O&M) expenses. The areas of need include direct gasifier temperature measurement, in situ measurement of slag viscosity, on-line (i.e., during operation) measurement of refractory wear, and real-time measurement of feed coal properties.

Perhaps the most important of these at the moment is the direct measurement of actual gasifier operating temperatures. Currently available gasifier instrumentation such as thermocouples have a very short life in commercial-scale coal gasifiers that operate at 2500°F-3000°F. As a result, peak temperatures in the gasification zone are inferred from the composition of the product gas. While these results are approximately correct, the result is the gasifiers are run at lower than optimum temperatures for carbon conversion to avoid excessive damage to refractories. This results in inefficiency.

DOE is sponsoring work at several locations: GE Energy; Albany Research Center in Oregon; Virginia Tech (with ConocoPhillips); Entertechnix; and the Gas Technology Institute (GTI). The work is focused on the development of temperature measuring systems based on direct observations of the flame in the gasification reactor and on processing the data from those observations to accurately and reliably infer temperature. The technical risks are as follows:

- It might not be possible to keep the sighting ports open and/or clean enough for long enough periods.
- The algorithms to process the data might not be accurate.
- The materials used in the systems might fail due to long-term exposure at high temperature.

Refractory Materials for Gasifier Walls

Current coal gasifiers (those of GE and E-Gas for example) that utilize an internal refractory to protect the metal walls of the vessel from exposure to high temperatures would benefit from this research, while gasifiers using a cooled membrane wall (those of Shell) would not. The chromium oxide-based refractory used in these high temperature gasifiers is attacked by the molten slag (derived from the mineral matter in the coal) that flows down their walls during gasification operations. Variations in operating temperature and slag composition result in different rates of attack. Typical intervals for planned refractory replacement, which require 3-4 weeks of unit downtime, are from 6 to 18 months. Unplanned outages are also required from time to time for minor refractory repairs. Increasing the replacement interval for an unspared gasifier from 18 months to 36 months, the target of the DOE program, would increase availability by 2.5 percent. Elimination of 50 percent of unplanned outages would also increase availability.

DOE has sponsored work at the Albany (Oregon) Research Center that resulted in new refractory formulations based on phosphate additions that are being tested in two currently operating commercial IGCC plants. Initial results are promising. The risks that must be faced are these:

- The refractions might not achieve their service life objectives.
- It might be too expensive to manufacture them in commercial quantities.
- They might be susceptible to premature failure when used with specific coals.

Hydrogen-Fired Gas Turbines

Most gas turbines were developed for use with methane as the primary fuel. Older models that operate at lower firing temperatures (1500°F-2000°F) were able to fire hydrogen-rich gases (up to 90 percent hydrogen) without downrating. However, newer models that operate at much higher firing temperatures (2300°F-2600°F) on methane must be fired at 200°F-300°F lower firing temperatures on hydrogen-rich fuels.

DOE has embarked on a major effort to develop gas turbines that can operate at the same high temperatures as methane-fired turbines and so achieve the same efficiency, with NO_x emissions of 2 ppm. The technical risks are as follows:

- It might not be possible to achieve the efficiency and NO_x emissions goals simultaneously.

- The combustor system required to achieve those goals simultaneously might be significantly more expensive.

Chemical Looping

Chemical looping involves a reactor system with two fluidized beds. This system can be utilized to achieve either total combustion or total gasification. Solid particles containing metal oxides are supplied to the reactor used for combustion or gasification to supply the oxygen. When the oxygen in those solids is utilized for combustion, the oxides are reduced to metals. The metals are circulated to a second reactor, where they are reconverted to oxides by reacting them with air. The CO_2-rich gas exiting the combustion reactor is diluted primarily with water, which can be removed by condensation, leaving an essentially pure CO_2 stream, which can then be captured.

DOE has funded experimental work on chemical looping in pilot plants at Alstom and GE. The oxides that have been tested for use in this cycle include both supported and unsupported materials. The solid oxide materials evaluated include calcium carbonate ($CaCO_3$), calcium sulfate ($CaSO_4$), and iron oxide (Fe_2O_3). Supported oxide materials included copper oxide (CuO), Fe_2O_3, manganese oxide (MnO_2), and nickel oxide (NiO) on various inert supports, including aluminum oxide (Al_2O_3), sepiolite, silicon oxide (SiO_2), titanium dioxide (TiO_2), and zirconium oxide (ZrO_2). The technical risks are as follows:

- The cycle efficiency might not be competitive;
- The capital costs might not be competitive;
- Gas cleanup to meet emission requirements (NO_x, hazardous air pollutants, etc.) might be difficult to achieve; and
- A solid that has appropriate chemical reactivity, chemical stability, and attrition resistance might not be available at the required cost level.

Market Risks

Market risks must be considered for the time frame associated with demonstration and commercialization of the IGCC technology developed by this program as well as by other, non-DOE programs. Following completion of R&D sponsored by the DOE in 2010, the panel estimates nominal requirements of 9 and 5 years, respectively, for demonstration and commercialization, which would mean a grid-ready plant in circa 2024.

Significant market risk factors can be broadly classified as follows: fuel prices, the regulatory environment, and competing technologies. Of these factors, that associated with the relative price of competing fuels is believed to carry the greatest risk. If the cost of natural gas were to drop below approximately $5 (in 2005 dollars) per million British thermal units (MMBtu), an IGCC power plant, even one resulting from achievement of the DOE program goals, could not compete with a natural gas combined cycle (NGCC) plant in terms of the levelized COE. For an IGCC plant to be cost competitive, today and in the future, the ratio of fuel costs (natural gas to coal) should exceed 3:1. Much will depend on growth of the nation's marginal needs for natural gas, the extent to which a liquefied natural gas (LNG) infrastructure is established to meet those needs, and global competition for natural gas supplies among net consumers.

Another significant risk would be the early implementation of an aggressive, mandatory CO_2 emission reduction regulation. Such a regulatory measure, in combination with low to moderate gas prices, would favor natural gas. Conversely, a more restrained approach to regulating CO_2 emissions, combined with high natural gas prices, would favor the commercialization of IGCC plants, which benefit from efficient means of CO_2 separation. However, if adequate (large-scale) CO_2 sequestration options fail to materialize, IGCC plants will lose the advantage afforded by efficient separation.

Competing technologies include ultra supercritical steam-pulverized coal (USC-PC) and circulating fluidized bed (CFB) combustion systems, nuclear power, fuel cells, and alternative/renewable sources such as wind, solar, and biomass. Advancements in the combustion (USC-PC and CFB) technologies, particularly if accompanied by cost-efficient means of carbon capture, represent the biggest technology threat to commercialization of IGCC plants. Table H-5 shows the significant differences between present costs and efficiencies for a range of coal-based and natural-gas-based power generation technologies and the DOE goals for IGCC. IGCC, utilizing a spare gasifier to bring its capacity factor up to 85 percent, has higher total plant costs and COE than alternative clean coal technologies. The total plant cost for NGCC is low, but the COE is high due to the high natural gas price and the resulting low capacity factor.[5]

Of the other technologies for generating electricity, nuclear power is encumbered by large capital costs, an extensive regulatory environment, and safety issues associated with radioactive waste disposal. Moreover, even with an eventual return to the construction of nuclear power plants, it is difficult to imagine any scenario other than one in which nuclear and coal options are concurrently developed. While fuel cells may one day provide a significant source of distributed power, progress will be slowed by related hydrogen supply and infrastructure requirements. However, sustained advancements in high-temperature (e.g., solid oxide) fuel cells may, in fact, enhance prospects for commercializing IGCCs by enabling hybrid systems with efficiencies exceeding 50 percent. Power generation from renewable forms of energy will continue to increase, particularly with the recent commitment of large corporations to the development and

---

[5]The calculation was made assuming a $5.00/MMBtu gas price, resulting in a capacity factor of less than 30 percent.

TABLE H-5 Costs for 500-MW Power Plants Using a Range of Technologies Without Carbon Capture and Storage

	PC		CFB	IGCC (E-Gas) w/ Spare	NGCC	
	Subcritical	Supercritical			85% CF	30% CF
Total plant cost ($/kW)	1,230	1,290	1,290	1,350	440	440
Total capital requirement ($/kW)	1,430	1,490	1,490	1,610	475	475
Avg. heat rate (HHV) (Btu/kWh)	9,310	8,690	9,800	8,630	7,200	7,200
η (HHV) (%)	36.7	39.2	34.8	39.5	47.4	47.4
η (LHV) (%)	38.6	41.3	36.7	41.6	50	50
COE (Levelized) ($/MWh)	46.5	46.6	46.0	49.9	47.3	56.5

NOTES: $/kW, dollars per kilowatt; $/MWh, dollars per megawatt-hour; Btu, British thermal unit; CF, capacity factor; CFB, circulating fluidized bed; COE, cost of electricity; E-gas, E-Gas coal gasification process; HHV, higher heating value; IGCC, integrated gasification combined cycle; LHV, lower heating value; NGCC, natural gas combined cycle; and PC, pulverized coal. Assumptions include book life of 20 years; commercial operation date of 2010; total plant cost includes engineering and contingencies; total capital requirement includes interest during construction and owner's cost; assumes EPRI's Technical Assessment Guide financial parameters; All costs in 2003 dollars.
SOURCE: National Coal Council (2004), Figure 2.6, p. 27.

implementation of related technologies, the growing application of renewable portfolio standards,[6] and the emergence of social values as a significant driver of national energy policies. However, the impact of these trends is likely to be small relative to the need for electricity associated with sustained economic growth.

Other risk factors are those associated with weak economic growth (reducing the need for new plants), constraints on the transmission system (the electricity grid), and insufficient human capital (i.e., engineers) to derive maximum benefits from available opportunities. However, these factors apply to all options, not just to IGCCs. There is also a risk associated with inadequate development of the transportation infrastructure for coal, which would affect USC-PC and CFB as well as IGCC systems.

In summary, the biggest market risks for IGCC commercialization are associated with reemergence of low to moderate costs for natural gas and with premature imposition of an aggressive CO_2 control regulation. Conversely, commercialization would benefit from high costs for natural gas and the gradual imposition of carbon constraints, as well as from opportunities for significant increases in efficiency and reductions in cost afforded by an immature technology such as IGCC.

RESULTS AND DISCUSSION

Overview

The committee's methodology asks that the panel evaluate the benefits of IGCC under each of three global scenarios representing three different possible future states of the world. It also suggests that the evaluation should explicitly consider the role of DOE funding; should present the technical risks and market risks that can affect the outcome and the value of the program's activities in a decision-tree format; and should include the net economic, environmental, and security benefits that will result from the portfolio. The panel considered each of these suggestions in its evaluation.

Consideration of Global Scenarios

The panel felt that the probability of technical success would not be influenced by the global scenarios, but that the probability of market success would be. As discussed above, two of the primary market risk factors are the cost of natural gas and the imposition of a carbon emissions restriction. The three global scenarios were defined specifically to consider the impact of changes in fuel prices and in the treatment of carbon emissions. These impacts are reflected in the probability estimates for market success, described below.

The Carbon Constrained global scenario was defined by the prospective benefits committee as a scenario wherein a $100/ton carbon tax is assumed to be in place in 2012. While the panel believes IGCC technologies could benefit significantly under this scenario, the benefits would depend critically on the success of research in carbon sequestration. Both IGCC R&D and sequestration R&D must make significant advancements for IGCC with sequestration to be a viable market choice; the panel does not believe it is possible to separate the benefits of IGCC R&D from the benefits of carbon sequestration R&D in this scenario. Thus, it did not evaluate the benefits of IGCC R&D in the Carbon Constrained scenario. The panel notes that another panel is using this methodology to evaluate DOE's carbon sequestration R&D and is focusing exclusively on Carbon Constrained scenarios.

---

[6]Renewable portfolio standards require a certain percentage of a utility's power plant capacity or generation to come from renewable sources by a given date.

Decision Tree and Probability Assessments

Role of DOE Funding

In defining the various levels of technical and market success described below, the panel considered the specific information provided by DOE about its program, as well as panel members' knowledge of the current performance of IGCC plants and of ongoing IGCC-related research that is being conducted independently of DOE's program, including research being carried out overseas. The panel assessed the probability of achieving various levels of technical performance and market success for IGCC technologies, first assuming the DOE R&D program is funded and then assuming the program is not funded.

Technical Success

As noted above, the panel considered the specific activities being funded by DOE and the potential impact of that research on both the capital cost and efficiency of IGCC plants. Based on that consideration, the panel identified four levels of technical success in terms of IGCC plant efficiency and capital costs. A secondary type of technical success was defined in terms of plant availability.

The four levels of technical success were defined by the estimated capital costs and efficiency for the nth plant, assuming R&D completed in 2012 and at least 80 percent plant availability:

- *Low success.* Efficiency of 38 to 40 percent on a higher heating value (HHV) basis at a capital cost of $1,400/kW, or approximately what is achievable with today's technology.
- *Moderate success.* Efficiency of 42 percent (HHV) at a capital cost of $1,265/kW.
- *High success.* Efficiency of 45 percent (HHV) at a capital cost of $1,135/kW.
- *Very high success.* Efficiency of 48 percent (HHV) at a capital cost of $1,040/kW. This level is approximately equal to DOE's stated goals for the IGCC program.

The panel then estimated the probability of achieving the specified levels of technical success, first assuming DOE's program is carried out and then assuming it is not.

Three levels of plant availability were also defined, and the panel estimated the probability of reaching each of these levels of availability both with and without DOE's R&D program. The estimated availability depended in part on the capital costs and efficiency for the plant.

- 80 percent availability,
- 85 percent availability, and
- 90 percent availability.

The panelists developed their probability estimates as a group during the second panel meeting. The consensus probabilities are shown in the decision tree in Figure H-1. Each path through the decision tree represents a level of overall technical success.

Market Success

As described above, the panel identified the key factors affecting the market success of IGCC technologies as fuel prices, the regulatory environment, and competing technologies. The impact of changes in fuel prices is captured by the High Oil and Gas Prices global scenario.

The primary factor that determines how many IGCC plants are built is how those plants compete with other technologies in terms of the COE. The panel identified the primary competing technologies as natural gas plants (if gas prices are low) and advanced coal combustion technologies (USC-PC and CFB). Other competing technologies include nuclear, advanced fuel cells, and renewables.

The panel also considered the results of DOE's FY04 benefits analysis when it identified the primary competing technologies. That analysis identified PC plants, natural-gas-fired turbines, fuel cells, and renewables (DOE, 2004, Figure 6.17) which is consistent with the panel's assessment.

Quantification of Benefits

Introduction

The economic, environmental, and security benefits of improvements in IGCC technologies depend on the degree of technical improvement and the resulting reduction in the COE from IGCC plants, the amount of IGCC capacity added, the technologies that would have been implemented absent IGCC, and the relative costs of electricity with those next-best alternatives.

The economic benefits that would be expected through technology advancements in IGCC technology include these:

- The reduction in capital cost of IGCC plants that would be realized by R&D.
- The increase in efficiency in IGCC plants that would be realized by R&D.
- The quantifiable impact on environmental benefits of IGCC plants that would be realized by R&D.
- The lower projected cost for removing CO_2 emissions compared with the cost of using currently available technology for other coal technologies.
- The impact on COE compared with the impacts of other options. COE is highly dependent on variables other than plant performance.

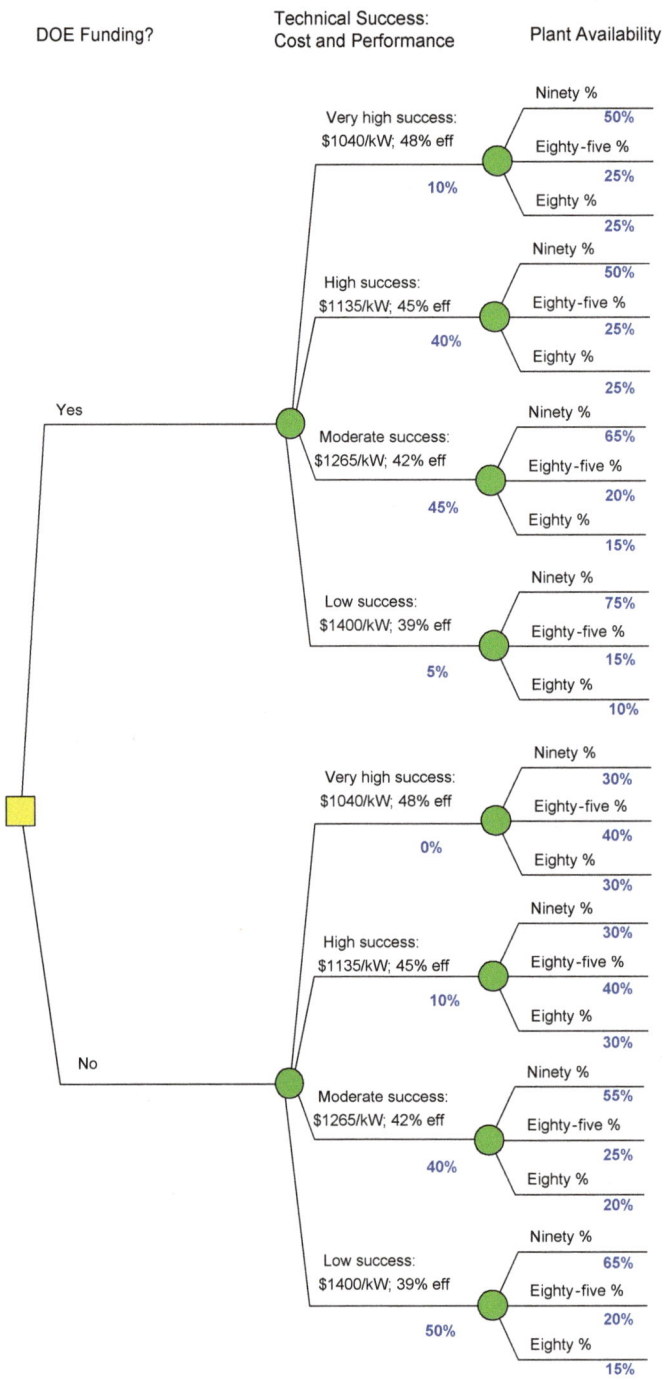

FIGURE H-1 Decision tree representing the panel's assessment of the likely technical outcomes of IGCC R&D.

The panel estimated the economic benefits by focusing on the COE for IGCC plants relative to other options. The economic benefit in any one year is the product of the amount of IGCC-generated electricity produced and the difference between the COE with IGCC generation and the COE with the alternative technology. The total benefits are calculated as the net present value of the annual benefits stream assuming a 20-year plant life. To implement this method of estimating benefits, the panel developed two models: one for estimating the COE from different generation technologies and one for estimating the amount of IGCC capacity that would be built.

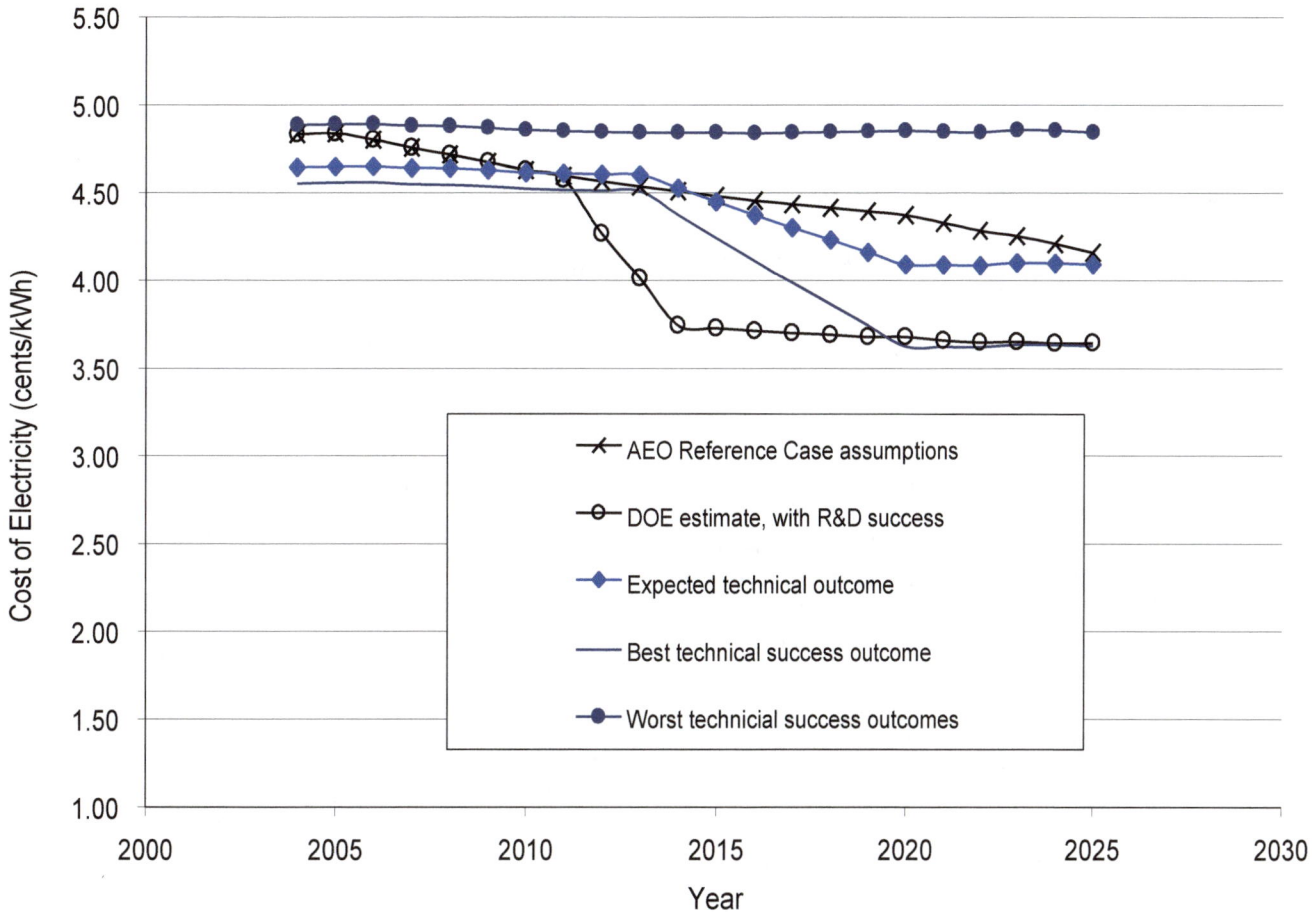

FIGURE H-2 Estimated COE for IGCC under different technical success assumptions for the AEO Reference Case global scenario.

Estimating the Cost of Electricity

The panel developed a simplified model for estimating the COE (i.e., the busbar costs) for IGCC and other technologies based on capital costs, plant efficiencies, operating and maintenance costs, and fuel costs.

The COEs for PC and NGCC plants—the presumed competitors to IGCC—were estimated based on plant characteristics taken directly from *Assumptions to the Annual Energy Outlook: Electricity Market Module* (EIA, 2005).[7]

The decision trees created by the panel specify a range of possible technical outcomes of IGCC research in terms of future capital costs, plant efficiencies, and plant availabilities, as described previously. For each of the 12 combinations of possible technical outcomes, a COE for IGCC plants can be calculated. Figure H-2 shows the range of the estimated COE over time for these different technical outcomes. The solid lines represent the COE for IGCC given the highest and lowest cost and performance outcomes from the panel's decision tree assessments. The COE projections for all 12 of the technical success scenarios lie between these two lines, and the line with the diamond markers represents the COE over time based on the probability-weighted average of all 12 outcomes. This "average" performance would correspond to a hypothetical IGCC plant with capital costs of $1,197/kW, efficiency of 43.65 percent, and availability of 87 percent. The figure also shows the estimated COE for an IGCC plant based on the costs and efficiency assumptions in the AEO 2005 reference case, and the estimated COE for a "DOE success" case, assuming the IGCC research meets the goals set out by DOE over time. Note that the panel estimated it would take much longer for those costs and efficiencies to be achieved through widespread commercial deployment than DOE estimated in its analyses.

To the extent that IGCC reduces the impact of fuel price fluctuations by reducing the need for electric power plants that use natural gas (e.g., NGCC), this will be captured as a security benefit.

---

[7]Especially Tables 38 and 48 of EIA (2005a).

Estimating the Amount of IGCC Capacity Built

The amount of IGCC built in any year is a function of the COE for IGCC and for competing technologies, as well as other macroeconomic factors. DOE uses NEMS to make annual forecasts of the amount of electricity generation that will be added in the nation between now and 2025. The panel decided to use these NEMS new-generation build forecasts as a starting point for estimating IGCC capacity additions. DOE's Office of Fossil Energy produces three build forecasts for IGCC based on different assumptions about the outcome of its R&D programs. There are forecasts for three cases:

- A no-R&D case, which applies the AEO Reference Case assumptions to all technologies, including IGCC.
- A case wherein the IGCC program is assumed to meet the technical goals established by DOE (e.g., $1,000/kW capital costs and 50 percent efficiency in 2014, increasing to 60 percent efficiency by 2024). This forecast for the case also assumes the success of all other DOE FE R&D programs, so that the IGCC technologies compete against conventional and other advanced technologies for market share.
- A case wherein all FE R&D programs except the gasification R&D are assumed to succeed.

To estimate the amount of IGCC that would be built under any one of the technical success scenarios defined by the panel, a scaling function was employed based on the relative COEs for the three cases: for IGCC in the technical success scenario, for IGCC based on meeting DOE goals, and for competing technologies, including both PC and advanced natural gas plants. Figure H-3 illustrates the capacity of IGCC built in each year under the AEO Reference Case and the High Oil and Gas Prices global scenario, assuming IGCC performs at a level reflecting the average of the panel's assessment of technical success. The figure also shows the quantities built in DOE's analyses with and without DOE's R&D programs.

The final step in estimating the benefit of IGCC advances is to establish what the next-best alternative technology would be; that is, to establish what type of generation will be built if IGCC is not cost-competitive. Based on the panel's assessment of market risks, IGCC technologies will compete with both NGCC plants and other advanced coal technologies.

Based on a review of DOE's FY04 benefits analysis and information presented by DOE to the panel, it appears that successful IGCC technologies compete with PC plants in some circumstances and with NGCC plants in others. Specifically, it appears that the success of other DOE FE research—for example, in distributed generation (DG)—affects what the next-best alternative to IGCC is: If DOE's FE programs on other power systems are assumed to succeed, NGCC is the next-best technology, and if those programs do not succeed, PC is the next-best technology.

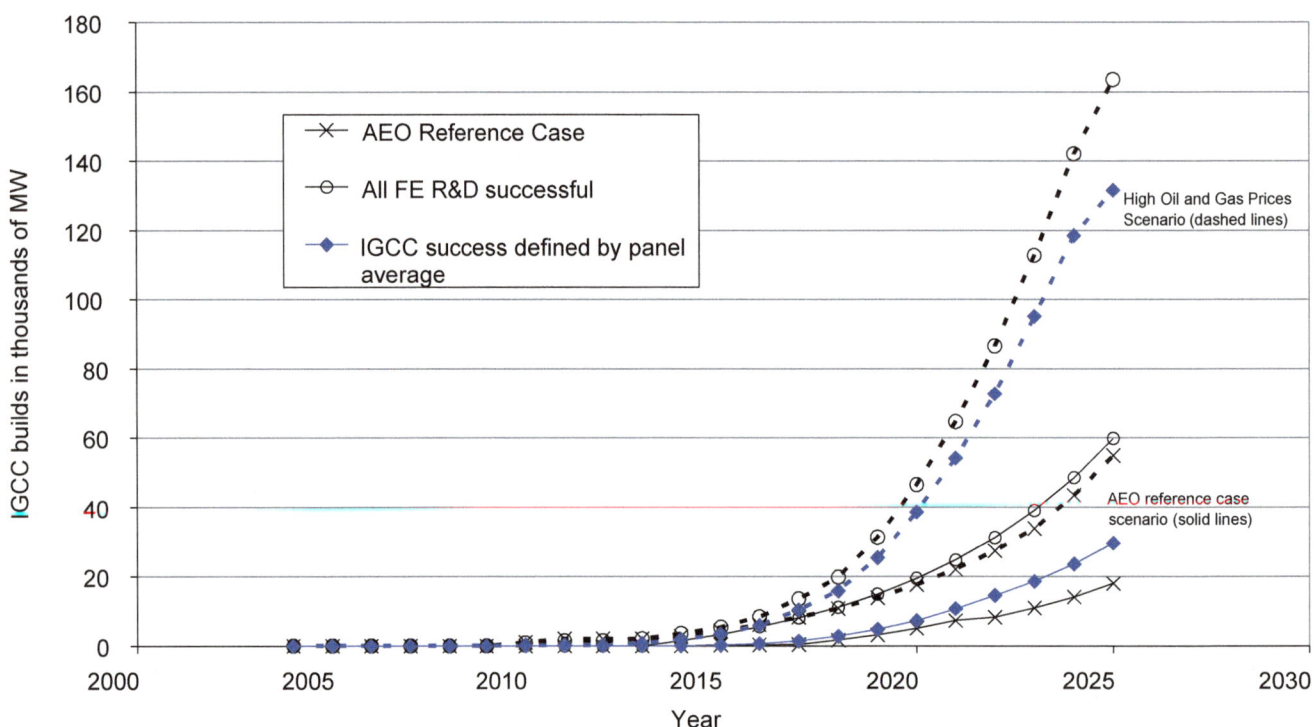

FIGURE H-3 Cumulative amount of IGCC built under three different technical success assumptions and two different global scenarios.

The panel also notes that other advanced coal technologies are currently under development (e.g., USC-PC) that could compete with IGCC. While assumptions about the success of DOE's other FE R&D programs or other advanced coal technologies have a significant impact on the benefits calculation, it is outside the scope of the panel's task to evaluate the chance of success. The panel's benefit estimates are therefore based on the assumption that IGCC becomes the most cost-effective coal-based technology. To the degree that other coal-based technologies turn out to be more promising than IGCC, the benefits estimated here would be reduced. To address the issue of what technologies successful IGCC would compete with, the panel estimated benefits two ways: once assuming (conventional) PC is the next-best technology, and a second time assuming NGCC is the next-best technology.

Economic Benefits Analysis Results

Using the simple models described above, it is possible to estimate the benefits of IGCC associated with each of the 12 technical success scenarios defined by the panel's decision tree. For the lowest level of technical success the benefits are zero—no IGCC is built because the technology is not competitive. For the highest level of technical success, the net present value of the benefits is $30 to $33 billion[8] for the AEO Reference Case global scenario, depending on whether IGCC is assumed to displace PC or NGCC.

Each of those technical success scenarios has two probabilities assigned to it by the panel: the probability of achieving that level of technical success without the DOE research program, and the probability of achieving that level of success with the DOE research program. The expected value of the IGCC improvements with or without DOE research is simply the probability-weighted average of the NPV for each technical success scenario using the appropriate probabilities. The value of the DOE research program is the difference between the expected value of IGCC research without the program and the expected value with the program.

Figure H-4 illustrates the expected economic benefits of the IGCC R&D, as well as the uncertainty surrounding those benefits assuming AEO Reference Case prices and that PC is the next-best alternative technology. The figure shows a cumulative distribution of net economic benefits with and without DOE support. The vertical lines represent the expected value of the benefits calculated as the probability-weighted average. The expected value of DOE's IGCC research program in this scenario is $6.4 billion, the difference between the expected value with the program ($8.6 billion) and the expected value without the program ($2.2 billion).

The expected benefits of the DOE program differ from one global scenario to another and from one next-best technology to the other. For the AEO Reference Case scenario but assuming NGCC is the next-best technology, the expected value of the benefits increases to $7.8 billion.

Under the High Oil and Gas Prices scenario, there is much larger difference in expected benefits based on what is assumed about the next-best technology: If IGCC replaces PC, the benefits of the program are $7 billion, if it replaces NGCC, the benefits are $47 billion.

Environmental Benefits Analysis

The environmental benefits that could be attributable to technology advances in IGCC include these:

• Reduction in regulated emissions, including sulfur oxides (SO_x), nitrogen oxides (NO_x), fine particulate matter ($PM_{2.5}$), and mercury.
• Reduction in CO_2 emissions due to higher efficiency than other coal technologies.
• Reduction in the amount of waste by-products compared to other coal-based technologies.
• Reduction in water consumption compared to other coal-based technologies.

As suggested by the full committee, emissions of criteria pollutants (mercury and so forth) are assumed to occur at the regulated level for all fossil-fuel-based technologies. If reaching those limits is less costly for IGCC than for PC, that benefit is captured as an economic one rather than an environmental one.

There will be quantifiable differences in carbon emissions if IGCC is deployed instead of either of the next-best alternatives. When IGCC plants are assumed to displace conventional PC plants, carbon emissions will be reduced because the former are more efficient. When IGCC plants are assumed to displace NGCC plants, carbon emissions will be higher. Using the build quantities estimated for IGCC as described above and emissions factors for different power plants from the *Annual Energy Outlook 2005* (EIA, 2005b), the panel was able to estimate the net change in carbon emissions attributable to the DOE's IGCC research program. If IGCC replaces PC, the net reduction in carbon emissions is 30 million tons (discounted at 3 percent) over the full 20-year life of all plants built between 2006 and 2025 in the AEO Reference Case scenario and 34 million tons in the High Oil and Gas Prices scenario. If IGCC replaces NGCC, there will be a net increase in carbon emissions of 90 million tons (discounted at 3 percent) in the AEO Reference Case scenario. However, in the High Oil and Gas Prices scenario, large quantities of IGCC are expected to be built with or without the DOE R&D program. The higher efficiencies of

---

[8]This value compares with the $30 billion in consumer savings from reduced electricity prices estimated by DOE as being a benefit of IGCC research that achieves DOE's stated goals for the program. Note that the level of performance the panel assumed at the highest level of technical success it evaluated is better than that assumed if DOE meets its technical goals. Further, the benefits estimated by the panel included the reduced COE for the full 20-year life of all plants built between 2006 and 2025.

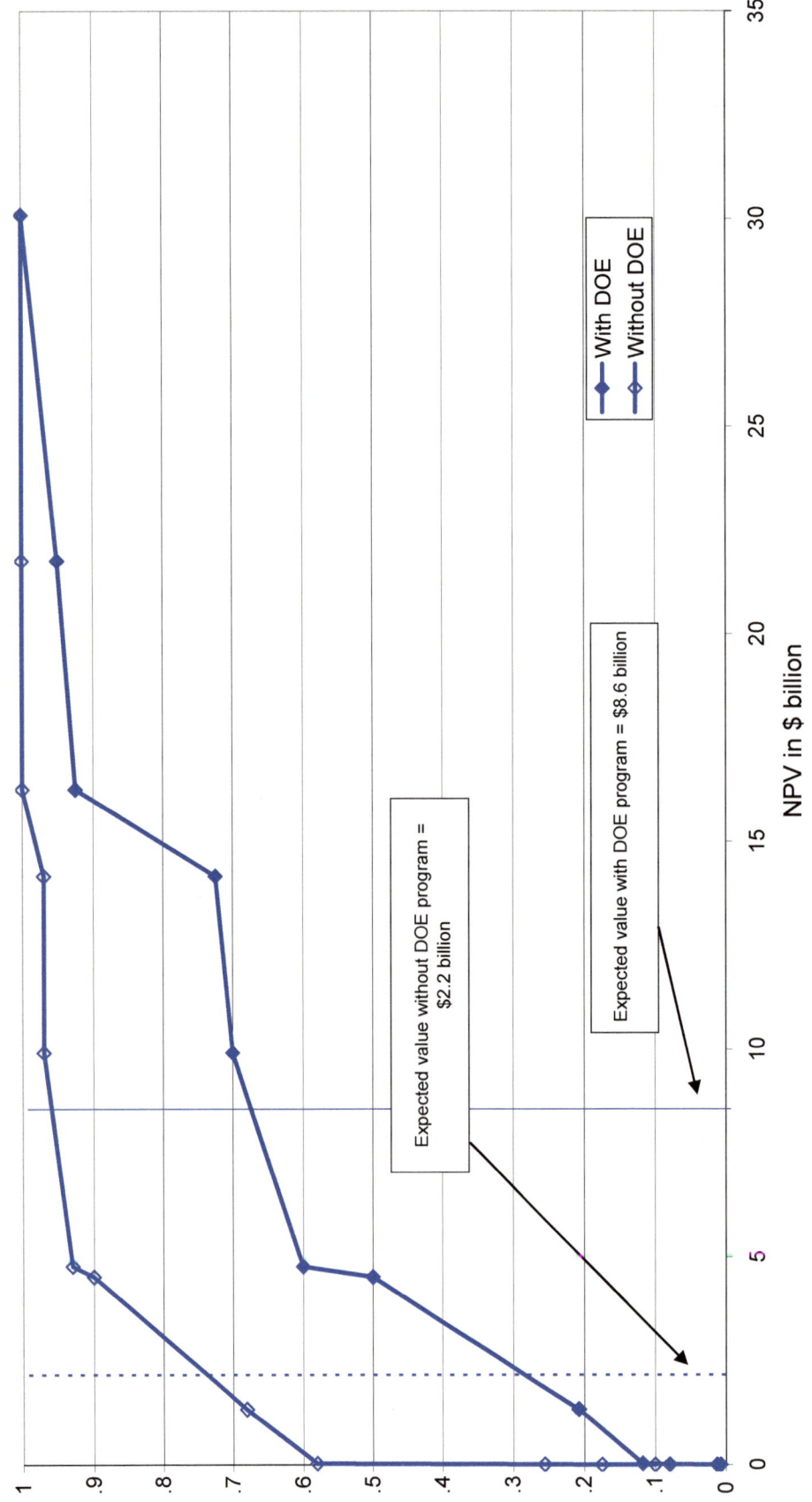

FIGURE H-4 Cumulative distribution on the net present value of IGCC research under the AEO Reference Case scenario, assuming IGCC replaces PC.

IGCC plants attained with the R&D program would decrease emissions by 36 million tons. Note that if IGCC plants were to replace USC-PC plants, the differences in carbon emissions would be negligible, because their thermal efficiencies are approximately the same.

Energy Security Benefits Analysis

The energy security benefits that could be seen through technology advancements in gasification technology because of the DOE IGCC program include the following:

- Increased use of domestic coal rather than imported natural gas for power generation.
- The ability to produce synthetic natural gas from coal.
- The ability to produce liquid fuels from coal.
- The ability to use coal rather than natural gas as a feedstock for the petrochemical industry.
- Maintaining the nation's technology leadership for both domestic and export applications as opposed to relying on importing technology from overseas.

If IGCC plants are built instead of natural gas plants, there are security benefits from the reduced usage of natural gas, especially to the degree that the reduced usage decreases the need for imports. Assuming IGCC plants replace NGCC plants, the net reduction in natural gas usage attributable to DOE's research is 4 to 5 quads. The reduction in the demand for natural gas for electricity generation also helps to reduce the cost and increase the availability of natural gas for other purposes, such as chemical manufacture and residential and commercial use.

Results Matrix

Figure H-5 summarizes the results of the panel's estimation of the benefits of DOE's IGCC R&D program. Benefits are calculated as described in Chapter 3 for two scenarios: the AEO Reference Case and the High Oil and Gas Prices scenario.

SUMMARY AND RECOMMENDATIONS

Summary

Coal gasification and the rigorous cleanup of the product syngas have made the use of coal in gas turbines possible, opening the way to efficient and environmentally clean coal-based power generation in combinations of gas turbines and steam turbines.

While its capital cost is presently higher than that of a comparable size pulverized coal supercritical (SC/PC) steam plant, IGCC lends itself more favorably to retrofit with capture and compression of CO_2 for sequestration. Success of DOE's advanced gasification research is a prerequisite for IGCC's timely commercialization.

Based on discussions with representatives of DOE, the panel extended the principal goals of the IGCC R&D program—efficiency and capital cost—to include availability, emissions, and COE. The panel evaluated the contributions of the major research projects under DOE's Advanced Gasification Research. The evaluation included projects that promise to improve IGCC availability, such as Durable Refractory for Gasifier Lining, and Gasifier Instrumentation and Diagnostics, and others that promise to reduce capital cost and increase efficiency—for example, Ion Transport Membrane Air Separation, Dry Coal Feeding, and Warm Gas Cleanup. Projects such as the Rocketdyne Gasifier, the Transport Gasifier, and Chemical Looping, which represent stepwise advances in gasification technology and can be expected to reduce costs and improve efficiency in the longer term, were also evaluated.

Technical risks for the individual projects have been identified; these are generally associated with the possibility that the researchers might not reach their goal, the results might not be scalable, or the capital cost and/or the parasitic energy requirement might be excessive. The panel quantified the technical risks and posited 12 future technical outcomes of IGCC research both with and without the impact of the DOE program. The 12 outcomes were defined in terms of capital costs, plant efficiencies, and plant availabilities. The prospects for cost and efficiency have been assessed as follows:

- 5 percent chance of achieving only minor improvements over today's technologies (39 percent efficiency and an nth plant capital cost of $1,400/kW) with the DOE program and a 50 percent chance of achieving only these minor improvements without it.
- 45 percent chance achieving $\eta = 42$ percent (HHV) efficiency and an nth plant capital cost of $1,265/kW with the DOE program and 40 percent without it.
- 40 percent chance of achieving $\eta = 45$ percent and an nth plant capital cost of $1,135/kW with the DOE program and 10 percent without it.
- 10 percent chance of achieving $\eta = 48$ percent and an nth plant capital cost of capital cost of $1,040/kW with the DOE program and 0 percent without it.[9]

For each of the above four cost and efficiency levels, plant availability was assessed at three levels—80, 85, and 90 percent. These comprise the 12 technical success scenarios.

The target date of 2010 given for the above goals is the time when the research will have been completed. This stage is to be followed by demonstration and commercial deployment—the latter to occur through operation of the nth

---

[9] Compare to the DOE goal of $1,000/kW and 50 percent efficiency in 2014.

		Global Scenario[a]	
		AEO Reference Case	High Oil and Gas Prices
Program Risks	Technical Risks	Technical success was defined by the cost and efficiency of future IGCC plants. The panel identified four levels of technical success: $1,400/kW and 39% efficiency; $1,265/kW and 42% efficiency; $1,135/kW and 45% efficiency; $1,040/kW and 48 percent efficiency. Estimated probability of achieving the highest level of technical success was 10% with the DOE program, 0% without the DOE program.	
	Market Risks	Technologies that will compete with IGCC are natural gas generation and advanced PC generation, including USC-PC. Primary market competition depends on the price of natural gas, on the progress of research in other coal-fired generation technologies, and on the results of other research programs such as fuel cells and distributed generation.	
Expected Program Benefits	Economic Benefits	Economic benefits depend on the next-best alternative technology, which depends in part on the results of other DOE R&D programs.[b]	
		If IGCC displaces PC, $6.4 billion at 3% $2.4 billion at 7% If IGCC displaces NGCC, $7.8 billion at 3% $3 billion at 7%	If IGCC displaces PC, $7 billion at 3% $2.7 billion at 7% If IGCC displaces NGCC, $47 billion at 3% $18 billion at 7%
	Environmental Benefits	Environmental impacts depend on the next-best alternative technology, which depends on the results of other DOE R&D programs.	
		If IGCC displaces PC, carbon emissions decrease by 30 million t If IGCC displaces NGCC, carbon emissions increase by 90 million t	If IGCC displaces PC, carbon emissions decrease by 34 million t If IGCC displaces NGCC, carbon emissions decrease by 36 million t[c]
	Security Benefits	If IGCC plants are built instead of NGCC plants, there are security benefits related to the decrease in consumption of natural gas. The total amount of natural gas displaced by IGCC, assuming a 20-year plant life, is as follows:	
		If IGCC displaces NGCC, natural gas consumption is reduced by 5 quad	If IGCC displaces NGCC, natural gas consumption is reduced by 4 quad

NOTE: /kW, per kilowatt; IGCC, integrated gasification combined cycle; NGCC, natural gas combined cycle; PC, pulverized coal; quad, quadrillion British thermal units; and USC-PC, ultrasupercritical pulverized coal.

[a]The panel determined that the benefits of the IGCC research program in the Carbon Constrained scenario depend critically on the success of carbon sequestration research. Specifically, in the Carbon Constrained scenario there would be no market for IGCC plants unless the carbon emissions can be captured and sequestered. A separate panel evaluated the carbon sequestration program.

[b]Economic benefit in this table is calculated as the difference in the discounted expected value with and without the DOE program of all IGCC plants built between 2006 and 2025. Economic benefits are discounted at both 3% and 7% real, while environmental and security benefits are discounted at 3%.

[c]In the High Oil and Gas Prices scenario, almost the same amount of IGCC is expected to be built with or without the program. The reduction in carbon emissions results from the increased efficiency of IGCC plants with the program.

FIGURE H-5 Results matrix of the Panel on DOE's Integrated Gasification Combined Cycle Technology R&D Program.

plant. The panel believes a grid-ready plant with the claimed improvements will be available by around 2024.

Market risks to the IGCC program are represented by reduced priced natural gas, competing technologies, and an unfavorable regulatory environment. IGCC deployment would benefit from the gradual imposition of carbon constraints, but the reemergence of moderate natural gas prices (<$5.00/MMBtu), combined with premature imposition of an aggressive mandatory control on CO_2, is considered to be the greatest risk to IGCC's market penetration.

Advancements in USC/PC and CFB combustion technologies in the near term—and, if accompanied by cost-effective oxy-combustion for CO_2 sequestration, also in the longer term—represent the biggest technology threats to IGCC deployment and commercialization.

The IGCC R&D program's benefits also come in the

economic, environmental, and energy security categories. IGCC is a technology that allows coal, an indigenous energy source, to be used cleanly, in compliance with environmental expectations. It is the route to zero-emissions power generation, to hybrid power plants with fuel cells, and to the economic capture and sequestration of CO_2. Its successful implementation will favorably affect public health, national security, the balance of payments, the competitiveness of energy-intensive industries, and jobs in the sector.

Conclusions

- IGCC has a favorable cost and efficiency outlook, but successful market penetration requires more operational experience to be gained in the near term while IGCC is still more expensive and is perceived to have lower availability than alternative coal technologies. Speeding up the research on topics that improve IGCC plant availability—such as durable refractory, dry coal feed, instrumentation and diagnostics, and improved availability of the syngas-fired gas turbine—would hasten implementation of IGCC technology.
- Efficiency is an important goal of the IGCC program. Increased efficiency saves fuel and also reduces emissions, but it can also result in a more complicated plant and, hence, reduced availability. When there is a conflict between efficiency and availability, as in the case of extensive subsystems integration, it is recommended that, at least for the near term, availability should be favored over efficiency increases.
- The date in DOE's program goals is the time when the research will have been completed to enable the construction of a demonstration plant. The claimed improvements would then be realized only after experience with the demonstration plant and the subsequent operation of commercial plants. A clearer definition of the timeline of this process would help in assessment of the time at which the benefits of the research can be expected to accrue.

ATTACHMENT A
PANEL MEMBERS' BIOGRAPHIES

Jack S. Siegel, *Chair,* is a principal with the consulting firm Energy Resources International, Inc., and president of its Technology and Markets Group. While at the U.S. Department of Energy (DOE), he held various positions of leadership, including deputy assistant secretary for coal technology and acting assistant secretary for fossil energy. Prior to DOE, he was at the U.S. Environmental Protection Agency and led efforts to regulate and enforce the Clean Air Act of 1970. Mr. Siegel has broad and extensive experience on energy and environmental issues and has recently been involved in studies on markets and barriers to clean coal technologies; conventional and advanced turbines; renewable energy systems; distributed power systems; the impact of electric power restructuring on fuel and technology choices in the energy sector; options for reductions of greenhouse gases; and energy and environmental analysis in support of a number of foreign countries, the World Bank, and the Global Environment Facility. He served as a member of the NRC Committee on Challenges, Opportunities, and Possibilities for Cooperation in the Energy Futures of China and the United States and was a member of the Committee on Benefits of DOE's R&D on Energy Efficiency and Fossil Energy and the Phase One committee. He has received the Presidential Award for Superior Achievement (1992) and the Secretary of Energy's Gold Medal for Outstanding Performance (1994). He has a B.S. in chemical engineering from Worcester Polytechnic Institute.

Rakesh Agrawal (NAE) is Winthrop Stone Distinguished Professor of Chemical Engineering at Purdue University, where he has worked since 2004. From 1980 until 2004, he was employed at Air Products and Chemicals, most recently as an Air Products fellow. His research interests include basic and applied research in gas separations, process development, synthesis of distillation column configurations, adsorption and membrane separation processes, novel separation processes, gas liquefaction processes, cryogenics, and thermodynamics. Dr. Agrawal has broad experience in hydrogen production and purification technologies. His current interest is in energy production issues especially from renewable sources such as solar. He holds 116 U.S. patents and more then 300 foreign patents. He has authored 64 technical papers and given many lectures and presentations. He chaired the Separations Division and the Chemical Technology Operating Council of the American Institute of Chemical Engineers and also a Gordon Research Conference on separations. He was a member of the NRC Committee on Alternatives and Strategies for Future Hydrogen Production and Use. Dr. Agrawal received a B.Tech. from the Indian Institute of Technology, in Kanpur, India; an M.Ch.E. from the University of Delaware; and an Sc.D. in chemical engineering from the Massachusetts Institute of Technology.

Janos Beer is currently professor emeritus of chemical and fuel engineering at the Massachusetts Institute of Technology. He is also a member of the National Coal Council, which provides guidance to the U.S. Secretary of Energy. Dr. Beer's current research interests include clean fossil energy electric power generation; turbulent combustion of gaseous, liquid, and solid fuels; and reduction of pollutant emissions from combustion processes. He has headed divisions of several prestigious research facilities, including the Combustion Section at the Budapest Heat Research Institute and The Netherlands Research Station of the International Flame Research Foundation. He served as dean of engineering from 1973 to 1976 at the University of Sheffield, England, and as director of the MIT Combustion Research Facility from 1976 to 1993. Dr. Beer earned his economics and engineering degrees at the Jozsef Nador University of

Technical and Economic Science in Budapest in the 1940s. He achieved his Ph.D. in 1960 and D.Sc. (Tech.) in 1968 at the University of Sheffield, England.

Francis P. Burke is vice president, research and development, CONSOL, Inc. Dr. Burke has been with CONSOL since 1975, engaging in a wide variety of coal-related R&D, including the development of technology for coal conversion and emissions control. His research activities have related to trace elements in coal and coal utilization processes; coal liquefaction process development; control technology development for oxides of sulfur and nitrogen; coal-related wasted management and utilization; and methanol reforming and synthesis. He has been involved in numerous national and international workshops and symposia on coal-related R&D needs. He has a B.S. in chemistry from Gonzaga University, a Ph.D. in physical chemistry from Iowa State University, and an executive education from the Darden School of Business, University of Virginia.

Linda R. Cohen is a professor of economics at the University of California, Irvine. She was previously chair of the Department of Economics, University of California, Irvine, where she taught in various capacities with increasing responsibility since 1987. Previously, Dr. Cohen was an economist associate at the Rand Corporation, a research associate for economics with the Brookings Institution, a senior economist at the California Institute of Technology's Environmental Quality Laboratory, and an assistant professor of public policy at Harvard University's Kennedy School of Government. She has been the Gilbert White Visiting Fellow at Resources for the Future, the Olin Visiting Professor in Law and Economics at the University of Southern California Law School in 1993 and 1998, a fellow of the California Council for Science and Technology in 1998, and a research fellow at the Brookings Institution in 1977. Dr. Cohen has written many articles and coauthored a book on federal research and technology policy. She is currently a member of the editorial board of *Public Choice* and a member of the California Energy Commission's Advisory Panel for the Public Interest Energy Research Program. She has served on a variety of panels and committees and was a member of the NRC Committee on Benefits of DOE's R&D on Energy Efficiency and Fossil Energy and of the Phase One committee. She has an A.B. degree in mathematics from the University of California at Berkeley and received her Ph.D. in social sciences from the California Institute of Technology.

Frank P. Incropera (NAE) is McCloskey Dean of Engineering and Brosey Professor of Mechanical Engineering, University of Notre Dame. His previous positions included head, School of Mechanical Engineering, Purdue University; visiting scholar, Mechanical Engineering Department, University of California, Berkeley; professor, School of Mechanical Engineering, Purdue University; and others. He has been a visiting scholar at a number of universities. He has been the recipient of numerous awards, including fellow, American Society of Mechanical Engineering (ASME); Senior United States Scientist Award of the Alexander von Humboldt Foundation (Bonn); Heat Transfer Memorial Award of the ASME; Worcester Reed Warner Award of the ASME; and has been named as one of the world's 100 most highly cited researchers in all fields of engineering. He has served on the NRC's Department of Energy Panel on Integrated Manufacturing and the Panel on Engineering, Applied Sciences, and Applied Mathematics. His expertise spans a wide range of heat transfer research, including all forms of convection, boiling, and two-phase flow, radiative transfer, and engineering applications, as well as manufacturing processes and materials engineering. He has an S.B. in mechanical engineering from the Massachusetts Institute of Technology and an M.S. and a Ph.D. in mechanical engineering from Stanford University.

Mike Mudd has over 30 years experience in the utility industry, with most of his career having been focused on coal-fired generation. He has been involved in the design, construction, start-up, and operation of large coal-fired power plants, including American Electric Power's 1,300-MW and 600-MW coal-fired power plants. He was responsible for several clean coal technology demonstration projects, including as project manager for the 70-MW Tidd pressurized, fluidized-bed combustion (PFBC) demonstration plant, the first such power plant in North America, built with the cooperation of the DOE and the Ohio Coal Development Office. In 1996, Mr. Mudd moved to AEP Resources, where he was a developer in the nonregulated utility business, responsible for the development of cogeneration projects in the United States and Canada and independent power producer projects in Mexico. In 2002, he returned to the R&D arena with AEP, where in his current position he is responsible for corporate R&D associated with energy supply technologies, including coal, gas, nuclear, and renewable energy technologies. Mr. Mudd is currently on the FutureGen Alliance board of directors and is playing a key role in the development of AEP's recently announced IGCC project. He is active in several industry associations, including participation in committees associated with the Coal Utilization Research Council, EPRI, and the National Coal Council. Mr. Mudd chaired a working group on a study for the National Coal Council on opportunities to expedite the construction of new coal-fired power plants that was published in 2005. He also serves on the IGCC Expert Working Group for the EPRI CoalFleet program. Mr. Mudd has a B.E. and postgraduate studies from Stevens Institute of Technology.

Ronald H. Wolk is principal, Wolk Integrated Technical Services. His previous positions include director, Advanced Fossil Power Systems Department, Electric Power Research Institute, and Associate Laboratory Director, Hydrocarbon Research, Inc. Mr. Wolk has extensive experience in assessing, developing, and commercializing advanced power generation and fuel conversion technologies, including fuel cell, gas turbine, distributed power generation, and integrated gasification combined-cycle technology systems. He served on the NRC Committee on R&D Opportunities for Advanced Fossil-Fueled Energy Complexes. He has a B.S. and an M.S. in chemical engineering from the Polytechnic Institute of Brooklyn (now Polytechnic University).

ATTACHMENT B
ESTIMATION DETAILS

Introduction

The benefits of IGCC depend ultimately on how many such plants are placed into operation and on the subsequent cost of electricity, the environmental consequences of electricity production, and the security implications of the industry structure. Demand for IGCC and its benefits depend on costs of alternative technologies, fuel cost and availability, demand for electricity, and constraints on the electrical system such as the need for peak load plants. Because of the important feedbacks between demand and prices and the interactions between different technologies on prices, costs, and demand for each, estimating IGCC benefits is a complex undertaking. Sophisticated computer modeling, such as the NEMS model employed by DOE, necessarily plays a critical role in benefit assessments. At the same time, the NEMS model is resource-intensive in itself, and given the many options that this study considers, the long time horizon and related uncertainly about the trajectory of fuel prices, the state of alternative technologies, and economy-wide factors that affect electricity demand, the simplifications that are employed here provide a ballpark estimate that is probably within the range of estimates that more sophisticated modeling efforts would yield. The NEMS estimates made for the four scenarios and two R&D cases,[10] which were used as a starting point for the panel's analysis, were modified in the ways discussed below. Some additional NEMS estimates, discussed below, would be useful for refining and checking the panel's estimates.

The analysis in this attachment is based on three sets of numbers from the NEMS output:

- Estimates of costs of electricity from IGCC and alternative technologies and of relative capacities of the different technologies deployed, given no DOE R&D program.
- Estimates of costs and deployment given successful—that is, programs attain all of the program goals related to cost, timing, and performance—DOE R&D for all FE components other than IGCC.
- Estimates of costs and deployment given successful DOE R&D for all FE program components, including IGCC.

The panel modified the NEMS estimates to account for variations in cost, timing, and performance for IGCC. It also modified them to correspond to a set of assumptions necessary for appropriate consideration of the technologies displaced by IGCC. Each of these is discussed here.

Cost and Performance

The capital cost, availability, and efficiency assumptions on each branch of the decision tree (Figure H-1) yield a levelized COE given the relative contributions to cost in the AEO ascribed to each electricity generation technology. When the estimated cost is smaller than that given by the AEO Reference Case, IGCC should fare better and the alternative worse; when the estimated cost is larger, the reverse situation holds. A scaling procedure that rests on a locally linear demand relationship for the different technologies allows an estimate to be made of the quantity of IGCC (i.e., the capacity) under the revised cost conditions. Let X_{ALT} be the average cost of the alternative technology; X_{IGCC} be the average cost of IGCC; b be the change in the average cost of IGCC; Q^0_{IGCC} be the original quantity of IGCC added during the year (the AEO estimate); Q^1_{IGCC} be the quantity of IGCC added given the change in average cost; and d be the difference in quantities in the two cases:

$$Q^1_{IGCC} = Q^0_{IGCC} + d$$

Then,

$$d = Q^0_{IGCC} * [b/(X_{ALT} - X_{IGCC})]$$

and

$$Q^1_{IGCC} = Q^0_{IGCC} [1 + b/(X_{ALT} - X_{IGCC})]$$

As is discussed above, the panel concluded that the DOE time goals for IGCC are optimistic. The panel thought a more likely time frame allowed completion of the R&D program in 2014, with prototypes and demonstration experience such that full competitive consideration of the technology would be possible in 2020. This time frame is in part reflected in the NEMS study, which allows only limited construction of IGCC in the years between 2012 and 2018 notwithstanding the formal completion of the R&D program. The panel's assessment is thus only somewhat more limited than that given by the NEMS model, although it departs formally from the DOE goal.

The panel's adjustment of the NEMS estimates of newly installed capacity relied on changing the capital cost estimates for IGCC, which was allowed to decline linearly from

---

[10]Julianne Klara, Senior Analyst, Office of Fossil Energy, DOE, "NEMS-based benefits of FE gasification R&D," Presentation to the panel. October 5, 2005.

$1,400/W in 2014—the current estimated capital cost—to the cost corresponding to the level of technical success achieved by 2020. This procedure allows a steady introduction of IGCC during the development and demonstration period, with full commercial credit given in 2020.

Competing Technologies

The NEMS runs made available to the panel by the Office of Fossil Energy provide an interesting illustration of the interaction among different DOE programs. According to the NEMS calculations, successful completion of other DOE program components leads to a large increase in baseload electricity generation in the United States. The amount of base (coal and NGCC) generation increases dramatically, the amount of distributed generation increases, and the contribution of gas turbines and diesel generators falls. These changes occur in both the Reference Case and the High Oil and Gas Prices scenario. According to the NEMS estimates, NGCC dominates additions to base capacity generation absent the DOE IGCC R&D program. With IGCC R&D, IGCC substitutes for an increasing share of NGCC capacity. Under the High Oil and Gas Prices scenario, it displaces NGCC in the later years. In the NEMS scenario, other coal technologies—in particular, advanced PC units—are dominated by both IGCC and NGCC given the reduction in gas prices that results from other DOE R&D activities.

In the absence of all DOE R&D, total baseload demand in the United States is much lower and natural gas prices relatively high. One consequence is that pulverized coal continues to compete with NGCC, but the combined demand for NGCC and PC is less than when DOE pursues R&D, either with or without an IGCC component. The panel did not have NEMS output for the case where only IGCC R&D is pursued. Given the other studies, however, it is possible to deduce that IGCC would further divide the baseload market between the coal and gas technologies. In this case, IGCC would compete initially with PC. In addition, if the cost of IGCC is low, it would also compete for some of the NGCC plants. The NEMS results imply that DOE's R&D program has complementary components: Success in DG and other distributed energy resources (DER) programs increases demand for IGCC. Similarly, success in the sequestration program increases IGCC demand in Carbon Constrained scenarios. Thus the economic benefits from the IGCC program are larger when the entire DOE program is successful than when only components of it are pursued to completion. The security and environmental benefits of IGCC change with success in the larger DOE program. If the entire program is pursued, IGCC replaces substantial gas generation, with positive security benefits and negative environmental consequences. If the remaining program is not undertaken, the environmental benefits of IGCC may be substantial, depending on emissions from PC plants, but the security benefits are negligible.

Estimating these options would, ideally, require a fourth set of results (Case D) from a NEMS run, corresponding to the case where IGCC R&D alone is pursued by the DOE program. Let p be the overall probability that the non-IGCC parts of the DOE R&D program succeed (of particular relevance are the components that allow more baseload power in the generation mix). Consider expected costs in four circumstances:

Cost (A) = cost when all DOE R&D is successful (A)
Cost (B) = cost given DOE R&D but no IGCC program (B)
Cost (C) = cost given the complete program
Cost (D) = cost given just the IGCC program (D)

The benefits of interest are

$$p^* [\text{Cost}(C) - \text{Cost}(B)] + (1 - p)^*[\text{Cost}(D) - \text{Cost}(A)]$$

Absent an estimate for the remaining program, which is beyond the purview of this panel, assessing benefits requires at a minimum a consideration of the two polar possibilities—presence or absence of the remaining DOE program. As the panel lacked a base NEMS run for the final option—IGCC research without the remaining program—it used an approximation, as described here. It is recommended that this approximation be replaced by an actual NEMS run.

Interim Approximation

The baseline IGCC capacities were derived for Case D assuming that IGCC could substitute for annual additions to baseload capacity only—that is, the amount of capacity represented in the no-DOE case by IGCC, PC, and NGCC additions (see Table H-6). With no program, the total base capacity additions amount to 213 gigawatts (GW). This figure can be compared to an addition of 200 GW with the DOE program and no IGCC and an addition of 240-256 GW, depending on source, for the DOE program with IGCC. Furthermore, it is assumed that gas prices are 5 percent higher in 2025 than they would have been had there been a comprehensive DOE R&D program, and represented in the NEMS FEBEN[11] estimate, or one-fifth the gas price increase given in the High Oil and Gas Prices scenario. IGCC thus competes with all of the PC plants, substituting for them as the technology becomes available, and also successfully competes with the NGCC plants to a degree calculated taking account of a higher gas price than with the all-DOE R&D program case. The 5 percent gas price penalty is introduced gradually in equal increments between 2016 and 2025 and an interpolation is then made between the share of IGCC in

---
[11]FEBEN is a calculation tool employed by DOE's Office of Fossil Energy to compare outputs from NEMS. (Julianne Klara, Senior Analyst, Office of Fossil Energy, DOE, "NEMS-Based Benefits of FE Gasification R&D," Presentation to the panel, October 5, 2005.)

TABLE H-6 Annual Additions of IGCC Capacity Used to Calculate Program Benefits

Year	Share of IGCC (%)			Annual Base Capacity Additions (GW)	Annual IGCC Capacity Additions (GW)
	High Oil and Gas Prices Scenario	Reference Case	Panel's Proposed Scenario		
2012	0.00	0.00	0.00	3.20	0.00
2013	15.97	0.00	0.00	3.64	0.00
2014	17.74	17.65	17.65	8.46	1.49
2015	17.74	17.65	17.65	10.82	1.91
2016	23.17	17.65	17.76	13.76	2.44
2017	30.14	17.65	18.15	17.21	3.12
2018	37.22	17.65	18.82	17.20	3.24
2019	71.49	23.23	27.09	16.13	4.37
2020	80.27	25.25	30.75	18.90	5.81
2021	82.62	23.58	30.66	22.04	6.76
2022	100.00	35.51	44.54	21.85	9.73
2023	100.00	43.59	52.61	26.22	13.79
2024	100.00	77.09	81.21	29.31	23.80
2025	100.00	79.96	83.97	21.49	18.04

NOTE: IGCC, integrated gasification combined cycle power plant; GW, gigawatt.

the High Oil and Gas Prices scenario and the Reference Case, whereby in 2025 the IGCC share is as follows:

Proposed IGCC share in 2025 = Reference Case share + (High Oil and Gas Prices share − Reference Case share)*0.20

In general,

Proposed share of IGCC = Reference Case share + (High Oil and Gas Prices share − Reference Case share)*0.02*(year − 2015)

for years between 2016 and 2025.

I

Report of the Panel on DOE's Carbon Sequestration Program

INTRODUCTION AND OBJECTIVE OF THE STUDY

The panel on DOE's Carbon Sequestration Program was formed by the National Research Council to examine the benefits of the U.S. Department of Energy's (DOE's) program on carbon sequestration as part of the activities of the Committee on Prospective Benefits of DOE's Energy Efficiency and Fossil Energy R&D Programs, Phase Two. The panel was charged with applying the method that the committee had developed for estimating the benefits of DOE's R&D. Although the panel was charged with estimating the likelihood that the program goals would be achieved within the budget and specified time period, the panel was not given detailed materials that would allow it to review individual projects to judge whether they would achieve their goals. Rather, it conducted a high-level program review, relying on members' knowledge of each area and the difficulties of achieving specific R&D goals.

The method developed by the committee asked the panel to come to a judgment concerning the likelihood that the DOE program as currently funded would achieve the goals within the specified time period. The panel was also asked to come to a judgment concerning the extent to which the technology would be deployed in the market. The committee's method outlined three scenarios for the panel and allowed the panel to add a fourth, which the panel believed would be of particular interest for this program.

The first scenario, the Reference Case, from the *Annual Energy Outlook (2005)*, assumed business as usual, the second assumed high oil and gas prices, and the third assumed that carbon emissions would be curtailed—namely, that a carbon tax would be imposed on emissions at $100 per ton of carbon.[1] The panel decided to evaluate another scenario wherein the carbon tax was assumed to be $300 per ton.

Since the panel did not think that a technology that separated and sequestered the carbon would be as inexpensive per megawatt-hour of generated electricity as a technology that did not, it concluded that carbon sequestration would not be implemented unless there were restrictions on carbon emissions. Thus, the panel concluded that the sequestration technology would not be implemented in the first two scenarios, even if DOE achieved its R&D goals. Thus, the panel focused its analysis on the scenarios with carbon taxes of $100 or $300 per ton of carbon emitted.

In evaluating the benefits of each scenario, the panel utilized the DOE NEMS model runs to provide baseline estimates of fuel costs and capacity additions in each scenario. Unfortunately, DOE was not able to make additional model runs for these two carbon tax scenarios within the time available, so previous NEMS runs for a carbon constrained scenario were adapted to provide the necessary estimates. The panel believes that the quantitative results are reasonable approximations to what new NEMS model runs would have given in these scenarios.

During Phase One of the prospective benefits study, the earlier panel estimated the benefits of the same DOE carbon sequestration program. Owing to differences in the extent to which two factors were considered, the Phase One panel calculated an expected economic benefit of $35 billion, whereas the current panel calculated benefits of $3.5 billion. The difference in results is primarily due to the current panel's more complete and rigorous application of the methodology outlined in Phase One. In particular, the current panel focused on what would happen without effort by DOE and the impacts of competing technologies. The earlier (Phase One) evaluation of the carbon sequestration program was done as part of the task of developing the methodology and consequently did not adequately consider these two factors.

---

[1]The tax is assumed to be imposed in 2012 and to increase at 3 percent per year thereafter.

APPENDIX I

SUMMARY OF DOE'S CARBON SEQUESTRATION PROGRAM

Carbon sequestration is the separation and storage of carbon dioxide CO_2 and other greenhouse gases (GHGs) that would otherwise be emitted to the atmosphere. GHGs can be captured at the point of emission or they can be removed from the air. The captured gases can be used, stored in underground reservoirs or possibly the deep oceans, or converted to rocklike mineral carbonates and other products. There is a wide range of sequestration possibilities to be explored, but a clear priority for near-term deployment is to capture a stream of CO_2 from a large, stationary emission point source and sequester it in an underground formation. Carbon sequestration holds the potential to provide deep reductions in greenhouse gas emissions since a little less than half of total U.S. GHG emissions are from large point sources of CO_2. Research is ongoing to develop a clearer picture of domestic geologic sequestration storage capacity, but it is likely that domestic formations have at least enough capacity to store several centuries' worth of point source emissions. Technologies aimed at capturing and utilizing methane emissions from energy production and conversion systems can be applied to carbon sequestration and will reduce an important GHG emission. Mobile and dispersed GHG emissions can be offset by enhanced carbon uptake in terrestrial ecosystems, and research into CO_2 conversion and other advanced sequestration concepts will expand the range of sequestration.

Program Goals

DOE established the carbon sequestration program in 1997.[2] The program, which is administered within the Office of Fossil Energy (FE) by the National Energy Technology Laboratory (NETL), seeks to move sequestration technologies forward so that their potential can be realized and they can play a major role in meeting any future needs for the reduction of GHG emissions. This program utilizes an annual Carbon Sequestration Technology Roadmap and Program Plan to identify research pathways that are expected to lead to commercially viable sequestration systems and sets forth a plan of action for sequestration research. Table I-1 is a top-level roadmap for core R&D and infrastructure development. The overarching program goal is 90 percent CO_2 capture with 99 percent storage permanence at no more than a 10 percent increase in the cost of energy services by 2012.

Core R&D

The goal of the core R&D program is to advance sequestration science and develop new sequestration technologies and approaches to the point of precommercial deployment.

The core program is a portfolio of work including cost-shared, industry-led technology development projects, research grants, and research conducted in-house at NETL. The core program is divided into the following six areas.

- *CO_2 capture.* CO_2 exhausted from fossil-fuel-fired energy systems is typically too dilute, at too low a pressure, or too contaminated with impurities to be directly stored or converted to a stable, carbon-based product. The aim of CO_2 capture research is to produce a CO_2-rich stream at high pressure. The research is categorized into three pathways: postcombustion, precombustion, and oxyfuels.
- *Carbon storage.* Carbon storage is defined as the placement of CO_2 into a repository in such a way that it will remain stored (or sequestered) permanently. It includes three distinct subareas: geologic sequestration, terrestrial sequestration, and ocean sequestration.
  —Trapping within a geologic formation is the primary method for storing CO_2. A layer, or cap, of impermeable rock overlies the porous rock into which the CO_2 is injected and prevents upward flow of CO_2.
  —Because the surface of sandstone and other rocks preferentially adheres to saline water in preference to CO_2, if there is enough saline water within a pore (75-90 percent of the pore volume), the water will form a capillary plug that traps the residual CO_2 within the pore space.
  —When CO_2 comes in contact with the saline water it dissolves into solution.
  —Over longer periods of time (thousands of years), dissolved CO_2 reacts with minerals to form solid carbonates. This process is known as mineralization.
  —Preferential adsorption of CO_2 onto coal and other organic-rich reservoirs takes place as a function of reservoir pressure.

Monitoring, Mitigation, and Verification (MM&V). Monitoring and verification for geologic sequestration has three components: (1) modeling, which facilitates the understanding of the forces that influence the behavior of CO_2 in a reservoir; (2) plume tracking, the ability to see the injected CO_2 and its behavior; and (3) leak detection systems, which serve as a backstop for modeling and plume tracking. MM&V for terrestrial ecosystems also has three components: organic matter measurement, soil carbon measurement, and modeling.

Non-CO_2 GHG Control. Because some non-CO_2 GHGs (e.g., methane, N_2O, and gases having high global warming potential) have significant economic value, they can often be captured or avoided at relatively low net cost. This area of the core sequestration program is focused on fugitive methane emissions, whereby non-CO_2 GHG abatement is integrated with energy production, conversion, and use. Landfill gas and coal mine methane are two top-priority opportunities.

Breakthrough Concepts. R&D on breakthrough concepts is

---

[2] Carbon Sequestration Technology Roadmap and Program Plan, U.S. Department of Energy, National Energy Technology Laboratory, May 2005.

TABLE I-1 Top-Level Carbon Sequestration Roadmap

	Pathways	Metrics for Success 2007	Metrics for Success 2012
CO_2 capture	Postcombustion Precombustion Oxy-fuel	Develop at least two capture technologies that each result in less than a 20% increase in cost of energy services.	Develop at least two capture technologies that each result in less than a 10% increase in cost of energy services.
Sequestration storage	Hydrocarbon-bearing geologic formations Saline formations Tree plantings, silvicultural practices, and soil reclamation Increased ocean uptake	Field tests provide improved understanding of the factors affecting permanence and capacity in a broad range of CO_2 storage reservoirs.	Demonstrate ability to predict CO_2 storage capacity with +/– 30% accuracy. Demonstrate enhanced CO_2 trapping at precommercial scale.
Monitoring, mitigation, and verification (MM&V)	Advanced soil carbon measurement Remote sensing of above-ground CO_2 storage and leaks Detection and measurement of CO_2 in geologic formations Fate and transport models for CO_2 in geologic formations	Demonstrate advanced CO_2 measurement and detection technologies at sequestration field tests and commercial deployments.	CO_2 material balance greater than 99%. MM&V protocols enable 95% of stored CO_2 to be credited as net emissions reduction.
Breakthrough concepts	Advanced CO_2 capture Advanced subsurface technologies Advanced geochemical sequestration Novel niches	Laboratory scale results from one or two of the current breakthrough concepts show promise to reach the goal of an increase of 10% or less in the cost of energy and are advanced to the pilot scale.	Technology from the program's portfolio revolutionizes the possibilities for CO_2 capture, storage, or conversion.
Non-CO_2 GHGs	Minemouth methane capture/combustion Landfill gas recovery	Deployment of cost-effective methane capture systems.	Commercial deployment of at least two technologies from the R&D program.
Infrastructure development	Sequestration atlases Project implementation plans Regulatory compliance Outreach and education	Phase II partnerships have pursued priority sequestration opportunities identified in Phase I and have conducted successful field tests.	Projects pursued by the Regional Partnerships contribute to the 2012 assessment under GCCI.

pursuing revolutionary and transformational sequestration approaches with potential for low cost, permanence, and large global capacity. These concepts are speculative but could offer performance and cost improvements that let them leapfrog existing technologies.

Field Projects. Because conditions in both terrestrial ecosystems and geologic formations are difficult to simulate, testing ideas in the field often enables significant learning and insight. Sequestration field tests serve as a test bed for CO_2 detection and measurement technologies and also present an opportunity to validate models.

Infrastructure Development

DOE initiated seven regional carbon sequestration partnerships (RCSPs) in 2003 with the goal of developing an infrastructure to support and enable future carbon sequestration field tests and deployments. The first stage of the RCSPs ended in June of 2005. The partnerships have established a national network of companies and professionals working to support sequestration deployments, created a carbon sequestration atlas for the United States, and identified and vetted priority opportunities for sequestration field tests. The primary and overarching objective of the second stage will be to move forward with the high-priority tests to validate sequestration technology that were identified in the first stage of the effort. In support of this primary objective will be the refining and implementing of MM&V protocols, improving the understanding of environmental and safety regulations; establishing protocols for project implementation, accounting, and contracts; and conducting public outreach and education. Also in the second stage, partnerships will seek to continue the characterization of the regions and to refine a national atlas of carbon sources and sinks. In FY 2009 DOE will consider an optional third stage effort for the RCSPs. A third stage, which would run through 2013, would be contingent on the continued importance of and synergies with the FutureGen initiative (partially funded by DOE), the need for validation of additional sequestration sites throughout the United States, and funding availability.

Program Budgets

The base sequestration program funding is roughly $55 million per year. DOE will provide approximately $100 million to support the RCSPs over the next 4 years. Each partnership will receive between $2 million and $4 million per year in DOE funding. At least 20 percent of project costs are covered by non-DOE funding. The total value of the projects exceeds $145 million over the next 4 years. The RCSPs are structured to become self-sustaining by 2013. The approximate actual and projected funding levels from 2001 to 2020 are shown in Figure I-1. Program costs through 2020 are expected to be $875 million.

TECHNICAL RISKS

While DOE has taken a portfolio approach for its CO_2 sequestration program, it has focused on developing components suitable for advanced integrated gasification combined cycle (IGCC) technology, with sequestration based on deep well injection. DOE sees advanced IGCC as the technology of choice to achieve the goal of 10 percent incremental COE for CO_2 sequestration beyond that achieved by IGCC units. (There is a goal of 20 percent increase in the COE for combustion-based systems. The cost of electricity generated from such systems, including carbon capture and storage (CCS), has been estimated to be about 10 percent greater than the cost of IGCC. A review of cost studies found that the cost of electricity generated using supercritical pulverized coal (SC-PC) technology with (post-combustion) amine-based CO_2 capture would be $77/MWh and, if using IGCC with a Selexol unit for carbon capture, $65/MWh. Both the cost estimates include $5 per ton CO_2 for geologic storage (Rao et al., 2005).

The carbon sequestration program is taking on a relatively high overall risk to create technologies for commercial demonstration by 2012 in that it relies heavily on the successful deployment of full-scale IGCC plants (more than 200-500 MW) in parallel with the sequestration program schedule. There are only a few IGCC plants operating worldwide, and advanced, commercial-scale IGCC units are only in the design stage and have no CO_2 sequestration. An end-to-end, full-sized plant demonstration of IGCC technology with sequestration will take longer.

The recent DOE systems analysis of the developments from the CO_2 capture program is framed in terms of four main components: (1) sorbent improvement associated with the Selexol process, (2) oxygen membrane separation to replace cryogenic separation, (3) membrane technology to facilitate the water gas shift reaction to produce hydrogen, and (4) storage of the CO_2 in deep geological formations. The storage component is said to be advanced, and based on

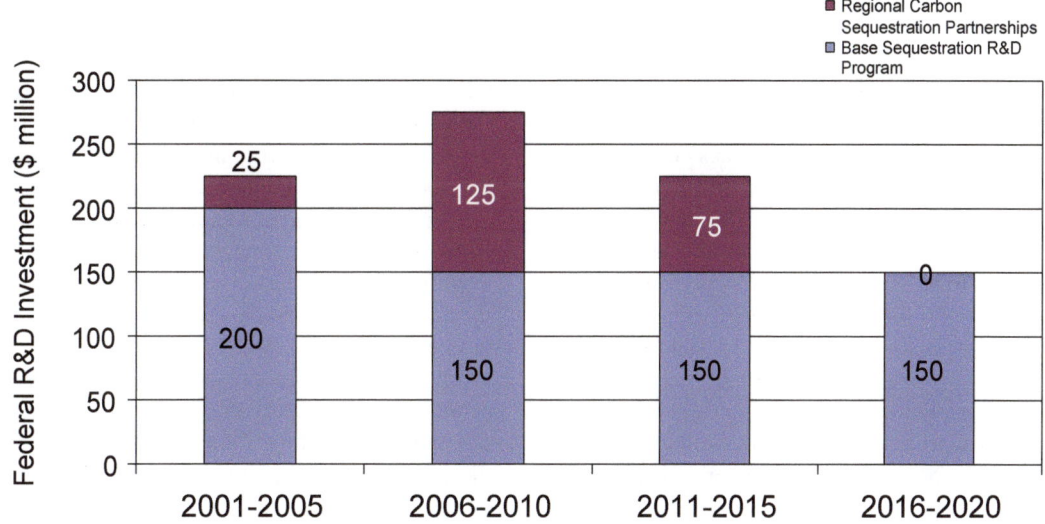

FIGURE I-1 Funding requirements for DOE's carbon sequestration program. RCSP, regional carbon sequestration partnerships.

the extensive commercial CO_2-enhanced oil recovery effort in place today. While this experience has shown the process to be viable and there have been no serious accidents, the storage time frame has been a few decades, compared with the centuries-long time frame needed for CO_2 storage. The cost reductions for electricity production with sequestration are more sensitive to increases in the efficiency of the IGCC system than to advances in CO_2 capture.

The results from new sorbent technologies are expected to improve the performance of CO_2 scrubbers greatly by increasing capture efficiency at higher temperature and pressure (DOE, 2005d). While sorbent research has shown a number of options for improved systems,[3] none of these have been tested in pilot or larger combustion systems relevant to power plant operations. The technologies need to be tested thoroughly for sorbent stability, operational reliability, and integrated performance to establish a cost-effective design basis. This is an ambitious task to be completed by the target date of 2012.

The second component envisaged for reducing IGCC power production costs, specific oxygen separation based on ion transport membranes (ITMs) for oxyfuels, is funded not within the DOE carbon sequestration program but within its advanced gasification research program. The technical risk of achieving the sequestration program's goals is increased by having this critical component controlled elsewhere. More important, this membrane technology is predicated on successful operation with temperatures of about 1000°C or even higher. While membrane materials for operation in this temperature range are well developed, the supporting equipment for operating a membrane-based system is problematic. Failure of this equipment could slow down practical applications and increase costs for the technology. Since none of the membrane technology has been tested beyond small pilot scale, the reliability and expected performance of these systems at larger scale for design engineering and costing remain an open question.

The water gas shift membrane technology is jointly funded by the DOE carbon sequestration and IGCC programs. While this technology appears to have a favorable future in the laboratory, the use of polymer membranes here depends on achieving the flux and membrane stability at ~300°C. This temperature is an ambitious goal for any polymer membrane; good membrane stability and performance at this temperature are yet to be demonstrated. The performance of this technology at the pilot scale and larger remains to be demonstrated at an aggressive pace, to provide an informed and confident basis for large-scale design and integration into IGCC technology, as planned.

The panel's perception, taken in toto, is that the DOE carbon sequestration program, which depends heavily on complementary work in fossil fuel technology, demands a highly ambitious, relatively high-risk effort to achieve its technical goals by 2012.

DOE is using systems analysis tools in a constructive way to evaluate past progress and future objectives. These tools have the potential to strengthen the program by guiding the choices of technologies to pursue most vigorously as well as the down-selection process. However, in the briefings the panel received about the DOE program, it observed the distorting effect of the program's "aspirational goals" on the systems analysis effort. The leadership of the program has set cost increment goals—sometimes 10 percent, sometimes 0 percent—for CCS. The systems analysis effort has been unduly influenced by an apparent need to show strategies that meet these goals. First, the difficulty of meeting them is hidden by comparing future (lower cost) IGCC systems with CCS to present systems without CCS, making it appear that the cost savings for advanced IGCC can be attributed to sequestration savings. (The savings, presumably, is that the COE from IGCC with carbon sequestration would be less than the cost of electricity from IGCC with venting and paying the tax.) The DOE goal is an increase of no more than 10 percent for IGCC with sequestration compared with IGCC without sequestration. The analysis thus makes a misleading comparison to arrive at a small increase (or no increase) due to sequestration. Second, in an effort to drive downward the apparent incremental cost of CCS for electricity production, systems analyses have built in large credits, sometimes for the sale of by-products of the carbon sequestration process (hydrogen or chemicals), sometimes for the sale of CO_2, and sometimes for the avoidance of sulfur management above ground via co-storage of sulfur (as H_2S or SO_2) with CO_2. All of these credits can be real in some situations but are zero in others. DOE should consider the ancillary benefits in scenarios whose assumptions are clear. Fortunately, the panel finds that the program as a whole has not been distorted to emphasize these secondary concerns. However, the value of the systems analysis effort to the program is considerably reduced by a focus on such credits. The panel recommends that the leadership of the program work harder to insulate the systems analysis group from pressure to produce results that conform to the program's aspirational goals, so as to get greater value from the expertise of the group.

The CO_2 storage component is smaller than the capture component as it currently relies on existing technologies needing relatively little innovation. The program emphasis is on small-scale demonstration in the field and partnership in one large project (Weyburn).

Once CO_2 is injected into the subsurface, there are two primary routes for leakage: through or around the reservoir seal (caprock) or through well bores that could be created for this purpose or that might have been drilled in the past for oil or gas exploration. The reservoir seal could be compromised by tectonic activity or by overfilling the reservoir, while the well bores could be attacked by carbonic acid formed when CO_2 dissolves in the formation water.

---

[3]See, for example, DOE (2005d) and Rao et al. (2005).

Industry has over 30 years experience with CO_2 injection for enhanced oil recovery, with no mishaps that would indicate the process has serious flaws. However, carbonic acid reacts with the Portland cement that is used in the construction of wells as well as with the tubular bores that communicate to the surface. These reactions can be evaluated in the laboratory over relatively short time spans, but there is no known way in the laboratory to evaluate the reactions that might degrade the well bore seals over hundreds to thousands of years. There may have to be some sort of protocol to monitor the wells periodically and make repairs as needed. The DOE program devotes little effort to remediation, assuming that the technology available in the industry is, or will be, adequate. The other major forms of carbon storage envisioned in this program are ocean and terrestrial. For ocean sequestration, environmental impacts may be more significant than concerns about safety, whereas the reverse is true of terrestrial sequestration in geologic formations (Herzog, 2001; Brewer, 2003; Orr, 2003).

While success of the capture program depends almost entirely on the ability to reduce the cost of the operation by technical means, the storage program cannot be successful if a significant fraction of the public views it as dangerous or unacceptable. Thus, the technologies must not only be safe and effective, they must be explainable to the public and the regulatory community in such a way as to instill confidence that they are in fact safe and effective. The federal government in general and the DOE in particular have not had a good track record in accomplishing this task in other programs, such as the Yucca Mountain nuclear waste repository, and the siting of terminals for unloading liquefied natural gas.

The cornerstone of the DOE program is the RCSPs, a collection of seven organizations run by respected entities and with a wide base of participation. These partnerships are in the second stage of their development and have developed work plans that include not only technical development and demonstration but also outreach. DOE holds meetings routinely to coordinate the efforts of the RCSPs and share results.

The RCSPs were told to develop demonstration projects relevant to their regions, and they hold storage field trials with significant monitoring and evaluation components. These projects, which will be completed over the next few years, will familiarize interested parties with the process. However, the RCSP program may not resolve uncertainties in extrapolating the volume scale and the time frame over which the demonstrations can operate.

MARKET RISKS

Both competing technologies and political factors will have an effect on the deployment of carbon sequestration technologies in the market. The primary driver for deployment is an incentive for reducing carbon emissions. The panel believes that only in a carbon-constrained scenario will any carbon sequestration technologies be implemented; accordingly, the benefits of DOE's carbon sequestration program were evaluated only for scenarios where a carbon tax exists.

Competing Technologies

A high carbon tax will make zero-emissions or very low carbon emissions electricity generating technologies more attractive. The panel believes IGCC with CCS is a promising technology. Other technologies that could potentially compete against IGCC with CCS are natural-gas-fired electricity generation technologies; technologies that transform coal into a noncarbon fuel, such as hydrogen, with carbon storage; high-efficiency combustion cycles with backend CCS; oxygen combustion with CCS; nuclear power systems; and renewables.

If only a modest reduction in carbon emissions is required, substituting natural gas (CH_4) for coal in a high-efficiency combined cycle generator can accomplish that reduction. However, natural gas prices have increased rapidly in recent years owing to high demand and static supply. At current prices, switching to natural gas would be a costly strategy with considerable doubt that the supply of natural gas would be sufficient through 2017.

Coal gasification can lead to a pure hydrogen stream with separation and sequestration of the CO_2. The resulting hydrogen could be burned in a turbine or used in fuel cells. This approach is a variant of IGCC and is attractive only if carbon separation and sequestration is an attractive, low-cost technology that effectively sequesters the carbon. If the DOE program were successful in creating an attractive IGCC technology with carbon sequestration, the hydrogen stream would be available for other applications.

Both higher efficiency combustion cycles (supercritical and ultrasupercritical) with backend CCS and oxycombustion systems are more expensive today than gasification with CCS,[4] and oxycombustion is in its early stages of development (Rao et al., 2005; Anderson et al., 2004). Whether these systems, which are being addressed in the DOE sequestration program, can provide a viable alternative remains to be seen. An as-yet- unresolved issue surrounding viable alternatives for coal remains the performance and cost of combustion or gasification with different types of coal. For example, lower rank coals such as lignite, when slurry-fed to the gasifier, bring in lower system efficiencies and net power outputs (Maurstad et al., 2006).

Several panel members believe that nuclear generation has significant market potential in the long run in a carbon-

---

[4]Based on an extensive literature review, Rubin (2006) has determined that a representative estimate of the cost of electricity, if generated using supercritical pulverized coal technology with CCS, would be $77/MWh and, if using IGCC with CCS, $65/MWh.

constrained scenario and could be a strong competitor for IGCC with sequestration. The relative attractiveness of the two technologies will depend on public acceptance and the cost of each technology, which will be influenced by DOE's fossil energy R&D program.

Political Risks and Other Market Factors

The panel identified several other potential barriers to the deployment of IGCC with carbon sequestration. Each of these barriers would make successful deployment of sequestration less likely and would tend to favor some of the competing technologies as a way to meet carbon constraints:

- *Public opposition based on the risk of sequestration.* It is not yet apparent whether the public would be receptive to carbon sequestration, and it is possible that people living near sequestration sites would have significant concerns that might lead them to oppose proposals to sequester CO_2 in their local environment. Strong public opposition could delay or even prevent the deployment of an IGCC plant with CCS. Some preliminary studies suggest that the public is not favorably disposed to carbon storage in the oceans or deep underground (Palmgren et al., 2004). To the panel's knowledge, there has not been a full risk assessment of carbon storage; such an assessment, could alleviate some public concerns.
- *Regulatory issues.* A variety of siting and permitting issues associated with carbon sequestration remains to be worked out, including jurisdictional issues that accompany the permitting process. Delays or problems in resolving these issues could significantly delay the deployment of sequestration technologies.
- *Physical siting requirements.* Storage in geological formations calls for sites having adequate capacity and injectivity, a confining unit (e.g., a caprock), and a geologically stable environment (IPCC, 2005). These requirements, along with regulatory requirements and public concern, could further limit potential sites and the penetration of IGCC with CCS. The location of generation away from load centers might raise costs to the consumer.
- *Competition from energy conservation and alternative energy sources.* In addition to public views and regulatory requirements, the competition will depend on the cost of electricity from each technology. This cost will be influenced by the regulatory requirements for each technology. For example, if regulators insisted that CO_2 had to be placed in areas where no oil or gas wells have been drilled, or below the depth to which wells have been drilled, IGCC with CCS could become less cost-competitive.

DECISION TREE MODEL AND PROBABILITY ASSESSMENT RESULTS

Rather than attempting to assess probabilities at a project level and somehow aggregate them, the panel decided to focus on an overall assessment of the effectiveness of the research program. The process and calculation methodology for this assessment[5] followed the recommended guidelines of the Committee on Prospective Benefits of DOE's Energy Efficiency and Fossil Energy R&D Programs, Phase Two. The impact of government support can be captured by considering the probabilities of various technical and market outcomes with and without government support. The decision tree developed by the panel is summarized in Figure I-2.

The main technological uncertainty considered was the increase in the COE associated with the capture and storage of carbon emissions from coal-fired power plants, specifically from advanced IGCC plants. DOE's R&D program assumes that IGCC plants without CCS will be the cheapest coal-based generation plants and that these plants will meet all EPA emissions requirements (aside from CO_2 emissions). Thus, the only significant difference between the two technologies is the COE and whether the carbon is sequestered. The panel considered COE in three time periods (2012, 2017, and 2022) and at four different levels of cost increase at each point in time. The probability assessments for costs in 2012 were conditional on the currently expected level of DOE funding for research on sequestration. The assessments for 2017 were made conditional on the 2012 results and the 2022 assessments were conditional on the 2012 and 2017 results as well as on the presence or absence of DOE support. Specifically, panelists were asked to assign probabilities that the COE increase associated with sequestration in 2012 would be 0 to 10 percent; 10 to 20 percent; 20 to 30 percent and more than 30 percent; four probabilities in total. For 2017, panelists were asked to assign conditional probabilities for the same ranges that depend on the cost increase in 2012. For example, if the cost increase in 2012 were in the 20-30 percent range, panelists were asked to specify probabilities that the costs in 2017 would 0-10 percent; 10 to 20 percent; 20-30 percent and more than 30 percent. In principal, there are four conditional probabilities for each of the four scenarios (16 in total), but many of these scenarios were judged to have zero probability: For example, panelists thought that there was no chance that the cost increases associated with sequestration would increase from 2012 and 2017. Thus, if the cost increase in 2012 were in the 20-30 percent range, there was no chance that the cost in 2017 would be more than 30 percent. The assessments for 2022 similarly depended on the outcomes in both 2012 and 2017; in principle there are a total of 16 scenarios requiring four probabilities each, but many of these scenarios were judged to have zero probability. To calculate expected costs and benefits, the 0-10 percent, 10-20 percent and 20-30 percent ranges were represented by their midpoints (5, 15, and 25 percent, respectively) and the over 30 percent range was represented by 40 percent. All of these probabilities were assessed assuming there would be

---
[5]A complete discussion of the methodology and process can be found in Chapters 3 and 4, respectively.

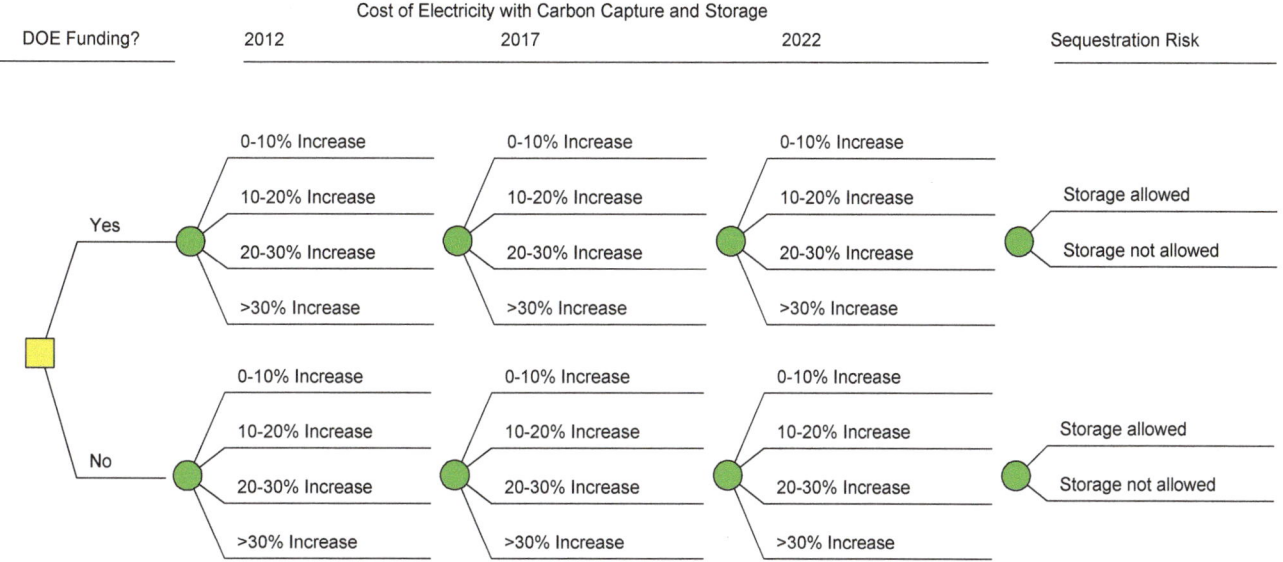

FIGURE I-2 Decision tree used by carbon sequestration panel.

a $100 per ton carbon tax beginning in 2012 with industry participants knowing well in advance of this impending tax. The same assessments were repeated assuming a $300 tax.

In the discussion of benefits below, the panel assumes that decisions about which technology to deploy are made with knowledge of the carbon tax level—$100 or $300 per ton carbon tax. However, when the panel discusses the COE for IGCC with or without carbon sequestration, it assumes the carbon tax is zero. In particular, if the COE of IGCC with carbon sequestration were 30-35 percent more expensive than for IGCC without sequestration, a $100 per ton carbon tax would make the COE about equal for the two plants. A $300 per ton carbon tax would make the COE for an IGCC plant with sequestration much cheaper than the COE for a plant without.

The results of these assessments are summarized in Figure I-3. Here are shown the expected costs by year, with and without DOE support, for the two different carbon taxes. The effect of a higher carbon tax is to induce greater near-term R&D efforts sooner to bring down the cost of IGCC with carbon sequestration. These expected COE increases are probability-weighted averages and were calculated from the probabilities the panelists provided. The costs expected by individual panelists are indicated by small crosses and the panel average is indicated by the larger diamonds. Reviewing these assessments, varying degrees of consensus among the panelists can be seen in the different scenarios. In the 2012 assessment with the $100 tax and no DOE support (the leftmost series shown in the figure), the panelists' expected cost increases average 35 percent and range from 32 percent to 39 percent. Estimates span a wider range for 2017 and 2022.

The panel's view of the effect of the DOE research support can be seen by comparing the expected COE increases with and without DOE research. For example in 2017 with the $100 tax, the panel's average expected cost increase without DOE support is 28 percent versus 24 percent with DOE support. These differences vary by year, with the impact of DOE research being smaller in 2012 and 2022 (by about 2 percent) than in 2017 (by about 4 percent). These results suggest that the panel believes that the impact of the DOE support is greatest in the medium-term. Comparing the low- and high-tax scenarios, it can be seen that the higher tax leads to lower expected costs both with and without DOE support, because higher tax would provide a greater incentive for the private sector to develop cost-effective CCS technologies. The estimated incremental effect of DOE support is approximately the same in the two tax scenarios.

These estimates are not compatible with the assumptions that DOE makes in its own benefit calculations. DOE assumes that it will succeed in developing a commercial design with only a 10 percent increase in the COE that will be available for demonstration by 2012 and for commercial deployment after 2016. The panel viewed this goal as very optimistic. In contrast, DOE's assumptions about the increased COE without DOE research funding were viewed as quite pessimistic: DOE's benefits calculations assume that without their sequestration research, there would be a 57 percent increase in the COE associated with carbon capture and storage in 2017 and a 50 percent increase in COE in 2022.[6]

---

[6]Julianne M. Klara, NETL, "NEMS-Based Benefits of FE Sequestration R&D," Presentation to the panel on September 29, 2005.

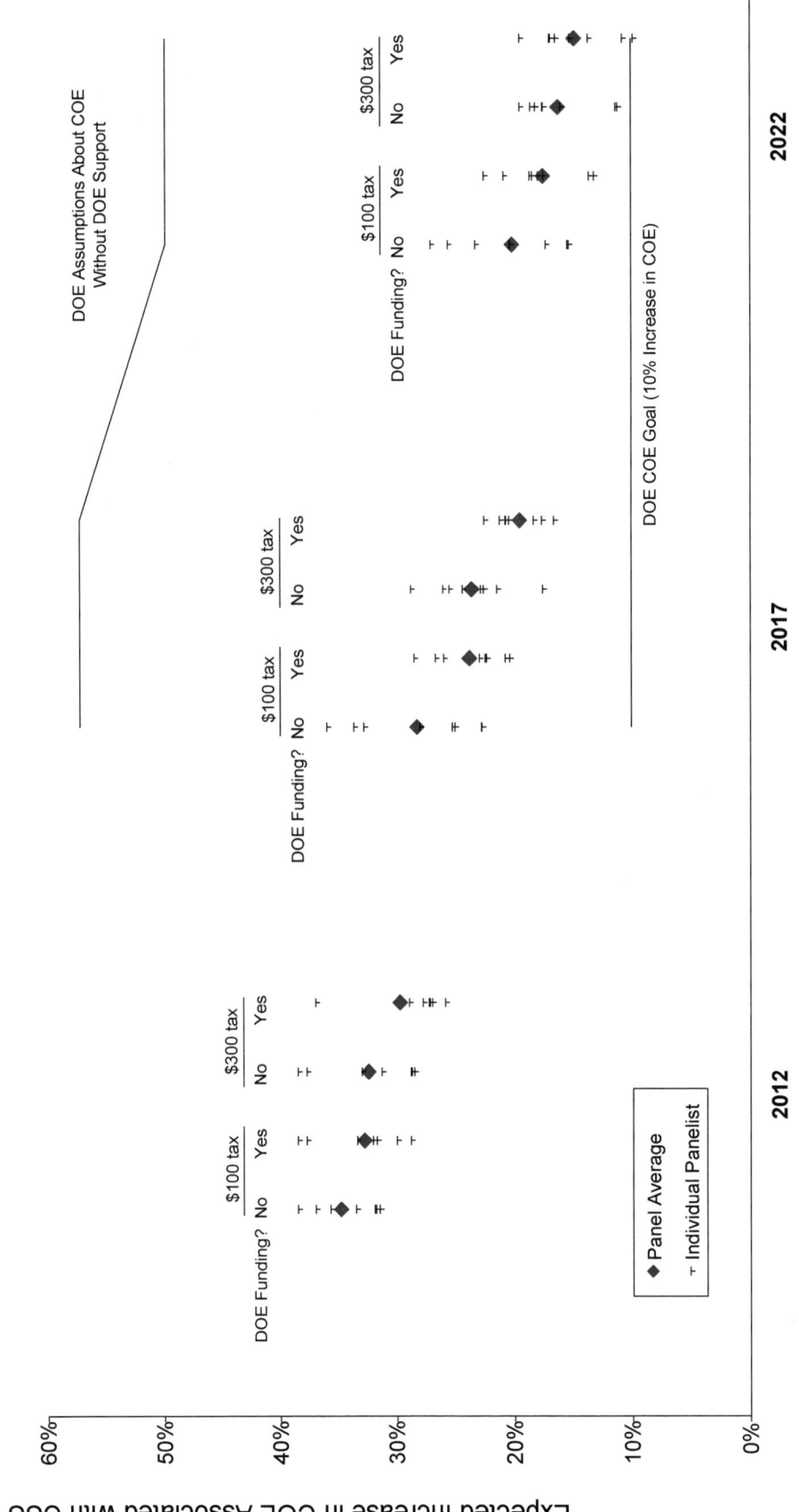

FIGURE I-3 Summary of probability assessment results.

The panel felt that the existence (or even the anticipation) of carbon taxes would lead to extensive private sector R&D that would reduce costs below these levels, even without DOE's research support. R&D activities overseas would probably increase as well if there were a U.S. commitment to reducing emissions. This combination of optimistic assumptions with DOE support and pessimistic assumptions without DOE support leads DOE to arrive at a much higher estimate of the benefits of its support than arrived at by the panel, although to be sure the panel assessments still show a high net payback.

In addition to the uncertainty about costs, the panel also considered a market acceptance uncertainty that focused on whether the public (and regulators) would allow large-scale underground storage of carbon. Without such acceptance, CCS technologies would not deployed. The panel's assessments of this uncertainty are summarized in Figure I-4. The average panel probability that the large-scale sequestration would be allowed is .66 without DOE's research support and increases to .77 with DOE's support. There was also a fair amount of disagreement about these probabilities, though the probabilities were all .5 or higher.

The panel considered competing technologies (e.g., nuclear power, natural gas with or without sequestration) in the benefits calculation, although without explicit modeling of the uncertainty about the costs of these competing technologies. If DOE's R&D programs in these competitive technologies progress rapidly, they could vitiate the benefits of IGCC with carbon sequestration.

QUANTIFYING THE BENEFITS OF THE DOE PROGRAM

The economic, environmental, and security benefits of improvements in carbon sequestration technologies depend on the degree of technical improvement, the amount of IGCC with carbon sequestration that is deployed, the technologies that would have been implemented absent carbon sequestration, and the COEs for IGCC with CCS and for the next-best alternative technology.

In assessing the benefits of DOE's carbon sequestration research program, the panel focused on the COE for IGCC plants with CCS, the COE from other technologies for generating electricity, and the COE for nonsequestering plants given either a $100 per ton or $300 per ton carbon tax. The panel concluded that carbon sequestration would add to the cost of IGCC within the time frame and that no carbon sequestration would be implemented absent some sort of limitation on carbon emissions.

The economic benefit of carbon sequestration improvements in any one year is the product of the amount of electricity produced by IGCC with sequestration and the difference between the costs of the IGCC with sequestration and the costs of the cheapest alternative technology. Since a generating plant lasts 30 or more years, a rational plant owner would

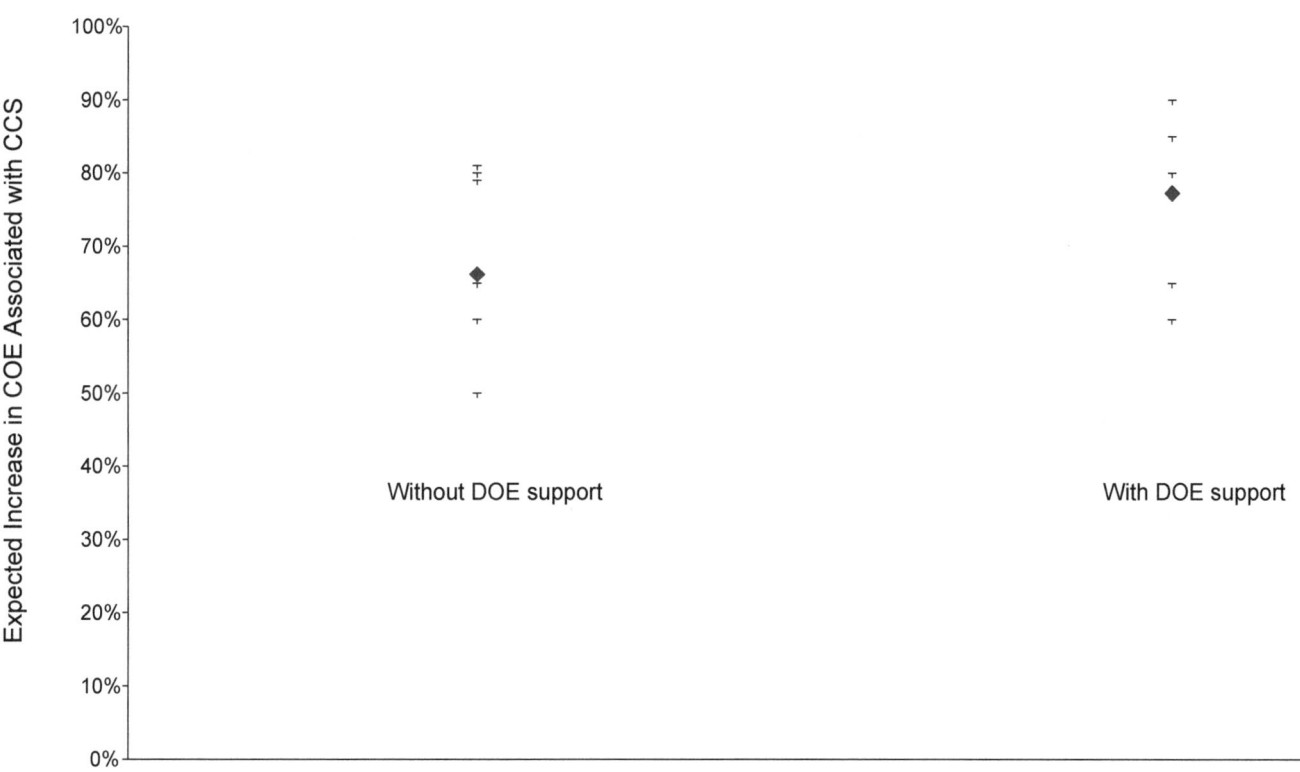

FIGURE I-4 Panel assessment of sequestration risks.

select the technology that is expected to be cheapest over the life of the plant. The total benefits can be calculated as the net present value (NPV) of the annual benefits stream. The carbon taxes affect the amount of sequestered IGCC capacity that is installed: Producers make their choices taking into account the taxes paid, as discussed below. However taxes are not considered in the COE calculations since the taxes net out from a societal perspective: Any carbon taxes paid by producers are receipts for the government. Thus, when comparing the COE for IGCC with sequestration and the COE for the cheapest alternative technology, the panel did not consider the taxes in either case. First, the panel developed a simple model for estimating COE with different generation technologies and next it estimated the amount of IGCC that would be built.

Estimating the COE for IGCC with Carbon Sequestration

The COE (busbar costs) for all electricity-generating technologies considered in the evaluation is based on capital costs, plant efficiencies, operating and maintenance costs, and fuel costs using plant characteristics taken directly from the Energy Information Administration's *Annual Energy Outlook 2005* (EIA, 2005b).[7] Fuel costs were taken to be the fuel cost projections in the AEO 2005 Reference Case, as suggested by the parent committee. Technologies considered explicitly included IGCC with and without sequestration, NGCC, nuclear, and several renewable sources (wind, biomass, and solar).

Baseline costs for IGCC without sequestration play an important role in estimating benefits. A separate panel, the NRC's Panel on DOE's Integrated Gasification Combined Cycle Program, evaluated the effect of DOE's R&D on IGCC technologies (Appendix H), and this panel (the "carbon sequestration" panel) used the results that panel's assessment of the future costs of IGCC as its baseline IGCC costs.

To estimate the COE for IGCC with carbon sequestration, the panel defined a range of possible technical outcomes of carbon sequestration research in 2012, 2017, and 2022, as described in the section on the decision tree model and probability assessment results. For each set of technical outcomes, a COE for IGCC with carbon sequestration can be calculated. Figure I-5 shows the estimated COE (including tax) over time using the baseline costs for IGCC as described above and the expected technical outcome of DOE's carbon sequestration research from the panel's probability assessments. The line with diamond markers corresponds to the expected increase in COE calculated from the probability-weighted averages shown in Figure I-3 for the $100 per ton carbon tax and assuming DOE funding of the research. Figure I-5 also shows the estimated COE for an IGCC plant without sequestration, with and without a $100 per ton carbon tax.

The abrupt rise in cost for IGCC without CCS reflects the tax being implemented in 2012. Thereafter, the change in COE is the sum of two contrary effects: a three percent per year rise in the carbon tax and a linear decrease in the capital cost that levels off in 2020. Finally, the smooth solid lines bound the range of estimates by panel members of the COE for IGCC with carbon sequestration. Thus, for a $100 per ton (or higher) carbon tax, under any cost scenario considered by the panel, the COE for IGCC with carbon sequestration is always less than the COE for IGCC with venting and the tax.

Estimating the Amount of IGCC with Carbon Sequestration That Will Be Built

To estimate the benefits of DOE's carbon sequestration research, we also need to know the amount of IGCC with sequestration that will be built. That amount is assumed to depend on the cost of IGCC with carbon sequestration and the costs of competing low- or zero-emissions technologies.

DOE has evaluated a global scenario with a carbon constraint that provides a starting basis for estimating how much IGCC with carbon sequestration will be built. In its analyses, DOE assumes a carbon cap (rather than a tax), and it assumes that the COE for IGCC with carbon sequestration will be 10 percent higher than for IGCC without sequestration. Under that scenario, about 70,000 MW of IGCC with sequestration is projected to be built by 2025. The panel took this as a reasonable upper bound estimate for the quantity that would be built under DOE's optimistic cost assumptions.

DOE's quantitative modeling is done with a U.S. energy model, perhaps because it would be so difficult to develop and implement a world energy model that would quantify the value of any energy technology. Such a model would have to account for the decisions of other governments regarding carbon emissions and the R&D in other nations.

To estimate the quantity of IGCC with carbon sequestration that would be built in each year under the cost scenarios identified by the panel, a simple cost comparison was made to determine which technology would be least costly for a utility making a decision about what to build. Whichever technology was least expensive was assumed to capture all of the possible low-emissions capacity added in that year.[8] The technologies are, in addition to those shown in Figure I-5, the following:

- NGCC with venting and paying the tax and
- Zero-emissions technologies: nuclear, wind, biomass, and solar.

---

[7] Assumptions to the *Annual Energy Outlook 2005* (EIA, 2005a), Electricity Market Module, especially Tables 38 and 48.

[8] This obviously is not a realistic assumption. Most years will see a combination of technologies built, and the relative costs will change with factors such as fuel resources, site availability, industrial supply capability, and many others. In the absence of detailed simulation, such as with the NEMS model, this approach still gives useful approximate results, which should be viewed as illustrative rather than as forecasts.

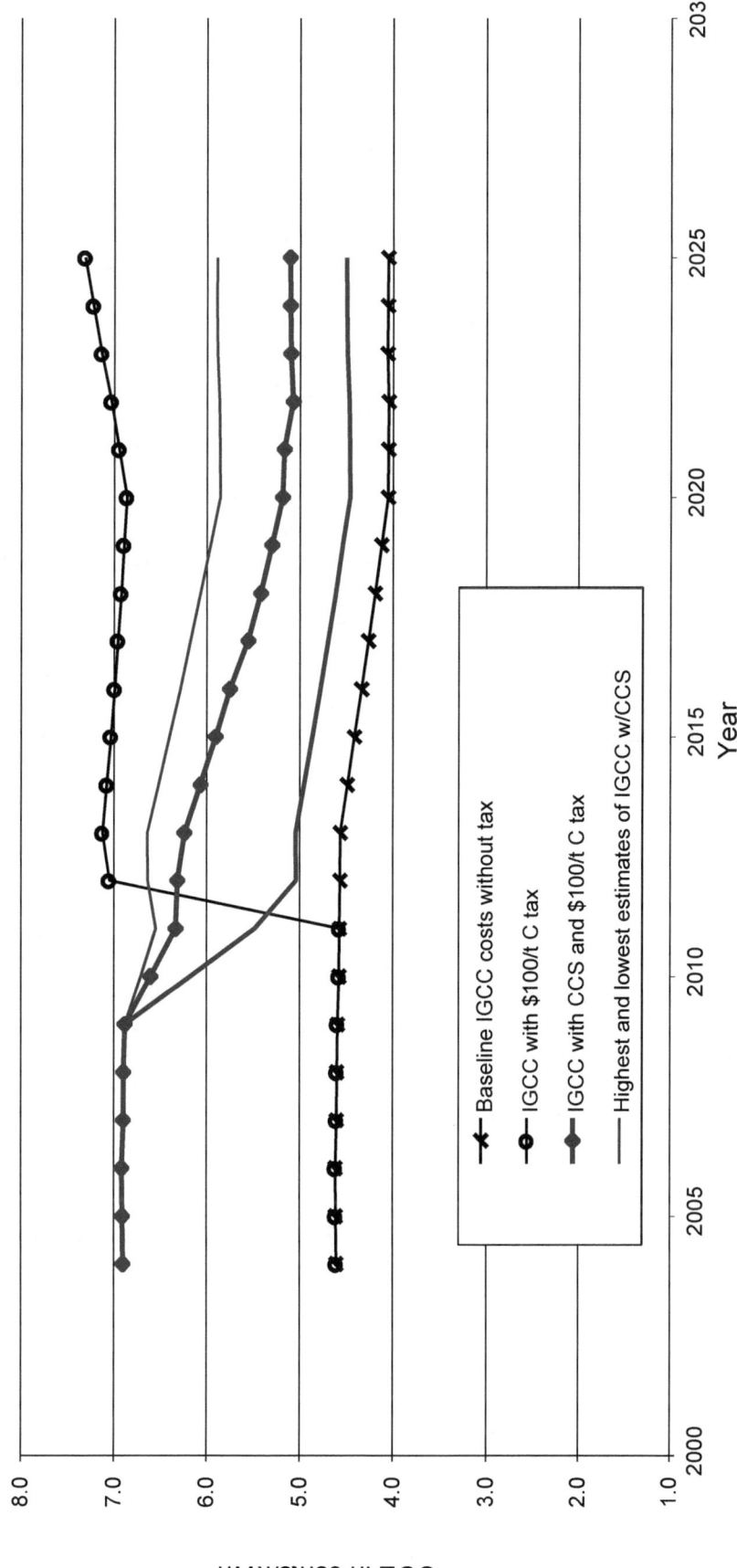

FIGURE I-5 Effect of carbon tax and incremental cost of CCS on COE for IGCC. The incremental cost of adding CCS was taken as the average of each panel members' individual estimates; the incremental cost was added to the baseline cost to determine COE for IGCC with CCS. Two additional lines are shown corresponding to the highest and lowest estimates made by individual panelists. The COE for baseline IGCC was taken from Appendix H, the report of the Panel on DOE's IGCC Technology R&D. C, carbon; CCS, carbon capture and storage; COE, cost of electricity; /t, per ton.

Costs are compared in Figure I-6 based on the net present value of the expected total costs over a 20-year life discounted at 14 percent.[9]

Assuming a winner-take-all competition among technologies, the panel estimated the amount of IGCC with CCS that would be built in each technology scenario. Figure I-7 shows the cumulative IGCC with carbon sequestration added, with and without the DOE program, based on the average probability assigned to each COE increase scenario identified by the panel. The figure also shows the maximum and the minimum IGCC with carbon sequestration added under any of the COE scenarios. The actual amount deployed varies by scenario. For example, the maximum amount will be deployed if there is a 0-10 percent or a 10-20 percent COE increase for carbon sequestration in 2012. However, no IGCC with sequestration will be deployed if the COE increase is always 20-30 percent or more. With intermediate costs, varying amounts of IGCC with carbon sequestration will be built.

Results of Expected Benefits Analysis

Using the approach described above, the panel estimated the benefits of carbon sequestration associated with each of the possible cost estimates defined by the panel (see Figure I-2). For the lowest level of technical success, the benefits are zero, and no IGCC with carbon sequestration is built because the technology is not cost-competitive. For the highest level of technical success considered, where the cost of IGCC with carbon sequestration is just 5 percent more than the cost without sequestration (starting in 2012), the net present value of the benefit is about $36 billion, assuming that large-scale carbon sequestration is allowed.

Each of those carbon sequestration cost scenarios has two probabilities assigned to it by the panel: the probability of achieving that level of technical success without the DOE research program, and the probability of achieving that level of success with it. The expected value of the carbon sequestration research with or without the DOE program is simply the probability-weighted average of the NPV for each technical success scenario using the appropriate probabilities, multiplied by the risk discussed in "Political Risk and Other Market Factors"—namely, that large-scale sequestration may not be allowed. The value of the DOE research program is the difference between the expected value of carbon sequestration research with the program and the expected value without the program.

Figure I-8 illustrates the expected economic benefits of carbon sequestration R&D, as well as the uncertainty surrounding those benefits. It shows a cumulative distribution on net economic benefits with and without DOE support. The net benefit of zero represents the panel's assessment of market acceptance in the case that large-scale carbon sequestration is not allowed by either the public or regulators. The vertical lines represent the expected value of the distribution of benefits: The expected value is calculated as the probability-weighted average of the benefits calculated for possible outcomes identified by the decision trees. The expected value of DOE's carbon sequestration research program is $3.5 billion, the difference between the expected value with the program and the expected value without the program (see Figure I-9).

With the carbon tax, the COE for IGCC with sequestration or for advanced nuclear or wind,[10] which release no carbon-dioxide to the atmosphere, is lower than for fossil fuel technologies without sequestration (either IGCC or NGCC), so none of the latter are built after 2015. Thus, it makes no difference to the environment from the standpoint of carbon emissions between IGCC with carbon sequestration and the viable alternatives, given a $100 per ton carbon tax, and the benefit of the DOE R&D program is simply the reduced cost of producing electricity.

The analysis illustrates that IGCC with carbon sequestration is likely to be such an important technology for generating electricity starting in 2012 that even a small reduction in the time required for the technology to become available, coupled with a small reduction in cost, would lead to a large benefit. The panel emphasizes that a DOE R&D project need not have a 100 percent chance of success or be focused on accomplishing something that could not have been achieved without DOE funding to make an important contribution. In an age of growing concern about GHG emissions, even a small contribution to the reduction of CO_2 emissions from fossil-fuel-based generation technology can be important. In the judgment of the current panel, DOE's R&D program is likely to attain these results only a few years ahead of when the private sector would have achieved the results without DOE funding. Thus, private sector R&D is effective here, and DOE should encourage it. Society would lose if DOE's actions discouraged private R&D or if DOE did not disseminate the results of its R&D to help make private R&D effective.

COMPARISON WITH THE PHASE ONE EVALUATION OF THE CARBON SEQUESTRATION PROGRAM

Although higher than government R&D expenditures, the expected economic benefit of $3.5 billion given by this analysis is substantially less than the expected benefit of $35 billion arrived at by the evaluation carried out in Phase One.[11] The difference in results is primarily due to the much

---

[9]This value was selected to represent what might be used by a utility or merchant generator. This is distinct from the discount rates of 3 and 7 percent that were applied to the benefits stream.

[10]Wind is not directly comparable with IGCC since it is an intermittent energy source while IGCC can run continuously. However, electric systems could utilize a much higher fraction of wind than they do now, and with improved storage and control, that fraction will increase.

[11]See Appendix G of the Phase One report, p. 97.

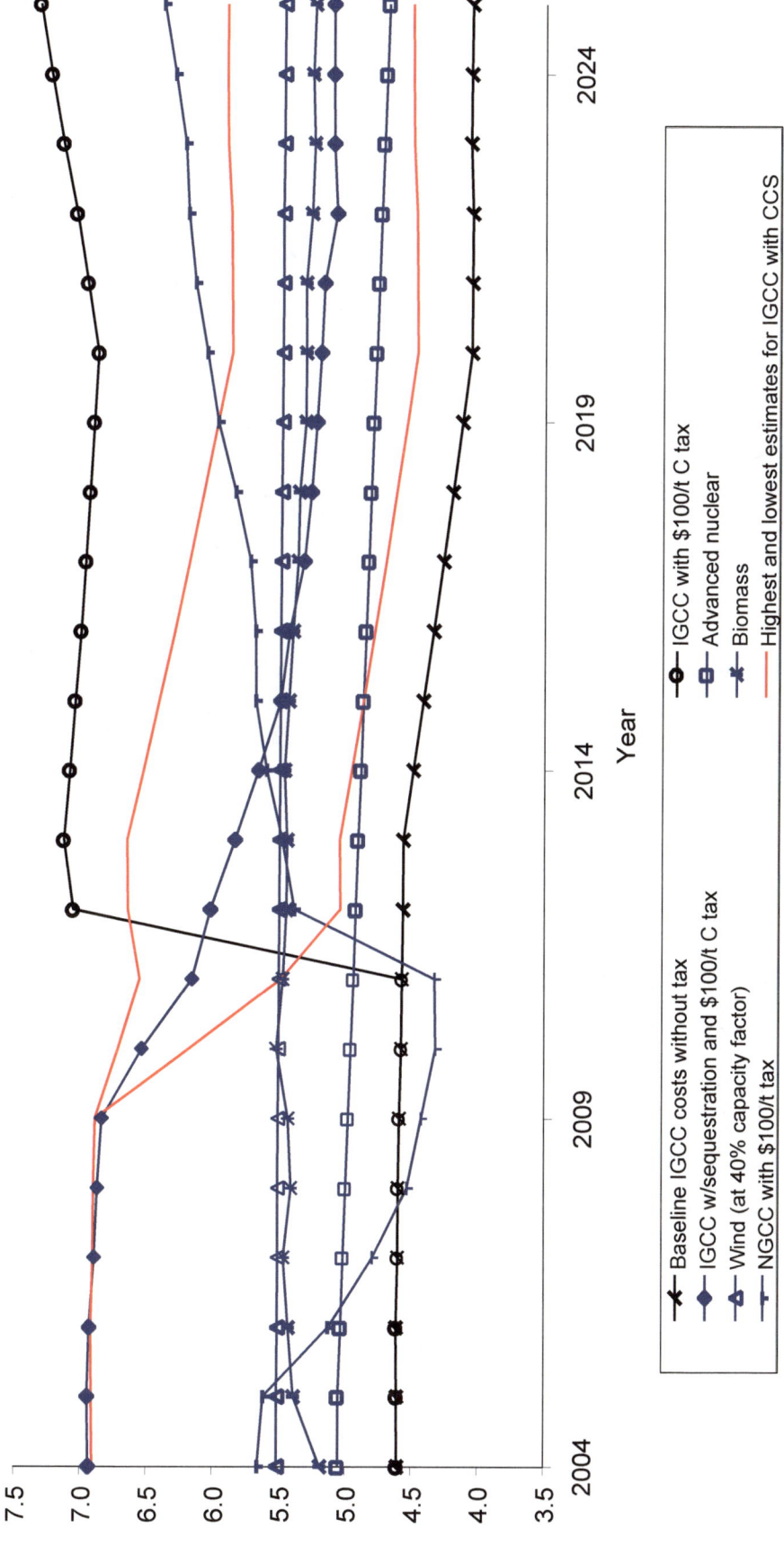

FIGURE I-6 Comparison of COE with competing technologies. COE for IGCC are as shown in Figure I-5. COE for other technologies is based on AEO technology forecasts. Carbon tax is assumed to be $100 per ton, increasing at 3 percent per year after 2012. C, carbon; CCS, carbon capture and storage; COE, cost of electricity; /t, per ton; IGCC, integrated gasification combined cycle.

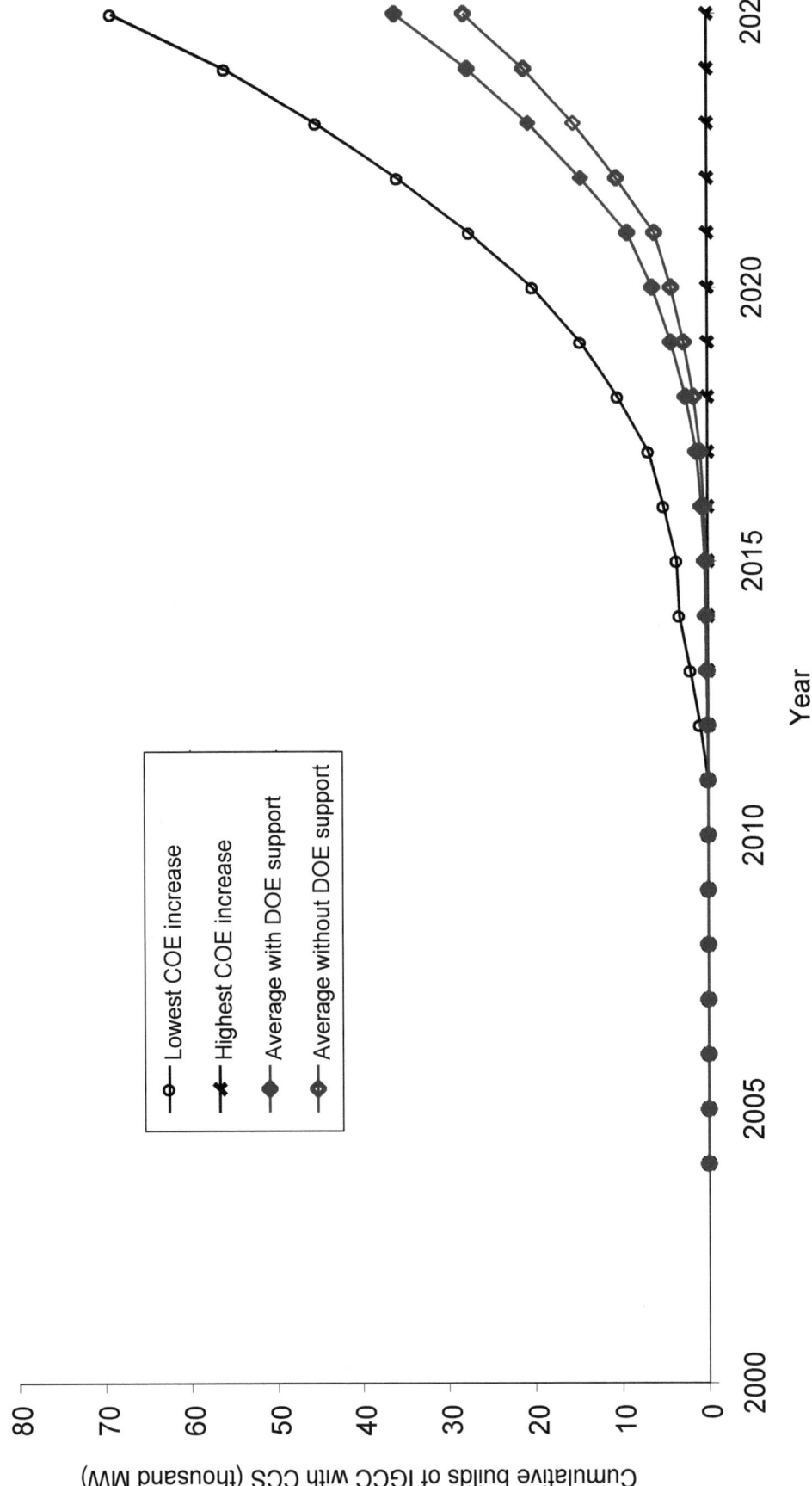

FIGURE I-7 Impact of DOE program funding on cumulative builds of IGCC with CCS, through 2025. The two lines, "with DOE support" and "without DOE support," were calculated by determining the number of builds for each scenario given in Figure I-2 and taking the probability-weighted average. Additional lines are shown corresponding to the number of builds estimated assuming the highest and lowest COE estimates from Figure I-5.

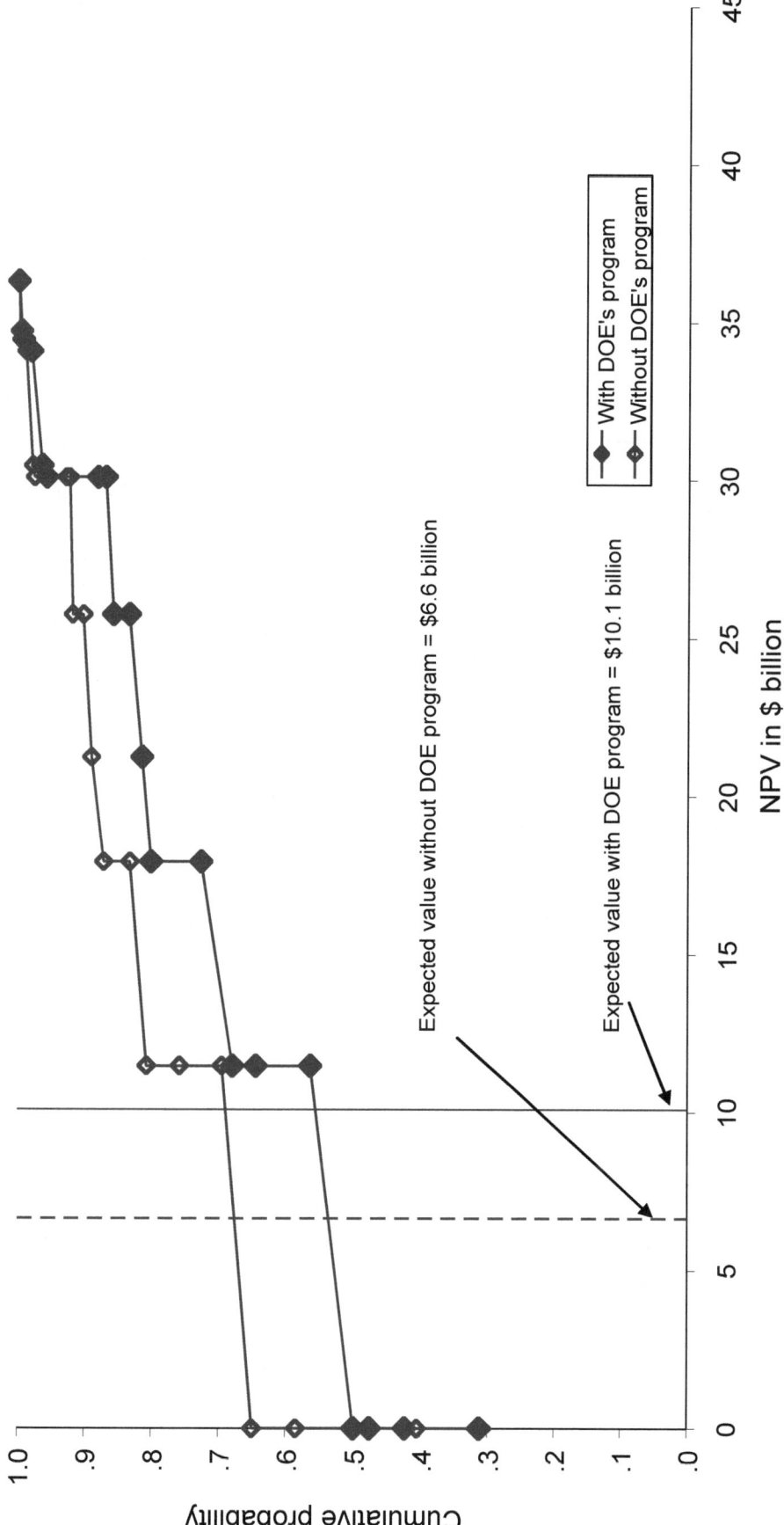

FIGURE I-8 Cumulative distribution on the NPV of the economic benefits of carbon sequestration research, with and without the DOE program. Expected values, with or without DOE research funding, were calculated as the probability-weighted average of the net present value of benefits in each cost scenario.

		Global Scenario[a]	
		Carbon Constrained	
		$100/ton Carbon Tax	$300/ton Carbon Tax
Program Risks	Technical Risks	Estimated as the probability of achieving specified impacts on the cost of electricity for IGCC plants with sequestration over those same plants without sequestration. Average of the panel assessments for the increase in cost of electricity (COE) for sequestration at three different times were as follows: 　　　　　　2012　2017　2022　　　　　　　2012　2017　2022 w/DOE program　33%　24%　18%　　w/DOE program　30%　20%　15% w/o program　　35%　28%　20%　　w/o program　　32%　24%　16%	
	Market Risks	Estimated as the probability large-scale carbon sequestration would be allowed by both the regulators and the public. Probabilities were assigned for the "with DOE program" case and the "without program" case and were assumed to be the same in the two global scenarios considered: with DOE program, 77%; without program, 66%.	
Program Benefits	Expected Economic Benefits[b]	$3.5 billion at 3% Range: $0-$36 billion $1.3 billion at 7% Range: $0-$13 billion	$3.9 billion at 3% Range: $0-$36 billion $1.5 billion at 7% Range: $0-$13 billion
	Environmental Benefits	The environmental benefit of carbon sequestration is reduced greenhouse gas emissions. The environmental benefit of DOE's carbon sequestration program depends on what technologies would be implemented if IGCC with sequestration does not become cost-competitive. Given the level of carbon tax in the scenarios evaluated, the least-cost alternatives to carbon sequestration are other zero-emissions technologies. Thus there is no quantifiable environmental benefit of the research program that is separate from the economic benefit of the reduced costs for very-low emissions generation. Under other assumptions, emissions would be reduced.	
	Security Benefits	The security benefit of carbon sequestration is the ability to continue to build electric generation plants that use coal, a domestic resource, minimizing our dependence on imported fuel resources. Given the level of carbon tax in the scenarios evaluated, however, the least-cost alternatives to carbon sequestration are a combination of nuclear generation and renewables; thus there are no quantifiable security benefits associated with the research program separate from the economic benefits. It is possible, however, that absent carbon sequestration, natural gas will be used instead. In such a case U.S. energy security would be decreased.	

[a]The panel judged that carbon sequestration technologies would not be implemented in global scenarios without a carbon constraint. They did not evaluate the program under the Reference Case or the High Oil and Gas Prices scenarios but evaluated it instead under the Carbon Constrained scenario and a fourth scenario defined by the panel to have a higher carbon tax than the Carbon Constrained scenario.

[b]Net economic benefits are calculated as the expected net present value (at 3% and 7% annual discount rates) of the reduction in the cost of electricity for zero- or very-low emissions generation over a 20-year plant life for all IGCC plants with carbon sequestration built between 2006 and 2025.

FIGURE I-9 Results matrix of the Panel on DOE's Carbon Sequestration Program.

more extensive analysis of during Phase Two what would happen without DOE. Two issues that had largely been neglected in Phase One were considered: the degree to which industry would develop sequestration technologies given adequate notice that they would be needed, and the impact of competing technologies. During Phase One, the methodology was still in an early stage of development, and the importance of examining the full range of options without DOE was not fully appreciated.

More specifically, in the Phase One evaluation of the carbon sequestration program, the panel assessed the likelihood of obtaining various costs of electricity with sequestration by various dates assuming DOE support. Although the assessments were framed differently (in Phase One they were stated in terms of costs; here, they are stated in terms of percentage increases in costs without sequestration), the structure of the earlier assessment was similar to the top branch of Figure I-2. The resulting forecasts for the COE with DOE support were similar. However the assessments without DOE support are quite different in the Phase One report and the current report. In the Phase One analysis, the panel adopted DOE's forecasts of costs without DOE support, which called for relatively modest decreases in the costs of electricity with sequestration. In the current study, panel explicitly considers the probabilities of achieving various cost levels without DOE support, as shown in the bottom branch of Figure I-2. The panel assumes that a CO_2 reduction program will be announced soon, giving 5 years to do R&D and commercialize the new technologies before the CO_2 emissions reductions are required. As shown in Figure I-3, the panel considers what the private sector is likely to do if a firm timetable is set for future abatement of CO_2 emissions. It concludes that private R&D will be increased greatly, thus coming closer to realizing the costs that DOE projects. Because the benefit of the DOE research program is taken as the difference between expected benefits with and without DOE support, improving the expected benefits of the technology without DOE support will decrease the estimated benefits of the program.

The second major difference in the two studies concerns the treatment of competing technologies. The panel that produced the Phase One report assumed that a fixed amount of carbon sequestration technology would be deployed regardless of its costs. The expected economic benefit of the program was then given by difference in expected costs with and without DOE support, multiplied by the fixed capacity that is deployed. The panel writing this report explicitly considers competition with other technologies, such as NGCC with venting and paying the tax as well as competition with other zero-emissions technologies such as nuclear, wind, and biomass. In scenarios with high costs of carbon sequestration, these competing technologies are cheaper than carbon sequestration; if they are cheaper, the carbon sequestration technology will not be deployed and hence would provide no benefit. Even when CCS technology is deployed, its economic benefits are measured relative to the costs of the competing technologies rather than to an assumed high-cost CCS technology. Recognizing competing technologies in this way reduces the benefits relative to those estimated by the Phase One study, which assumed a fixed capacity would be deployed and that benefits would be measured relative to an assumed high-cost alternative.

Although the models used in this study are still approximations, the difference in results between the Phase One evaluation and this Phase Two evaluation of the carbon sequestration program highlights the importance of thinking carefully through the "without DOE support" scenario and capturing, at least in a rough way, the impact of competing technologies.

CONCLUSIONS AND RECOMMENDATIONS

This panel found that the method developed by the parent committee for estimating the benefits of DOE R&D worked satisfactorily in this case. It was frustrated, however, by not having been charged with examining the R&D and not having been given the data to do that. Members thought that they could have given somewhat better estimates of the likelihood of the R&D projects achieving their goals had they had the detailed data. Nonetheless, they found that they were able to implement the method proposed by the parent committee. While individual members had different judgments about the likelihood of achieving the R&D goals and the extent of market penetration for the resulting technology, there was general agreement on these conclusions:

- Carbon sequestration technology will not be implemented commercially without carbon emissions constraints.
- A carbon tax of $100 per ton is sufficient to make carbon sequestration competitive with IGCC plants that vent their carbon.
- DOE's R&D program will speed the attainment of the carbon sequestration program's R&D goals by about 3 years because there is so much private sector interest and R&D in these technologies.
- If the technology is demonstrated to be reliable and cost-effective, IGCC with carbon sequestration could be widely deployed following the implementation of carbon emissions constraints.
- The expected benefit of the DOE program is large, roughly four times the R&D costs incurred by the federal government.
- DOE's CCS R&D can make a contribution to society of about $3.5 billion in spite of the panel's view that it will accelerate the attainment of the program goals by only a few years. Setting aside DOE's overly optimistic assumptions about the contribution of its R&D program and recognizing the private sector R&D the panel finds that even attaining the national goals only a few years sooner is important and would amply repay the R&D investment.

Recommendation: DOE should encourage private sector R&D in conducting its program. The DOE R&D results should be made available quickly to the private sector.

Recommendation: The panel recommends that the leadership of the DOE sequestration program work harder to insulate its systems analysis group from pressure to produce results that conform to the program's aspirational goals, so as to get greater value from the expertise of the group.

ATTACHMENT A
PANEL MEMBERS' BIOGRAPHIES

Lester B. Lave (IOM), *Chair,* is the Harry B. and James H. Higgins Professor of Economics and University Professor; Director, Carnegie Mellon Green Design Initiative; and Co-Director, Carnegie Mellon Electricity Industry Center. His teaching and research interests include applied economics, political economy, quantitative risk assessment, safety standards, modeling the effects of global climate change, public policy concerning greenhouse gas emissions, and understanding the issues surrounding the electric transmission and distribution system. He is a member of the National Academies' Institute of Medicine and a recipient of the Distinguished Achievement Award of the Society for Risk Analysis. He has a B.S. in economics, Reed College, and a Ph.D. in economics, Harvard University.

Charles Christopher is a project manager in the Exploration and Production Technology Group of BP Americas in Houston. He is an internationally recognized expert in improved oil recovery and greenhouse gas issues. He is the co-lead of the storage, monitoring and verification team of the CO_2 Capture Project, a $25 million joint industry project sponsored by 8 energy companies and three governments. The purpose of the project is to identify and develop technologies to allow CO_2 to be effectively and economically captured and stored in the subsurface. Mr. Christopher is also the subsurface technical liaison for BP to the Princeton Carbon Mitigation Initiative, and principal BP representative for the Weyburn Joint Industry Project, the Mt. Simon project, and the Frio CO_2 Injection Demonstration. He helped organize several DOE-funded regional CO_2 sequestration centers and is BP's North American representative for greenhouse gas technology issues.

George M. Hidy is retired Alabama Industries Professor of Environmental Engineering at the University of Alabama, where he was also professor of environmental health science in the School of Public Health. From 1987 to 1994, he was technical vice president of the Electric Power Research Institute (EPRI), where he managed the Environmental Division and was a member of the Management Council. From 1984 to 1987, he was president of the Desert Research Institute of the University of Nevada. He has held a variety of other scientific positions in universities and industry and has made significant contributions to research on the environmental impacts of energy use, including atmospheric diffusion and mass transfer, aerosol dynamics, and chemistry. He is the author of many articles and books on these and related topics. Dr. Hidy received a B.S. in chemistry and chemical engineering from Columbia University; an M.S.E. in chemical engineering from Princeton University; and a D.Eng. in chemical engineering from the Johns Hopkins University.

W.S. Winston Ho (NAE) is a University Scholar Professor in the Department of Chemical and Biomolecular Engineering at the Ohio State University. His research interests include molecularly based membrane separations, fuel-cell and fuel processing and membranes, transport phenomena in membranes, and separations with chemical reaction. Dr. Ho holds a B.S. from Taiwan National University and an M.S. and a Ph.D. from the University of Illinois at Urbana-Champaign.

David Keith is an assistant professor in the Department of Engineering and Public Policy, Carnegie Mellon University. Dr. Keith's policy work addresses the uncertainty in climate change predictions, geoengineering, and carbon management. He has been a collaborator in research on climate-related public policy at Carnegie Mellon since 1991 and an investigator in the Center for the Integrated Study of the Human Dimensions of Global Change since its inception. His current research involves an analysis of the use of fossil fuels without atmospheric emissions of carbon dioxide by means of carbon sequestration. This research aims to understand the economic and regulatory implications of this rapidly evolving technology. Questions range from near-term technology-based cost estimation to attempts to understand the path dependency of technical evolution; for example, how would entry of carbon management into the electric sector change prospects for hydrogen as a secondary energy carrier? In addition, Dr. Keith is working on a study of geoengineering that explores its historical roots and its ethical implications. As an atmospheric scientist, he collaborates with Professor James Anderson's group at Harvard on observations of water vapor, cirrus clouds, and stratosphere-troposphere exchange. He was the senior scientist for INTESA, a new Fourier-transform spectrometer that flies on the NASA U-2, and he worked as project scientist on Arrhenius, a proposed satellite aimed at establishing an accurate benchmark of infrared radiance observations for the purpose of detecting climate change. He has a B.Sc. in physics from the University of Toronto and a Ph.D. in experimental physics from MIT.

Larry W. Lake (NAE) is a professional engineer (Texas) and the W.A. "Monty" Moncrief Centennial Endowed Chair for the Department of Petroleum and Geosystems Engineering at the University of Texas, Austin, where he has served on the faculty since 1978. He has 5 years of industrial experience and has authored one book and more than 50 technical articles and reports. His research interests are in the areas of enhanced oil recovery, geochemical flow processes, and petrophysics, all of which involve numerical simulation in one form or another, and flow through permeable media. In addition, Dr. Lake has been most involved in finding ways to model geologically realistic reservoir properties—primarily permeability quantitatively—with the hopes of improving the ability to predict hydrocarbon recovery better. This has led to efforts that seek to merge sedimentary concepts with the discipline of geostatistics. Dr. Lake holds a Ph.D. in chemical engineering from Rice University and was elected to the National Academy of Engineering in 1997.

Michael E. Q. Pilson is professor emeritus of Oceanography at the University of Rhode Island (URI). He was director of the Marine Ecosystems Research Laboratory at URI for 20 years. His current research interests include the chemistry of seawater, biochemistry and physiology of marine organisms, and nutrient cycling. He received a B.Sc. in chemistry-biology from Bishop's University, in Canada, an M.Sc in Agricultural Biochemistry from McGill University, and a Ph.D in marine biology from the University of California, San Diego. He is a member of the American Association for the Advancement of Science; Sigma Xi; the American Geophysical Union; the American Society of Mammalogists; the American Society of Limnology and Oceanography; and the Oceanography Society. He has published extensively, including the text book *An Introduction to the Chemistry of the Sea*.

Jeffrey J. Siirola (NAE) is a research fellow in the Chemical Process Research Laboratory at Eastman Chemical Company in Kingsport, Tenn. He received his B.S. degree in chemical engineering from the University of Utah in 1967 and his Ph.D. in chemical engineering from the University of Wisconsin-Madison in 1970. His research centers on chemical processing, including chemical process synthesis, computer-aided conceptual process engineering, engineering design theory and methodology, chemical technology, assessment, resource conservation and recovery, artificial intelligence, nonnumeric (symbolic) computer programming, and chemical engineering education. He is a member of the National Academy of Engineering.

James E. Smith is professor of decision sciences at the Fuqua School of Business at Duke University. He teaches courses in probability and statistics and decision modeling. Dr. Smith's research interests lie primarily in the areas of decision analysis and real options, focusing on developing methods for formulating and solving dynamic decision problems and valuing risky investments. His research has been supported by grants from the National Science Foundation, Chevron, and the Eli Lilly Foundation. Dr. Smith received B.S. and M.S. degrees in electrical engineering from Stanford University (in 1984 and 1986) and worked as a management consultant prior to earning his Ph.D. in engineering-economic systems at Stanford in 1990. He has been at Fuqua since the fall of 1990 and received the Outstanding Faculty Award from the daytime MBA students in 1993 and 2000. He served as associate dean for the Duke MBA Program from 2000-2003. He has been a member of the Advisory Panel for the National Science Foundation's Decision Risk and Management Science program and has been departmental editor for decision analysis at the journal *Management Science*.

Robert H. Socolow is a professor of mechanical and aerospace engineering at Princeton University, where he has been on the faculty since 1971. He was previously an assistant professor of physics at Yale University. Professor Socolow is a fellow of the American Physical Society and the American Association for the Advancement of Science. He currently codirects Princeton University's Carbon Mitigation Initiative, a multidisciplinary investigation of fossil fuels in a future carbon-constrained world. From 1979 to 1997, Professor Socolow directed Princeton University's Center for Energy and Environmental Studies. He has served on many NRC boards and committees, including the Committee on R&D Opportunities for Advanced Fossil-Fueled Energy Complexes, the Committee on Review of DOE's Vision 21 R&D Program, and the Board on Energy and Environmental Systems. He has a B.A., an M.A., and a Ph.D. in physics from Harvard University.

John M. Wootten is retired vice president, Environment and Technology, Peabody Energy. He spent most of his professional career with Peabody Holding Company, Inc., the largest producer and marketer of coal in the United States. His positions at Peabody and its subsidiaries included that of director of environmental services, director of research and technology, vice president for engineering and operations services, and president of Coal Services Corporation (COALSERV). His areas of expertise include the environmental and combustion aspects of coal utilization, clean coal technologies, and environmental control technologies for coal combustion. He has served on a number of NRC committees, including the Committee on R&D Opportunities for Advanced Fossil-Fueled Energy Complexes and the Committee to Review DOE's Vision 21 R&D Program. He received a B.S. (mechanical engineering) from the University of Missouri-Columbia and an M.S. (civil engineering, environmental and sanitary engineering curriculum) from the University of Missouri-Rolla.

J

Report of the Panel on DOE's Natural Gas Exploration and Production R&D Program

INTRODUCTION

As part of the study by the Committee on Prospective Benefits of DOE's Energy Efficiency and Fossil Energy R&D Programs, Phase Two (the committee), the Panel on Benefits of DOE's Natural Gas Exploration and Production Program (the panel; see Attachment A) was appointed by the National Research Council in September 2005. The primary focus of the panel was to apply the committee's prospective benefits methodology to R&D activities for natural gas exploration and production (E&P) in the Office of Oil and Natural Gas, which is part of DOE's Office of Fossil Energy.

As noted in the next section, "Overview of the Natural Gas Exploration and Production Program," the Office of Oil and Natural Gas will be impacted by the Energy Policy Act of 2005 (EPACT-2005, P.L. 109-58), Section J, Ultra-Deepwater and Unconventional Natural Gas and Other Petroleum Resources. Under EPACT-2005, royalties of $50 million per year will fund the E&P program, and a private consortium will be formed that will select R&D projects, which may result in a portfolio of projects different from those currently in the DOE program and being reviewed here by the panel. In addition, EPACT-2005 authorizes appropriations of up to $100 million per year. Nevertheless, Section J includes unconventional natural gas resource E&P technology, as well as the technology challenges of small producers. Both of these areas are covered in the existing E&P program. As a result of discussions among the committee chairperson, the panel chairperson, and DOE, the committee and panel chose to focus on four key subprograms of the Office of Oil and Natural Gas E&P program, which encompass DOE's unconventional natural gas R&D projects: (1) existing fields; (2) drilling, completion, and stimulation; (3) Deep Trek; and (4) advanced diagnostics and imaging. The committee and panel believe that even with the changes that are expected to occur under EPACT-2005, a portion of the program will still focus on the areas addressed by the panel, so that any insights provided by the panel could help the Office of Oil and Natural Gas even as it transitions under EPACT-2005. It reviewed DOE's estimates of the benefits of its program, reviewed projects in the portfolio, made judgments about technical risks and market risks, and worked with the committee's consultant to apply the committee's methodology to estimate overall technical and market risks and prospective net benefits to the nation for the E&P program as a whole. The four subprograms are discussed in more detail in the next section. In addition, like the other panels formed under the committee, the panel beta tested the committee's methodology, and it offered comments to the committee about the efficacy of the methodology, noting what works well, what its limitations are, and what improvements may be necessary.[1]

OVERVIEW OF THE NATURAL GAS EXPLORATION AND PRODUCTION PROGRAM

As noted in the Office of Fossil Energy's Natural Gas Technologies Program Plan (DOE, 2004b), the mission of the program is to develop environment-friendly technologies through R&D and policy options that will diversify natural gas supply options and steadily expand the nation's economically recoverable gas resource base. The program, broadly speaking, includes three main areas: domestic supply, supply from global resources, and delivering America's energy. The panel focused its review and evaluation efforts on DOE's E&P activities to increase domestic supply but did not include methane hydrates;[2] it also does not address "supply from global resources" or "delivering America's en-

---

[1]The panel chair also interacted with and provided feedback to the committee chair, the consultant, and NRC staff during the study.

[2]Early in the committee's study, the committee chair discussed with DOE the possible formation of a separate panel to evaluate the benefits of the methane hydrates program. Since the methane hydrates program is in a research phase, the committee decided that the methodology would best be applied when that program's activities are further into technology development and possible applications.

TABLE J-1 DOE's Performance Targets for Expanding Domestic ERR by 50 Tcf Through 2015

Year	Performance Milestones	Contribution to Target (Tcf)
2006	Develop technologies to increase gas finding efficiency, increase well productivity, reduce well abandonment, and address excessive water production from existing fields	7 by 2015
2008	Develop technologies to reduce the cost of drilling for unconventional and other gas by 5%	13 by 2015
2008	Provide technologies for hydrate avoidance or seafloor stability mitigation to assure the safety of ongoing deepwater hydrocarbon exploration	2 by 2015
2011	Develop improved reservoir imaging systems to increase finding efficiency in unconventional gas reservoirs	11 by 2015
2013	Develop reliable E&P systems for gas located 20,000 ft below the earth's surface	5 by 2015
2015	Develop integrated deep drilling system that reduces drilling cost by 30%	10 by 2015
2007-2015	Develop environmentally sound approaches that minimize the gas E&P footprint, enabling expanded access to gas in environmentally sensitive areas	2 by 2015

SOURCE: DOE. 2004b. Natural Gas Technologies Program Plan. Washington, D.C.: Office of Fossil Energy.

ergy." The objective shown in the Program Plan is to develop technologies by 2015 that expand the nation's economically recoverable resources (ERR) by 50 trillion cubic feet (Tcf) while minimizing environmental impact.[3] Table J-1 lists the performance targets for expanding the domestic ERR by 50 Tcf through 2015.

A variety of R&D projects are carried out in the three areas, which range widely in funding levels and duration; their descriptions are available on the National Energy Technology Laboratory (NETL) Web site.[4] As noted in the "Introduction," even though the E&P program may change as a result of Section J of the EPACT-2005, it is likely that R&D will continue in a number of the areas covered in the panel's evaluation.[5] In the President's Budget Request (PBR) for 2007, the administration proposes to cancel the requirements in Section J of the Energy Policy Act of 2005 by means of a future legislative proposal.

The FY05 budgets for the key subprograms for enhancing domestic supply in DOE's program were as follows:

- Existing fields subprogram, $1.6 million;
- Drilling, completion, and stimulation, $7.3 million;
- Advanced diagnostics and imaging, $3.8 million; and
- Deep Trek, $1.5 million.

These subprograms are described in more detail in the following sections. The PBR for FY06 did not ask for money for subprograms assessed by the panel; it requested $10 million for closeout of the natural gas technologies program. Nevertheless, the Congress appropriated a total of $33 million to the program. For the subprograms addressed by the panel, these appropriations include $9 million for advanced drilling, completion, and stimulation, including Deep Trek; $4 million to continue work aimed at expanding the recoverability of natural gas from low-permeability formations; and $2 million for stripper wells and technology transfer.

ASSESSMENT OF THE NATURAL GAS EXPLORATION AND PRODUCTION PROGRAM

The committee's methodology suggests that an assessment of the benefits of a specific subprogram should explicitly consider the role of DOE funding and the technical risks and market risks that can affect the outcome and the value of that subprogram's activities. The methodology also requires that benefits be estimated under three different global scenarios representing possible future states of the world.

Role of DOE Funding

DOE defines the goals for its natural gas exploration and production R&D program in terms of additions to ERR that can be attributed to the success of DOE-funded research.

---

[3]Economically recoverable resources are resources, both discovered and undiscovered, that are economically extractable under a given set of price-to-cost relationships and technological assumptions. DOE defines ERR as resources that can be found and produced profitably at a given point in the future at prevailing prices estimated by the Energy Information Administration (EIA) in its *Annual Energy Outlook*.

[4]See <http://www.netl.doe.gov/scngo/NaturalGas/index.html>.

[5]The $50 million per year of funding from federal oil and natural gas royalties starting in FY07 will be divided as follows: (1) conduct R&D for ultradeepwater E&P and integrated systems (35 percent); (2) conduct R&D for unconventional oil and gas resources (32.5 percent); (3) assist small producers with production problems of complex and unconventional resources (7.5 percent); and (4) conduct complementary R&D at NETL (25 percent). Technology transfer will be included, and DOE will award a consortium contract by May 2006.

The panel adopted this definition as the basic metric to use in assessing technical and market success of the research program, and it developed estimates of the likely outcomes in each of the four subprograms in terms of the probability of adding specific amounts of ERR as a direct result of DOE's R&D program. This framing of the goal created several complexities:

- What is determined to be "economically recoverable" depends on assumptions about future gas prices. The panel addressed this by making separate estimates of the probability of reaching different levels of increased ERR for the three different global scenarios.
- Several panelists felt that their estimate of DOE's contribution to increases in ERR would be more informed if they had estimates of the total increase in ERR attributable to all R&D, but such estimates are not available nor are they readily attainable. After discussion, the panelists agreed to estimate the net benefit of DOE's research directly rather than attempt to estimate separately the increase in ERR from all R&D (including DOE's) and from all R&D except for DOE's. DOE personnel provided an estimate that 20-25 percent of the total increase in production and reserves estimated by the Energy Information Agency (EIA) in its National Energy Modeling Systems (NEMS) analyses could be attributed to the DOE program. The panel interpreted that as meaning that the total anticipated increase in ERR is about four times the increase DOE estimates from its program.

Consideration of Global Scenarios

The panel considered the three global scenarios (defined in Appendix F) in estimating the probability of technical and market success: the AEO Reference Case scenario; the High Oil and Gas Prices scenario, where gas prices are assumed to be twice those in the AEO Reference Case; and a Carbon Constrained scenario, where a $100 per ton carbon emissions tax is assumed to be put in place in 2012. The scenarios affect the probability of technical success, in that higher gas prices and constraints in carbon emissions are believed to impact the quality and focus of the research, as well as the definition of what is economically recoverable. The scenarios affect the economic benefits by virtue of their impact on natural gas prices: at higher gas prices, more gas is economically recoverable.

The next four subsections cover the four main subprograms in the natural gas R&D program and present the panel's assessment of the activities, issues, and technical and market risks in each subprogram.

Existing Fields Subprogram

Near-term efforts by DOE focus on maximizing the efficiency of recovery from existing wells and fields that are operating near the bottom margin of profitability. Generally speaking, these are low-volume stripper gas wells, defined by the Interstate Oil and Gas Compact Commission (IOGCC) as a natural gas well that produces 60 thousand cubic feet (Mcf) per day or less. This amount of gas is approximately the energy equivalent of the better known oil stripper wells, which produce fewer than 10 barrels of oil per day.

The IOGCC statistics indicate that marginal gas wells account for 8 percent of the total natural gas produced in the United States. This amount is approximately equal to coal-bed methane production and is therefore an important component of the nation's domestic gas supply. An estimated 271,856 stripper wells produced from them 1.54 Tcf in 2004. The number of marginal wells and gas production from them have both increased each year from 2001 to 2004. Current marginal wells also represent a significant increase over 1995 figures of 159,669 wells and a production of 0.92 Tcf.

Although the number of stripper wells has increased, the average daily production per well, and therefore the production baseline that DOE R&D has to build up from, has stayed steady at 15+ Mcf/day for the last 10 years. Abandoned natural gas wells, like abandoned marginal oil wells, are those that have been permanently plugged. The IOGCC statistics show a significant trend: The total number of plugged stripper wells in 2004 increased for the fourth consecutive year, while demand for natural gas continued to rise.

DOE efforts have progressed: Previous research in secondary gas recovery in conventional fields has become advanced diagnostics research on optimal infill drilling practices to maximize recovery in tight and fractured accumulations. Additionally, an ultra-short-radius composite drill pipe is being developed that can be used to efficiently reenter existing wells and drill horizontally to maximize recovery. It is unclear to the panel how much of this research on infill drilling and composite drill pipe is directed at marginal wells.

Ongoing efforts focused at the well level (through NETL's Stripper Well Consortium) are expected by DOE to result in the commercialization of an array of technologies in 2006 that will significantly reduce the incidence of premature plugging and abandonment of producing wells in existing fields. According to the Program Plan, p. 3, the performance target for existing fields is to "develop technologies to increase gas finding efficiency, increase well productivity, reduce well abandonment and address excessive water production." These efforts are forecast to contribute approximately 7 Tcf of additional ERR by 2015. The 7 Tcf target is to be reached with a budget of $7.4 million over 5 years for the Stripper Well Consortium (with an additional $1.85 million dollars cost-share). An earlier project, Advanced Technologies for Stripper Wells, was funded at $245,714, with an additional cost-share of $141,000.

The benefits expected by DOE appear to the panel to be unrealistically high, given the relatively small research budgets, high market risks, and—but to a lesser extent—the technical risks. The technical risks are small compared to those of other programs that emphasize exploration or drilling. The marginal wells exist and are currently productive. Some subset of the wells, however, could have problems with well bore stability, casing, and tubing because of age, corrosion, and lack of maintenance. In some instances, these problems may cost more to repair than could be realized by applying newly developed technology. Innovations and breakthroughs will require R&D that allows significant, inexpensive, incremental improvements in the life and/or production rates of the existing wells. It is not clear to the panel that the DOE Stripper Well Consortium is the most effective approach to funding the bulk of this R&D.

Marginal gas wells are operated (for the most part) by small independent operators, not major oil and gas companies. These operators number in the thousands and may have little access to research funds or even, perhaps, to technical literature. Market risks are primarily affected by (1) the unavailability of capital for some of the producers, (2) the lack of a mechanism to effectively transfer information on improved/advanced technology to the thousands of producers, and (3) no way to make small producers aware of the benefits their investment would bring.

The panel was asked to evaluate the probability of DOE achieving technical success defined as the increase in ERR by 2015 made possible by DOE's research in stripper well technologies. It identified four possible outcomes, illustrated in the decision tree in Figure J-1. The panel defined market success as the fraction of the potential market that would implement the technologies at a given level of technical success. Combining these assessments yields an estimate of the expected increase in ERR attributable to the existing fields subprogram, as shown in the first row of Table J-2. The table shows estimates of the increase in ERR for each of the subprograms—first DOE's estimates and then the average of the panel's estimates under each global scenario. A range of individual opinions was offered by panel members, with the table also showing the lowest and the highest ERR estimates. The panel discussed the range of assessments and felt that an average value would best represent the expertise of the group and should be the basis for the benefits calculations described below.

Drilling, Completion, and Stimulation Subprogram

Drilling is the most costly item in producing unconventional gas resources and makes it uneconomical to develop many unconventional gas fields. DOE is working on new tools that will drill faster and instruments that help avoid expensive drilling problems, including laser-, percussion-, and hydraulically pulsed drills that have the potential to drill faster and to significantly reduce drilling costs.

Work is also under way on improving surface and downhole instruments and on speeding up data transmission from the bottom of the well. These tools will allow drillers to optimize drilling operations, to drill into higher temperature environments, and to detect well problems before they become major.

Lightweight, flexible composite drill pipe will increase the depth to which vertical and horizontal wells can be drilled and allow the drilling of horizontal wells of smaller

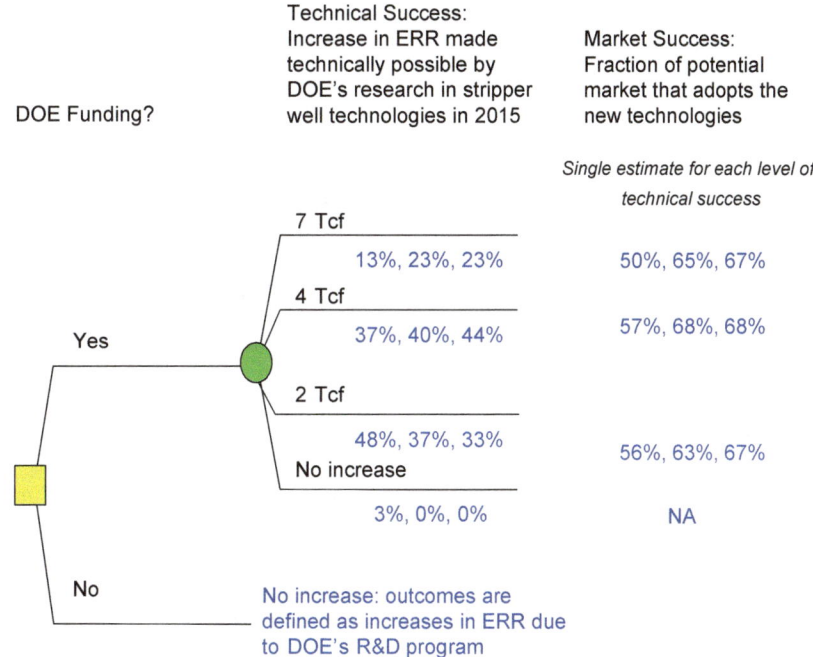

FIGURE J-1 Decision tree for the existing fields subprogram. Values under each branch are the average of the panelists' estimates of the probability of that technical outcome for the Reference Case, the High Oil and Gas Prices, and the Carbon Constrained scenarios, respectively.

TABLE J-2 Panel Assessments of Technical and Market Risks for Each Subprogram (trillion cubic feet)

Subprogram	DOE Estimate	Results of Panel Assessments		
		AEO Reference Case Scenario	High Oil and Gas Prices Scenario	Carbon Constrained Scenario
Existing fields	7	1.8 (0.7 to 3.2)	2.6 (1 to 4.5)	2.7 (1.1 to 5.3)
Drilling, completion, and stimulation	13	3.9 (1.2 to 6.2)	5 (1.1 to 7.6)	4.9 (1.1 to 7.6)
Advanced diagnostics and imaging	11	3.4 (1.4 to 5.2)	4.1 (2 to 5.8)	4 (2 to 5.8)
Deep Trek	15	4 (0.9 to 8.3)	5.7 (2.5 to 9)	5.6 (1.9 to 9)
Total	46	13 (4.1 to 22)	17 (7 to 26)	17 (7.3 to 26.5)

NOTE: Technical success is defined as the estimated increase in ERR attributable to DOE's research based on the average of the panelists' technical and market risk assessments. Values in parentheses show the range of estimates as calculated from each panelist's assessments. The totals are rounded to two significant figures.

radius to intersect natural fractures and thereby increase gas production. Work is also under way on a horizontal drilling system to improve gas production from highly fractured and faulted complex gas reservoirs that are difficult to produce economically with current drilling systems.

DOE is participating in a joint industry project to improve ultra-deep-water drilling through the development of better subsea data processing, composite production risers, and deepwater casing drilling systems. DOE partners on this project include two large operators and four large service companies, so this technology should be quickly applied once developed. The potential payout for this project is very high because of the high costs and high risks associated with ultra-deep-water drilling.

This novel drilling R&D has the potential to make breakthrough improvements rather than the incremental improvements typically made by oilfield service companies. Drilling improvements made by DOE are also applicable to conventional oil and gas wells, so the potential payouts are much larger than improvements applicable to unconventional gas wells alone.

The technical risk for any one of these breakthrough drilling projects is high, but the probability of succeeding on one or two of these high-payout projects is also high. On the other hand, the market risk for such projects is low, because new drilling technology is quickly implemented throughout the industry. The fact that operators jointly develop many offshore fields leads to widespread technical interchange throughout the industry.

Well completions are extremely important, because improper completions can damage well bores and reduce gas production by up to 50 percent. This is especially important with unconventional gas reservoirs, because many of them have very low permeability, resulting in low gas production rates and marginal economics.

DOE is conducting R&D on the use of Aphron drilling and completion fluids that will temporarily seal off fractures and pores so that gas pay zones are not permanently damaged while drilling. These fluids allow the fractures and pores to open up once the wells have been drilled and put into production.

Essentially all of the unconventional gas wells are in tight reservoirs where the permeability of the rock is low, limiting gas flow toward the well bore. In many cases these wells cannot produce natural gas economically unless there are natural fractures that allow the gas to flow more readily to the well bore. DOE has a project to improve mapping natural fractures so that horizontal wells can be drilled through the most productive parts of the reservoirs.

Work is also under way on a down-hole power generation and wireless communications system for intelligent completions that will allow gas production to be optimized throughout the life of wells. This wireless system continuously measures down-hole temperatures and eliminates the electrical cables used with existing systems, reducing their cost and improving reliability.

A down-hole fluid analyzer is being developed that will measure the fluid fractions (water, oil, gas) produced downhole in real time as the well is produced so that the well's production can be optimized and remedial actions taken if problems develop. This system transmits data from the well bottom to the surface of the well using a fiber-optic cable for data transmission.

Stimulation of unconventional gas reservoirs is important because of the low permeability of many of these reservoirs. The fracture mapping systems described in the foregoing discussion on completions are a key to drilling horizontal wells into the most productive parts of reservoirs to maximize gas production. These stimulation projects relate primarily to the detection and mapping of natural fractures. The panel encourages the DOE to initiate projects to develop improved hydraulic fracturing techniques or novel drilling techniques to connect existing natural fractures to well bores.

As shown in Table J-1, the goals defined by DOE for drilling, completion, and stimulation for unconventional resources are to reduce the cost of drilling for unconventional and other gas by 5 percent, resulting in a 13 Tcf increase in ERR by 2015. The panel characterized uncertainty about technical success for this program in terms of the decrease in drilling costs and the increase in ERR that would result from specific decreases in drilling costs, as shown in the decision tree in Figure J-2. The panel does not believe there are any significant barriers to market adoption of these technologies, so no market risks were evaluated. The second row of Table J-2 shows the results of the panel's assessments.

Advanced Diagnostics and Imaging Subprogram

Seismic imaging of the subsurface geologic structures is widely used for oil and gas E&P. Most of the imaging is done with arrays of seismic sources and receivers at the surface and highly advanced processing of the data to obtain a 3-D image of the subsurface. Through diagnostic analysis of seismic phases (e.g., attributes), the seismic imaging not only detects potential reservoirs but also determines their fluid (oil, gas, brine) content. By repeating the imaging of the same reservoir at successive times (called 4-D seismic for the three dimensions of space and one dimension of time), it has been possible to monitor the changes in oil-water contact and other fluid properties in the reservoir. In that sense, 4-D seismic has become an important part of reservoir monitoring.

Most advanced applications of seismic imaging have been used for conventional oil and E&P. Offshore prospects have been the favored targets because it is relatively easy to acquire seismic data in the marine environment. Seismic imaging on land has lagged because of the cost and the difficulties arising from the topographic and near-surface geologic heterogeneities. In addition, unconventional gas E&P made up a smaller market share than the conventional oil and gas E&P. As a result, geophysical service companies have not moved aggressively to develop imaging and diagnostic techniques for the unconventional gas prospects.

The DOE natural gas R&D program has been playing a very important role in applying advanced seismic imaging methods to unconventional gas fields. The DOE program directed its limited resources wisely to get the most for its investment. First, it supported meetings and consortia to bring new developments and technologies to the attention of the small producers without large R&D budgets. Second, it cofinanced with industry some new technologies relevant to unconventional gas and helped to implement them. Third, they financed some well-chosen advanced R&D efforts directly relevant to unconventional gas. Without the DOE funding it was unlikely that this R&D would have started. The projects were chosen well and assigned to competent groups. Overall, they were very successful technically.

The performance milestone for this program is to increase ERR by 11 Tcf by 2015, as shown in Table J-1. The panel characterized uncertainty about technical success for this program directly in terms of the increase in ERR that would result from the research sponsored by the DOE, as shown in the decision tree in Figure J-3. In the past, industry was eager for technological innovations to improve productivity, and since the industry was directly involved in most DOE-sponsored imaging projects, new developments were adopted naturally. The panel does not believe there are any significant barriers to market adoption of the technologies currently under development, and so it did not evaluate market risks.

The third row of Table J-2 shows the panel's assessment of the ERR increase attributable to the DOE program in advanced diagnostics and imaging.

The panel commends DOE for its subprogram on advanced diagnostics and imaging.

Deep Trek

The goal of the DOE/NETL Deep Trek subprogram is to develop technologies that lower the cost and improve the efficiency of drilling and completing deep wells. New tools and technologies that help operators safely drill faster, deeper, cheaper, and cleaner will help ensure an adequate supply of clean-burning natural gas for the nation.

According to the Office of Fossil Energy's *Natural Gas Technologies Program Plan* (DOE, 2004b, p. 5), targets for Deep Trek are as follows:

> to provide fundamental advances in high-temperature, high-pressure materials and electronics (target 2007) that will enable the construction of durable deep drilling tools (2010). These tools will then be integrated into a field-tested and demonstrated deep drilling system (target 2015) that will result in major reductions in the cost . . . and risks . . . of deep drilling (target 2013). In addition, a new initiative in improved diagnostic and imaging technologies tailored for deep gas exploration and development . . . further improving deep gas economics. Together, these two initiatives will result in the expansion of the ERR [economically recoverable reserves] by approximately 15 TCF through 2015.

Currently budgeted at $1.5 million for FY05, Deep Trek will benefit industry with a diverse number of program areas that enable access to resources below 20,000 feet, including but not limited to

- Imaging superdeep gas plays across the Gulf of Mexico shelf,
- High-temperature down-hole electronics,

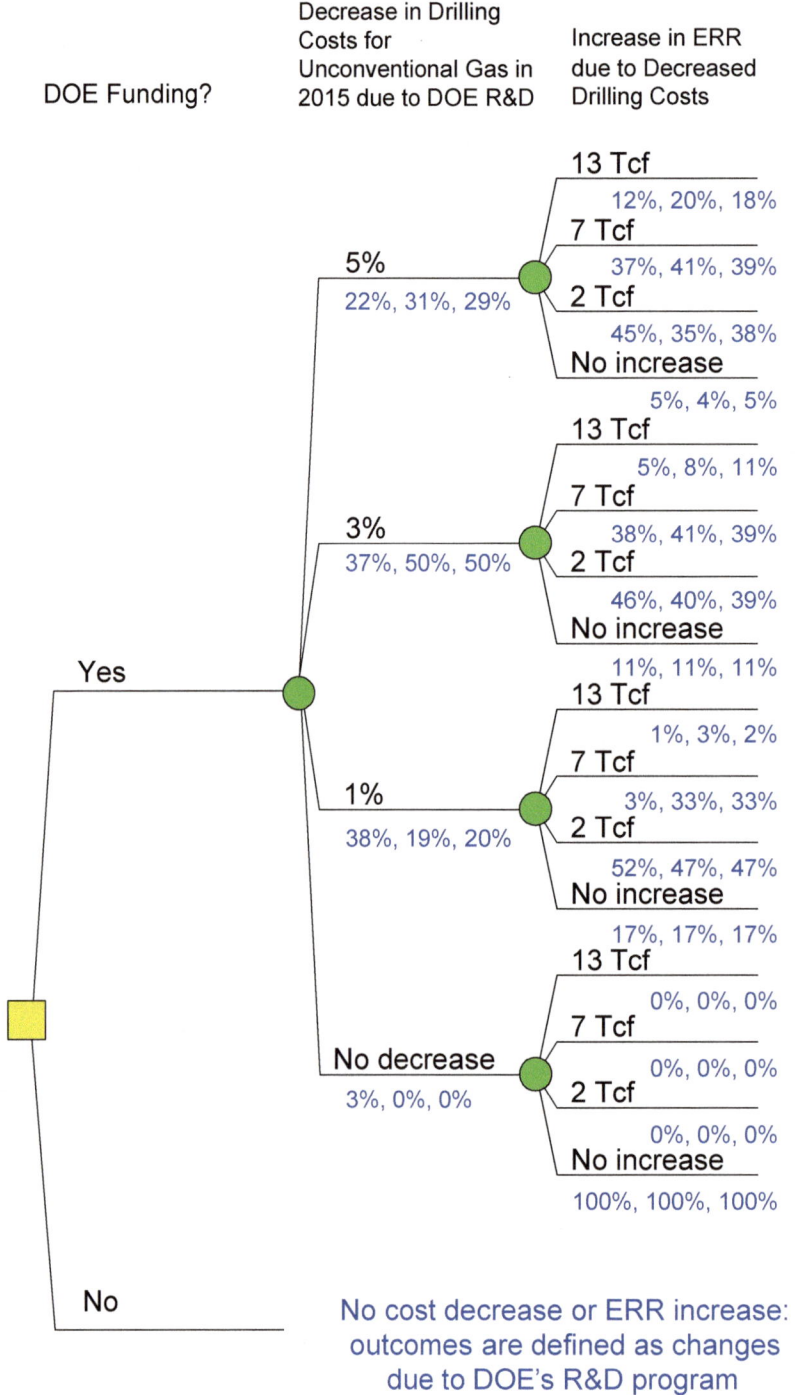

FIGURE J-2 Decision tree for the drilling, completion, and stimulation subprogram. Values under each branch are the average of the panelist's estimates of the probability of that technical outcome for the Reference Case, the High Oil and Gas Prices, and the Carbon Constrained scenarios, respectively.

- High-temperature/high-pressure measurement while drilling (MWD) tool,
  - Supercement,
  - Down-hole vibration monitoring and control, and
  - In-house high-temperature drilling laboratory.

Funding to date for the Deep Trek program has totaled over $16 million, with nearly $9 million contributed by research partners.

All of these technology areas are priority items for private industry and would probably eventually be developed anyway, but DOE's R&D program could still have a significant

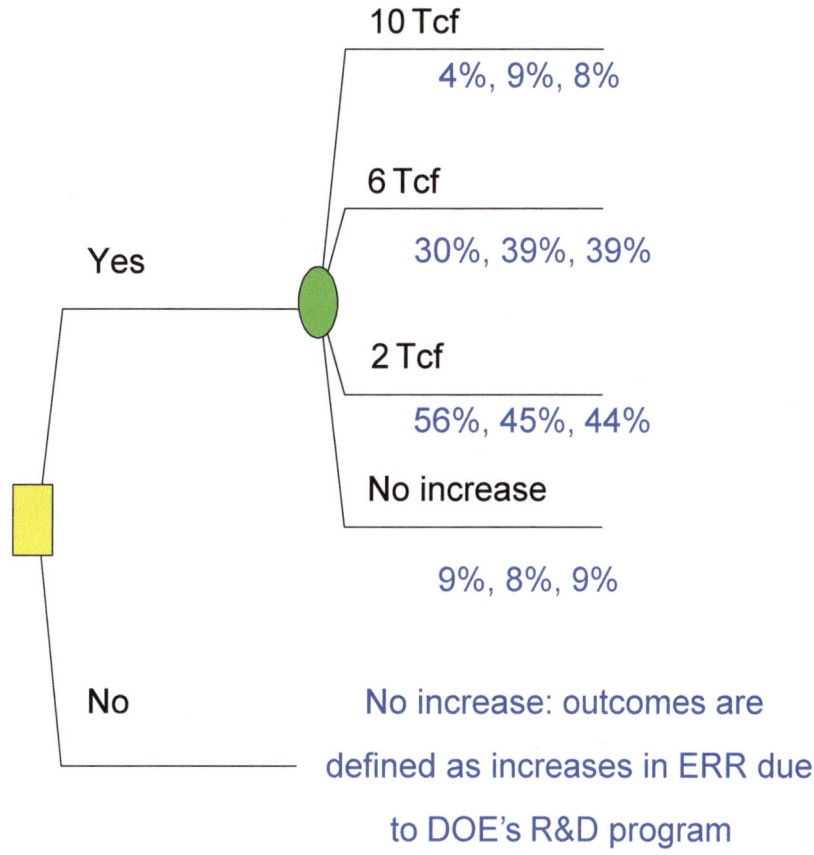

FIGURE J-3 Decision tree for the advanced diagnostics and imaging subprogram. Values under each branch are the average of the panelist's estimates of the probability of that technical outcome for the Reference Case, the High Oil and Gas Prices, and the Carbon Constrained scenarios, respectively.

impact. Though industry projects long-term economic value for development of the technology, the present-day value of expenditures may be an even greater consideration. When prices are high, drilling becomes more economical. Companies tend to accept more drilling risk when they expect the return on investment to be greater. Though R&D continues during such times, drilling outpaces it. In low-price environments, drilling activity decreases, and funds available for R&D typically decrease as well. R&D by DOE to achieve long-term technological advances during low-price environments would bear fruit during future high price environments, when drilling activity increases. Assuming the resource does exist in substantial amounts, the question is whether DOE-funded research will be both successful and well-timed relative to industry-funded research to claim credit for a significant proportion of the resources that are ultimately discovered.

Though its R&D will certainly have a beneficial impact

on recovering resources at depths greater than 20,000 feet (and more likely down to 30,000 feet in the Gulf of Mexico shelf), DOE probably takes more credit for the future success of this program than it deserves. On the other hand, DOE's investment in Deep Trek encourages deep gas exploration by keeping it up to speed and facilitating R&D with industry through technology transfer and exchange. It is the opinion of the panel that DOE's policy role is just as important as the dollars it spends. DOE should keep up its investment in this important R&D area.

The potential benefits of Deep Trek depend on whether a substantial gas resource exists in the United States at depths below 20,000 feet. Currently, there are only two substantial very deep gas plays in the nation: (1) the deep Norphlet play around Mobile Bay in the eastern Gulf of Mexico, with 7 Tcf, and (2) the deep Madison in Wyoming's Wind River Basin (Madden field), with 2 Tcf. Both were discovered in the 1980s. Uncertainty about the existence of very deep gas is caused by two factors: the adverse effects of the high pressures and high temperatures at great depths on reservoir quality and the adverse effects of high temperatures on gas quality. Gas below 20,000 feet is likely to be found only where the thermal gradient is low and where early migration and the characteristics of the reservoir matrix permit porosity and preserve permeability. That such resources are limited was indicated by drilling activity on the Gulf of Mexico Outer Continental Shelf (OCS) in 2004. Of 1,050 total shelf completions in what is considered the most promising area in the country for very deep exploration, there were only 15 completions between 16,000 and 18,000 feet true vertical depth, 3 between 18,000 and 20,000 feet, and none below 20,000 feet.

The EIA estimated that 7 percent of all U.S. natural gas production came from deep formations between 15,000 and 20,000 feet in 1999 and said that was expected to increase to 14 percent by 2010. The new production will come from the Rocky Mountains, the Gulf Coast, and the Gulf of Mexico. Seismic imaging techniques utilize ocean-bottom-cable technology for data acquisition. The benefit is that sensors can be deployed close to platform legs, wellheads, and other obstacles and receivers can be extended to greater offset distances. This allows flexibility in placing receiver stations at optimal positions for imaging superdeep geologic targets. Data are being collected to demonstrate both the capabilities and limitations of the technology. This project is making considerable progress and has been tested to 30,000 feet, at which depth it provides excellent images.

In many of the deep drilling applications progress has been made toward developing prototypes of wireless communications, stimulation, and completions, or of other technologies. These prototypes have been tested in the laboratory and, in some instances, in the field. For example, a harsh-environment solid-state gamma detector for down-hole gas and oil exploration was tested in the first phase and a decision is pending on whether to proceed with the second phase.

DOE R&D promotes shared development of technology. Primarily, industry relies on service companies for drilling technologies in hazardous high-temperature, high-pressure environments. When provided by service companies as a product, the technology is available to all consumers. However, the cost may be so high that only large companies can afford to utilize the technology. Technologies that are developed by DOE could be made available to a greater portion of the market at a lower cost, enabling more industry competitors to utilize the technology. Technologies developed in-house, held as proprietary, and not shared would give their owners a competitive advantage, and fewer companies would be able to participate in drilling for deeper gas resources without DOE R&D, fewer resources would be developed for consumers, and gas prices would probably be higher.

Industry is concentrating new efforts on drilling deep gas plays both onshore and across the Gulf of Mexico OCS. A multitude of challenges are encountered in this newly evolving prospective arena. As very few wells have been drilled to 30,000 feet, seismic data are the primary tool for exploration, and seismic imaging techniques become critical. Already, industry has taken advantage of partnering with universities in seismic imaging techniques. This partnership also should be a driver supporting university geoscience department enrollment, a critical but declining skill set. The mean age of geoscientists in industry is 47. By 2008, 80 percent of the oil and gas industry's workforce will be eligible for retirement. Geoscience enrollment in universities has steadily declined since 1985 from approximately 12,000 graduates per year to fewer than 8,000, of whom only 2,000 or so are enrolled in graduate research programs, where the R&D partnership with universities is focused (Hill, 2002). DOE partnering with universities in R&D projects will help develop and sustain a workforce with the critical skills required for continued exploration and development of oil and gas resources.

As described previously, the performance milestone for the program is an addition of 15 Tcf to ERR by 2015. The panel identified both technical and market risks associated with this goal. Uncertainty about technical success was characterized directly in terms of the estimated increase in ERR that would result from the program. Four different levels of ERR addition were identified, as illustrated in the decision tree in Figure J-4, and the panelists estimated the likelihood that DOE's R&D program would result in each of the specified levels of ERR increase. Owing to the highly exploratory nature of the Deep Trek program and the anticipated difficulties of accessing deepwater resources even if the Deep Trek program is successful, the panel believes market penetration of the technologies is a significant uncertainty. The panel defined market success as the fraction of the technically achievable ERR that will ultimately be achieved.

The fourth row of Table J-2 shows the panel's assessment of the technical and market risks associated with the Deep Trek program.

APPENDIX J

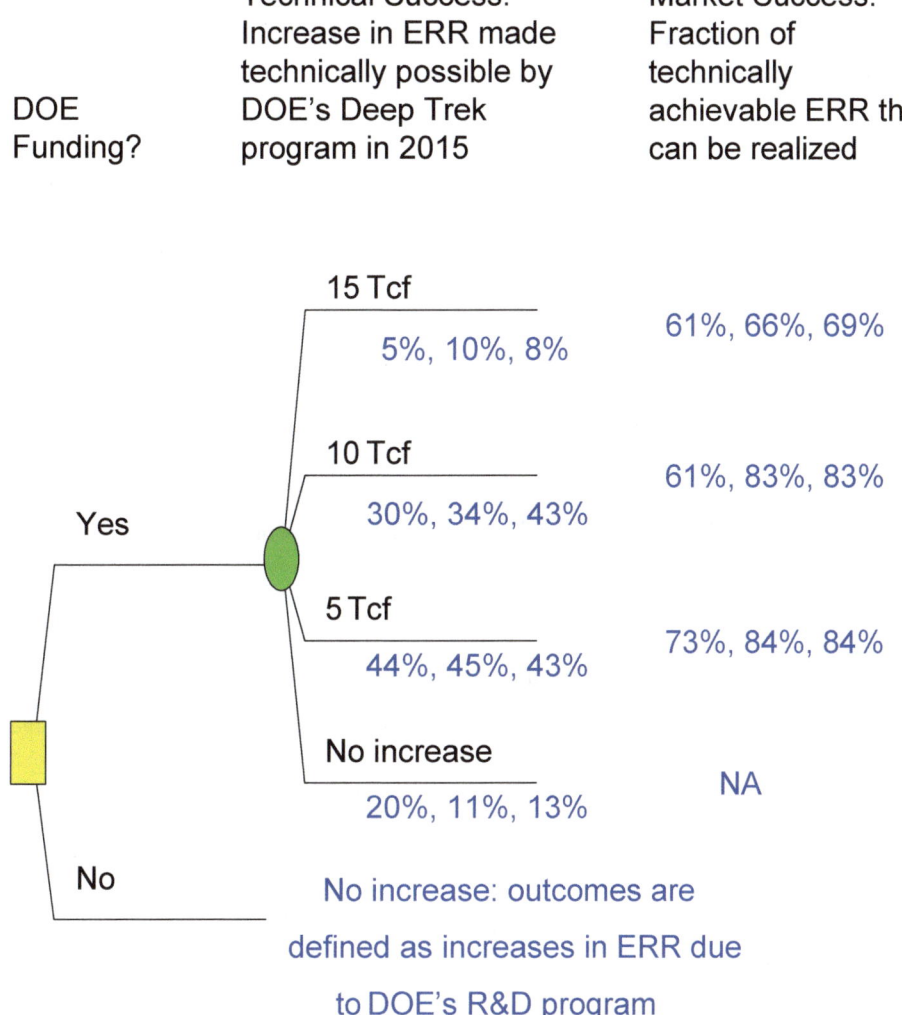

FIGURE J-4 Decision tree for the Deep Trek subprogram. Values under each branch are the average of the panelist's estimates of the probability of that technical outcome for the Reference Case, the High Oil and the Gas Prices, and the Carbon Constrained scenarios, respectively.

RESULTS AND DISCUSSION

The assessments described above and illustrated in the decision trees result in an estimate (with uncertainty) of the increase in ERR attributable to each specific subprogram. Because the benefits of the R&D program will be estimated for the program as a whole (see following section), these subprogram-level assessments were combined to generate an estimate of the total increase in ERR attributable to the R&D program.

Table J-2 summarizes the results of all the technical and market success assessments at the subprogram level. Each of the three global scenarios is represented, with the expected value (the probability weighted average) of the increase in ERR attributable to the R&D program. The final row of Table J-2 shows both the average and the highest and lowest individual estimates of increase in ERR attributable to the program.

Benefits Estimation

Increases in ERR for natural gas resources result directly in economic, environmental, and security benefits for the nation. The panel relied heavily on DOE's own evaluation and quantification of the benefits associated with its natural gas R&D program in estimating those benefits[6] but also made several important modifications.

---

[6]DOE, *Estimating the DOE Office of Fossil Energy's R&D Program Benefits*, FY2004 Final Report, Volume 1, and DOE presentations to the panel.

Economic Benefits

The primary economic benefits that result from increases in ERR come in the form of reduced costs of production and increased domestic natural gas production. Based on DOE's benefits analysis, about 23 percent of the total increase in ERR over the 2003 to 2025 time period translates into increases in expected production. Of the total estimated increase in expected production, roughly 58 percent is realized as increased production and the remainder is added to proved reserves (DOE, 2004a, pp. 41 and 102). Using these scaling factors and the average of the panel's estimates of the ERR increase due to the R&D program (13 Tcf by 2015) results in an estimated increase in cumulative domestic natural gas production of about 1.7 Tcf by 2015 in the AEO Reference Case scenario. The panel also notes that DOE incorporates estimates of risk into its benefits calculations based on the assessments of its program managers in the various technology areas.

DOE estimates the economic benefits of its R&D program as the total consumer energy savings resulting from the increased domestic natural gas production attributed to the program. For each year, DOE estimates a change in natural gas price that results from the increase in supply and multiplies the reduction in gas price by the total gas consumption for that year to yield the reduction in consumer expenditures on natural gas.

As pointed out in the Phase One report, these benefits—savings to consumers—do not measure the economic benefits to the nation of the program. If the research being evaluated changes the cost of production only for the natural gas that would not otherwise be produced—that is, if the cost of production for the vast majority of domestic gas produced in a given year is unchanged by the research—then the consumer savings from reduced prices are offset by the reduced revenues to producers. This is a transfer payment rather than an economic benefit to the nation.

Both the Phase One committee and DOE have noted that a more appropriate economic benefits measure would include not only the cost savings to consumers but also the reduced revenues of the producers—that is, the economic benefit measure should be the net consumer and producer surpluses. DOE has developed a framework for evaluating the net surplus associated with a change in the supply curve for natural gas and presented this method both to the panel and to the Phase Two committee.[7] DOE's net surplus evaluation requires multiple evaluations with NEMS, which the panel was not able to implement independently, and existing preliminary analyses by DOE could not be modified to extract the benefits of the natural gas E&P program alone. The panel implemented a simplified version of this approach to estimate economic benefits.

The details of the economic basis and the calculations are included in Attachment B. To implement this approach for estimating economic benefit, the panel estimated two quantities: (1) the amount of domestic natural gas production (and the fraction of total domestic natural gas production) in each year that is produced at reduced costs due to DOE's E&P research and (2) the reduction in the costs of production for that natural gas.

First, because the research being evaluated addresses only a small fraction of the total domestic natural gas production (stripper wells and unconventional resources), the results of such research are assumed to reduce the costs of production for only a small fraction of the total domestic natural gas produced in any given year—just the portion that would not have been produced otherwise. Figure J-5, derived from DOE (2004a), illustrates DOE's and the panel's estimates of the annual increase in domestic natural gas production attributed to the natural gas R&D program. The panel's estimates are a simple scaling of DOE's estimates, based on the ratio of DOE's estimate of increased ERR to the panel's estimate of increased ERR attributable to the program, and the fraction of increased ERR assumed to become increased production from DOE's benefits analysis. The Reference Case and High Oil and Gas Prices scenarios are shown. DOE does not have a projection for a scenario analogous to the Carbon Constrained scenario, so the Reference Case projections for increased production were used to estimate benefits for that scenario.

The change in the costs of production for natural gas that is produced as a result of advances from the drilling, completion, and stimulation subprogram was estimated directly by the panel, as described previously and shown in the decision tree in Figure J-2. Cost reductions due to the other subprograms were not estimated directly, and simplifying assumptions were made to derive an estimate of those cost reductions. Noting that the magnitude of ERR in each of the programs is reasonably close to that of the drilling, completion, and stimulation subprogram, a reasonable estimate of the cost reduction was assumed to be the middle value provided for that program, or 3 percent for all incremental production attributed to the other three subprograms.

The resulting estimated economic benefits of the program from 2006 through 2025 are $220 million, $590 million, and $300 million in the Reference Case, High Oil and Gas Prices, and Carbon Constrained scenarios, respectively, with annual benefits discounted at 3 percent as recommended by the Phase Two committee. DOE's estimate of the discounted cumulative economic savings through 2025 from reduced natural gas prices for the Reference Case scenario is about $50 billion, assuming the program goals are met (DOE, 2004a, p. 105), but the panel notes again that this is an estimate of consumer cost savings rather than an economic benefit to the nation and thus is not directly comparable with the panel's estimate.

Figure J-6 summarizes the panel's estimates of the risks

---

[7] John Pyrdol, economist, DOE Office of Oil and Gas, "Assessing the benefits of R&D: A framework of analysis and an application using NEMS," Presentation to the panel on October 12, 2005.

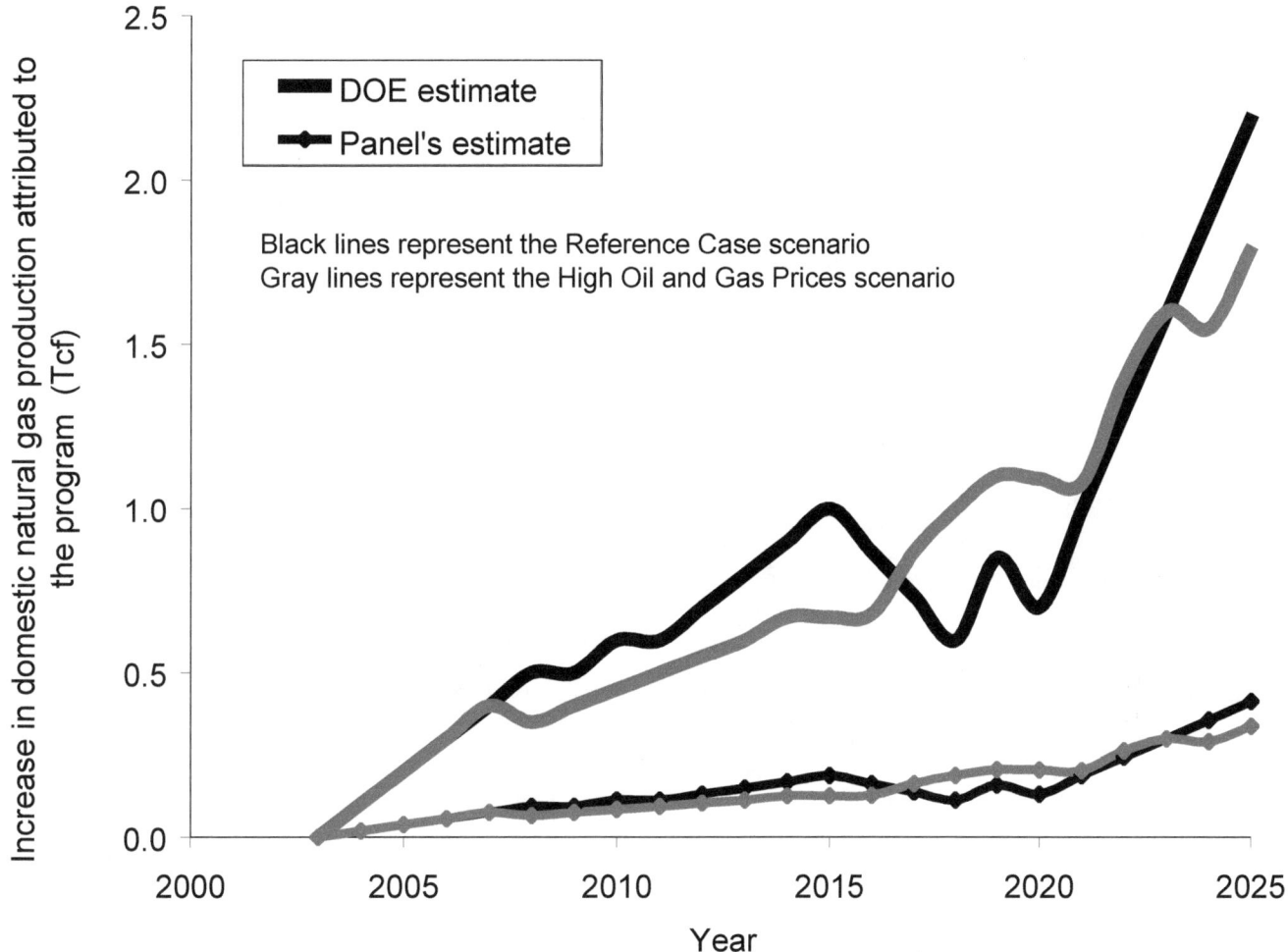

FIGURE J-5 Estimated increase in domestic natural gas attributed to the program by year.

and benefits of DOE's natural gas exploration R&D program. Benefits are calculated as described in Chapter 3 and are reported for all three global scenarios.

Environmental Benefits

In many ways natural gas is the ideal fuel for residential and commercial heating and for generation of electricity. It is also a very important feedstock for the chemical industry. Natural gas is the cleanest burning fossil fuel; of all fossil fuels produces the least carbon dioxide per unit of energy generated and the least amount of other noxious by-products. It also requires less capital investment for combustion equipment than coal or oil. The distribution system is well developed in most parts of the country and relatively safe. Whenever natural gas can be substituted for coal or oil, the result is less emissions of greenhouse gases and cleaner air and generally a cleaner and safer environment. Truck traffic and rail traffic are reduced. Every increase in the use of natural gas at the expense of other fossil fuels results in environmental benefits.

Improvements in the quality of geophysical exploration and in drilling technology, and in the technologies that enhance recovery, result in more gas produced per well and, overall, in a smaller footprint for the production activity, reducing the total impact on the environment. This is somewhat offset by the increased capacity to produce from more remote regions and by the consequent increase in wells in such regions.

Security Benefits

According to a report by the NRC (2002b), which included attention to the security of energy systems in the United States, the U.S. economy and quality of life require a plentiful and continuous supply of energy. Though energy accounts for less than 10 percent of the GNP, much of the remaining economy will not function without energy assets.

		Global Scenario		
		AEO Reference Case	High Oil and Gas Prices	Carbon Constrained
Program Risks	Technical Risks	Technical risks were evaluated at the subprogram level and were defined as the likelihood of achieving specific increases in economically recoverable resources (ERR) by 2015 as a result of DOE's R&D program. Aggregated at the program level, the expected increase in ERR made technically possible by the DOE program in each scenario is shown below. Numbers in parentheses represent the 10th and the 90th percentiles of the uncertainty and assume no market risks.		
		17 (9, 25) Tcf	20 (11, 28) Tcf	20 (11, 29) Tcf
	Market Risks	Market risks were identified for two of the four subprograms evaluated and were defined as the fraction of the total potential addition to ERR that would be realized. The effect of including market risks is to decrease the estimated net increase in ERR in 2015. The expected value and the 10th and 90th percentiles of ERR increase given both technical and market risks are shown.		
		13 (6.8, 19) Tcf	17 (9.5, 25) Tcf	17 (9.5, 25) Tcf
Program Benefits	Expected Economic Benefits[a]	$220 million at 3% ($100-$280 million) $145 million at 7% ($68-$180 million)	$590 million at 3% ($275-$675 million) $380 million at 7% ($180-$440 million)	$300 million at 3% ($150-$370 million) $200 million at 7% ($100-$250 million)
	Environmental Benefits	Several elements of the program have the potential to lead to generally smaller footprints for natural gas production, although this benefit is at least partially offset by the increase in producibility from previously marginal regions. The net impact on the environment in terms of disturbed area has not been estimated. The other potential environmental benefit would be a reduction in carbon emissions if natural gas substitutes for other fossil fuels. At the relatively modest level of increased production envisioned for this program alone, environmental benefits are expected to be modest.		
	Security Benefits	Increased domestic natural gas production partially offsets gas imports. Based on DOE's estimates, about 50% of the increased production results directly in reductions in imports under Reference Case prices (25% in a High Gas Prices scenario). Natural gas imports avoided (2005 to 2025).		
		1.2 Tcf	0.6 Tcf	1.2 Tcf

[a]Benefits are expected values, with 10th and 90th percentiles shown in parentheses. Economic benefits are present values discounted at both 3% and 7% real. Environmental and security benefits are discounted at 3%. The net economic benefits were calculated based on the assumption that the technology from the DOE program affects no existing gas production and that there will be a decrease in the cost of gas that would have been uneconomical to produce without the DOE program. DOE (2004a) estimates the discounted, cumulative economic savings through 2025 to be $50 billion for the Reference Case, although the committee notes that this includes consumer cost savings largely offset by reduced revenue to producers and, as such, is not indicative of the net economic benefit to the nation.

FIGURE J-6 Results matrix of the Panel on DOE's Natural Gas Exploration and Production Program.

Oil products provide 97 percent of the energy used in the transportation sector, while natural gas provides over 25 percent of residential and industrial energy needs. A significant disruption to either of these basic resources for more than a few days would have serious consequences for the U.S. economy and the health and well being of the population (NRC, 2002b). (Interruptions in oil supply can be mitigated for a time by the Strategic Petroleum Reserve, but there is no such reserve to mitigate interruptions to natural gas supplies.) At the present time the United States still imports a relatively small amount of natural gas, but projections call for this to increase. To the extent that domestic production can be increased, imports can be lessened. Every additional cubic foot that is produced domestically means a cubic foot less imported from overseas. Reducing the number of LNG tankers delivering natural gas improves the balance of trade and reduces the security risk of disruptions in the supply from abroad, as well as security risks to LNG facilities.

The vulnerability of key infrastructure components is an important component of the energy security issue. Industry is not capable of handling extensive organized acts of terrorism with weapons or explosions, cyberattacks on control systems via the Internet, or natural disasters that impact key elements of the energy system. Each of the natural gas R&D subprograms examined by the panel is designed to enhance exploration for and development of additional resources. Though DOE R&D could enhance national security, none of these subprograms has a goal of decreasing the vulnerability of the U.S. natural gas economy vulnerability to disruptions in supply. Although it did not have the resources and task to investigate the area of vulnerability and infrastructure security, the panel believes there is a need for government (probably through the Department of Homeland Security) and industry to share the cost and execution of the needed R&D, with the government contributing, as appropriate, its expertise in this matter.

Increased domestic natural gas production partially offsets natural gas imports. Based on DOE's estimates, about 50 percent of the increased production directly reduces imports under Reference Case prices, and about 25 percent of increased production offsets imports in the High Oil and Gas Prices scenario.

SUMMARY AND RECOMMENDATIONS

The panel reviewed four subprograms of DOE's natural gas activities: existing fields; drilling, completion and stimulation; advanced diagnostics and imaging; and Deep Trek. It considered three global scenarios in estimating the probability of technical and market success: the AEO Reference Case scenario; a High Oil and Gas Prices scenario, where gas was assumed to cost twice as much as in the AEO Reference Case; and a Carbon Constrained scenario, where a $100 per ton carbon tax was assumed to be in place in 2012.

DOE natural gas R&D budgets are small compared to industry natural gas R&D budgets and are likely to consist of early development and seed funding efforts as opposed to complete R&D through to final development and commercialization. This makes assessment of the ultimate benefit of the research efforts of DOE relative to that of private sector efforts particularly difficult, in that numerous additional steps must be funded and taken beyond the R&D projects being evaluated before benefits will actually be realized. For this program, it is also difficult to separate gas from oil R&D in many cases: For example, drilling is relatively similar whether for oil or for gas, and gas drilling R&D has the potential to reduce drilling costs for oil resources as well. Nevertheless, the panel found that the decision tree probabilistic assessment approach was a well-organized way to convert the expert knowledge of the panel members into quantifiable measures. The utility of the decision tree approach was that (1) it provided a logical template for experts in quantifying their evaluations and (2) the results for different scenarios could be compared since they are arrived at following a uniform methodology.

While the panel's estimates of ERR and the resulting economic benefits were substantially lower than DOE's estimate, the panel's estimates show, nonetheless, a significant and justifiable return on the modest R&D investments of DOE, both in terms of ERR added to the nation's resources base and the economic benefits resulting therefrom. This is especially true in the High Oil and Gas Prices scenario, which seems the most likely scenario for the United States. In addition, although it is difficult to value, the DOE program sponsors of university research in the natural gas E&P area, which in turn fosters the development of a high-level technical workforce.

The panel recognized the substantial benefits of increased consumption of natural gas relative to that of other fossil fuels and notes that the United States, and indeed the world, is at the threshold of a methane economy, where natural gas, both as a clean, efficient fuel and as a source of hydrogen, is the transition fuel to a carbon-free fuel economy. In the United States the bulk of future additions of natural gas from the resource base will come from so-called unconventional natural gas resources. Whether these resources can be available at an affordable price will depend heavily on the development and application of advanced technology. This is particularly significant in North America, where production will be less and less able to meet needs, and more and more natural gas will be purchased by the United States. Much of this LNG will be landed in the Gulf Coast area, and the hurricanes of 2005 have had an impact. Operational dates for much LNG landing capacity in the Gulf Coast area are likely to be as much as 5 years as new locations, designs, regulations, and hurricane requirements are included. This may increase the cumulative price of North American gas over the next decade.

In North America, the only two large, yet (relatively) untapped gas resources are low-permeability shales and ultradeep gas. Both have major uncertainties. For low-permeability shales, industry estimates are currently that 8-12 percent of the gas in place can practically be recovered in the next decade or two. If this were 2-3 percent, the quantity of gas recovered would be very small; if it were 25-30 percent, the quantity would be very large. For the recovery of ultradeep gas, how to create and maintain permeability for gas recovery under the extremely high stresses that prevail remains an unsolved problem. Therefore, gas R&D opportunities are significant, so successful R&D would also be significant. The panel concludes that the DOE program now in place is geared to such development and application.

The panel is mindful that as the Energy Policy Act of 2005 makes itself felt, the future funding structure for federally supported R&D in natural gas will likely differ from the structure in place and evaluated by the panel, but it judges that the program for developing unconventional natural gas will follow the same basic thrust.

Finally, the matter of the accrued value of DOE's R&D continues to be the subject of many studies, panel discussions, technical associations, and other groups. None of these efforts—even those looking back on completed R&D and technology successes that have occurred since the R&D was performed—can quantify precisely the dollar return on the R&D investment. Nevertheless, because they seem to be the best way to grasp the benefits, such attempts at quantification continue and their value seems to be well recognized, as is demonstrated by the large voluntary investments that continue to be made by industry. At a minimum, industry apparently believes that investment in R&D returns at least as much as capital spending on ongoing or new business development—otherwise resources would go to capital spending as opposed to R&D investments. The panel commends the DOE natural gas program on including risk in its estimate of benefits but urges DOE to continue to pursue its proposed approach to evaluating the net benefits to the nation. Following up on and refining what the panel and the Phase Two committee have done will help in focusing the new efforts that come out of the EPACT-2005, Section J, initiative.

Recommendation: The DOE Office of Natural Gas should continue to pursue its proposed methodology for evaluating the "net benefits to the nation" of their R&D programs, so as to develop more accurate estimates of R&D benefits.

Recommendation: As the natural gas program changes under the EPACT-2005, DOE, in following up on the methodology of the full committee, should use its more accurate method of estimating R&D benefits to help focus its new R&D efforts.

ATTACHMENT A
PANEL MEMBERS' BIOGRAPHIES

William L. Fisher (NAE), *Chair*, holds the Leonidas Barrow Chair in Mineral Resources, Department of Geological Sciences, University of Texas at Austin. His previous positions at the University of Texas at Austin have included director and state geologist of Texas, Bureau of Economic Geology; director, Geology Foundation; chairman, Department of Geological Sciences; and Morgan J. Davis Centennial Professor of Petroleum Geology. He has been assistant secretary, energy and minerals, U.S. Department of the Interior and deputy assistant secretary, Energy, U.S. Department of the Interior. He is a fellow of the Geological Society of America, Fellow of the Texas Academy of Science, Fellow of the Society of Economic Geologists, and member of the National Academy of Engineering. He has served on numerous federal government committees and councils and NRC committees. He served on the NRC Committee on Benefits of DOE's R&D in Energy Efficiency and Fossil Energy, a precursor to the current study. He has expertise in energy policy, oil and gas resources and recovery, fossil fuel exploitation and technology, geology, and mineral resource policy. He has a Ph.D. in geology from the University of Kansas.

John B. Curtis is a professor in the Department of Geology and Geological Engineering and director, Petroleum Exploration and Production Center/Potential Gas Agency at the Colorado School of Mines. Dr. Curtis has been at the Colorado School of Mines since July 1990. Before that, he had 15 years experience in the petroleum industry with Texaco, Inc., SAIC, Columbia Gas, and Exlog/Baker-Hughes. Dr. Curtis serves on and has chaired several professional society and natural gas industry committees, which included the Supply Panel, the Research Coordination Council, and the Science and Technology Committee of the Gas Technology Institute (Gas Research Institute). He co-chaired the American Association of Petroleum Geologists (AAPG) Committee on Unconventional Petroleum Systems from 1999-2004 and is an invited member of the AAPG Committee on Resource Evaluation and its Committee on Research. He was a counselor to the Rocky Mountain Association of Geologists from 2002 to 2004. He is an associate editor of the *AAPG Bulletin* and *The Mountain Geologist*. He has published studies and given numerous invited talks on hydrocarbon source rocks; exploration for unconventional reservoirs; and the size and distribution of U.S.; Canadian and Mexican natural gas resources, and comparisons of methodologies for resource assessment. As director of the Potential Gas Agency, he directs a team of 145 geologists, geophysicists and petroleum engineers in their biennial assessment of remaining U.S. natural gas resources. He teaches petroleum geology, petroleum geochemistry, and petroleum design and stratigraphy at the Colorado School of Mines, where he also supervises graduate student research. He received a B.A. (1970) and M.Sc. (1972) in geology from Miami University and a Ph.D. (1989) in geology from the Ohio State University.

Sidney J. Green (NAE) is chairman and chief executive officer of TerraTek in Salt Lake City, a geotechnical research and services firm focused on natural resource recovery, civil engineering, and defense problems. Previously, he worked at General Motors and at the Westinghouse Research Laboratory. He has an extensive background in mechanical engineering, applied mechanics, materials science, and geoscience applications. He has extensive experience in rock mechanics, particularly as related to oil and gas and other natural resource recovery. Expertise includes research and development, multidisciplinary problem solving, management, and corporate structure organization. He has experience in taking research to commercial application and in entrepreneurial high-technology business development. He is a former member of the NRC Geotechnical Research Board. He was named Outstanding Professional Engineer of Utah and received the ASME Gold Medallion Award and the Society of Experimental Mechanics Lazan Award. Mr. Green received a Degree of Engineer from Stanford University in engineering mechanics, an M.S. degree from the University of Pittsburgh, and a B.S. degree from the University of Missouri at Rolla, both in mechanical engineering.

Patricia M. Hall is currently BPs geoscience recruiting manager and early development program coordinator. Earlier she was a senior geologist for BP, where her primary assignment was evaluating complex subsalt regional basins and the frontal fold belt in the Gulf of Mexico Deep Water Exploration Group. Ms. Hall has held numerous positions as a geologist with Shell, Amoco, and the Gulf Exploration and Preproduction Co. Ms. Hall is the chairperson of the National Association of Black Geologists and Geophysicists and a member of NASA's Minority Education Awareness Committee.

Previously she was co-chair of the annual conference of the National Association for Black Geologists and Geophysicists in 2001 and chair of the Geological Society of America's Committee for Women and Minority Geoscientists in 1996. Ms. Hall received her B.S. in earth sciences and her M.S. in geology from the University of New Orleans.

Martha A. Krebs is director, Energy R&D Division, California Energy Commission. Prior to that she was a consultant with Science Strategies. She was a senior fellow at the Institute for Defense Analyses (IDA), where she led studies in R&D management, planning and budgeting. She has extensive experience in DOE's basic and applied energy programs. Dr. Krebs also served as DOE assistant secretary and director, Office of Science, where she was responsible for the $3 billion basic research programs that underlay DOE's energy, environmental, and national security missions. She also had the statutory responsibility for advising the secretary on the broad R&D portfolio of DOE and the institutional health of its national laboratories. She has been associate director for planning and development, Lawrence Berkeley National Laboratory, where she was responsible for strategic planning for research and facilities, laboratory technology transfer, and science education and outreach. She also served on the House Committee on Science, first as a professional staff member and then as subcommittee staff director, responsible for authorizing DOE non-nuclear energy technologies and energy science programs. She is a member of Phi Beta Kappa, a fellow of the American Association for the Advancement of Science, a fellow of the Association of Women in Science, and received the Secretary of Energy Gold Medal for Distinguished Service (1999). She is a member of the National Academies Committee on Scientific and Engineering Personnel and the Navy Research Advisory Committee. She is also a member of the Committee on Prospective Benefits of DOE's Energy Efficiency and Fossil Energy R&D Programs (Phase Two). She received a bachelor's degree and a Ph.D. in physics from the Catholic University of America.

William C. Maurer (NAE) is president of Maurer Enterprises Inc. His previous positions include senior research specialist for Jersey Production Research Company and senior research specialist for Exxon Production Research. His primary area of interest is advanced drilling, including novel drills such as lasers, electric beams, and high-pressure water jet drills. Other areas of interest include oil-field horizontal drilling techniques, which significantly increase oil and gas production, and advanced drilling software, which improves drilling operations and reduces drilling costs and environmental impact. Another area of interest is applying Russian oil-field technology in the United States to allow more economical production of marginal oil and gas fields. This technology includes electrodrills, percussion drills, geared turbodrills, and retractable drilling systems that can be removed from deep wells without removal of the drill pipe from the well. He holds 32 U.S. patents and 6 foreign patents and is author or co-author of over 50 publications and two books. He is a member of the National Academy of Engineering and a fellow of the American Society of Mechanical Engineers. He has extensive knowledge of market risk, having started 18 successful oilfield service companies. He has a Ph.D. in mining engineering and an M.S. in engineering from the Colorado School of Mines and a B.S. in mining engineering from the University of Wisconsin, Platteville.

Richard Nehring is president of NehRinG Associates (aka NRG Associates), in Colorado Springs. Prior to that, he was a project director in the Energy Policy Program of the Rand Corporation in Santa Monica, where he led or participated in numerous projects studying various fossil fuel supply issues. Since 1983 at NRG Associates, he designed the Significant Oil and Gas Fields of the United States Database and its subsequent expansions, directed the initial development and subsequent updates, upgrades, and expansions of the database; planned the three editions of the *Oil and Gas Plays of the United States* (the play description book accompanying the database) and directed their preparation; composed all database documentation; designed and directed the development of *Significant Oil and Gas Pools of Canada Database*; and marketed both databases and handled most customer relations. Mr. Nehring has served on several national committees dealing with oil and gas resource and supply issues, including the Resource Subgroup, Supply Task Group, NPC Committee on Natural Gas (2002-2003); the Committee on Resource Evaluation, American Association Of Petroleum Geologists; and the NRC Committee on Undiscovered Oil and Gas Resources. At Rand, he consulted in the energy supply area, for clients such as Atlantic Richfield, General Motors, McDonnell Douglas, the Office of Technology Assessment, Paine Webber, and Sohio. He has a B.A. in history, Valparaiso University; a B.A. in philosophy, politics and economics (as a Rhodes scholar), Oxford University, and was a Danforth fellow, Stanford University, in political science and economics for 4 years.

Michael E. Q. Pilson is professor emeritus of oceanography at the University of Rhode Island (URI). He was director of the Marine Ecosystems Research Laboratory at URI for 20 years. His current research interests include the chemistry of seawater, biochemistry and physiology of marine organisms, and nutrient cycling. He received a B.Sc. in chemistry-biology from Bishop's University, in Canada, an M.Sc. in agricultural biochemistry from McGill University, and a Ph.D in marine biology from the University of California, San Diego. He is a member of the American Association for the Advancement of Science; Sigma Xi; the American Geophysical Union; the American Society of Mammalogists; the American Society of Limnology and Oceanography; and the Oceanography Society. He has published extensively, including the text book *An Introduction to the Chemistry of the Sea*.

Reginal Spiller is executive vice president, exploration and production, Frontera Resources Corporation. Mr. Spiller has been a senior executive of the Company since May 1996 and has been responsible for Frontera's exploration and production activities. Mr. Spiller has over 25 years' of experience working in the United States and international oil and gas industries. From 1993 until joining the company, Mr. Spiller was deputy assistant secretary for Gas and Petroleum Technologies at DOE. For 5 years before that, he was the international exploration manager for Maxus Energy Corporation, which held properties in Bolivia, Bulgaria, Czechoslovakia, and Indonesia. He is a member of the NRC Committee on Earth Resources. He also serves as a director of Osyka Corporation and is an active member of the American Association of Petroleum Geologists and the National Association of Black Geologists and Geophysicists. Mr. Spiller is a graduate of Penn State University with an M.S. in geology and of the State University of New York with a B.S. degree in geology.

M. Nafi Toksöz is professor of geophysics and director, George R. Wallace, Jr., Geophysical Observatory, Department of Earth, Atmospheric, and Planetary Sciences, Massachusetts Institute of Technology. His interests and expertise are in the seismology and tectonics of the eastern Mediterranean caused by the collision of the Arabian and Eurasian Plates; and seismic tomography for characterization of Earth's crust and petroleum reservoirs. He has a Ph.D. in geophysics from the California Institute of Technology.

ATTACHMENT B
ESTIMATING ECONOMIC BENEFITS FOR NATURAL GAS EXPLORATION AND PRODUCTION

The DOE natural gas E&P program aims to increase the supply and production of domestic natural gas. DOE reports its goals and predicts the outcomes of the program in terms of increases in economically recoverable resources (ERR). In its benefits analyses, DOE translates the changes in ERR into changes in expected production, as described in the main text of this panel report. Economic benefits accrue from changes in the supply curve for domestic natural gas.

This attachment describes the simplified approach used by the panel to evaluate the net consumer and producer surplus that results from the change in the supply curve. The approach is consistent with recommendations from the full committee and with the preliminary approach being proposed by DOE.[8]

---

[8]John Pyrdol, economist, DOE Office of Oil and Gas, "Assessing the benefits of R&D: A framework of analysis and an application using NEMS," Presented to the panel on October 12, 2005.

Economic Basis

Figure J-7 shows a demand curve and two supply curves for natural gas, where Q is the quantity of gas produced and P is the market prices of the gas. The initial supply curve represents the supply of gas absent any technical improvements from the program. The modified supply curve represents the impact of making *some* natural gas (ΔQ) less costly to produce. The lower portion of the supply curve is unchanged: It represents all natural gas production that cannot or does not benefit from the technological advances being evaluated in the program—that is, gas for which production costs are unchanged. The supply curve bends, as represented by the dotted line, at some point, representing the amount of natural gas production that does benefit from the technologies being evaluated—that is, gas for which production costs have decreased.

P_1 and P_2 represent the market prices of gas given the original and modified supply curves. The area between the P_1 and P_2 curves represents the benefits calculated by DOE for the natural gas program: the change in the market price of gas multiplied by the quantity consumed. To the extent that the costs of production for most of the gas have not decreased, this quantity represents a transfer from producers to consumers rather then a net economic benefit to the nation.

The net economic benefit is the difference in total surplus (the area between the supply and demand curves) with and without the program. The program causes a change in the supply curve, increasing the total surplus. The additional surplus is represented in the Figure J-8 by the area of the dark gray triangle formed by the demand curve and the initial and modified supply curves. The area of the dark gray triangle is approximated by the area of triangle shown in light gray in Figure J-8. The area of the triangle depends on (1) the amount of gas produced at lower cost due to technical advances from the program (ΔQ) and (2) the reduction in the cost of production for that gas (ΔC_p). The area of the blue triangle is $0.5 * \Delta Q * \Delta C_p$. The area of the light gray triangle represents an upper bound on the area of the dark gray triangle: The two areas would be equal if the market price of gas were insensitive to the small change in supply.[9]

So for each year t, the economic benefits can be approximated as

---

[9]Equality follows from construction: The change in the supply curve is due to reduced cost of producing gas that was previously uneconomic. The pre-DOE program cost of this marginal gas is shown in Figure J-8 on the supply curve to the right of the preprogram equilibrium price (P_1), and its total area is given by the light gray triangle. The cost reduction shifts the location of this gas to the left, and the net change in total consumer and producer surplus is equal to the area of the dark gray triangle. Note that the area of the shaded triangles will not necessarily be equal, differing by the area of the triangle formed by the demand curve, the modified supply curve, and the horizontal line corresponding to the price P_1. Note further that if there is no change in price—that is, the demand curve is horizontal—the shaded triangles are equal.

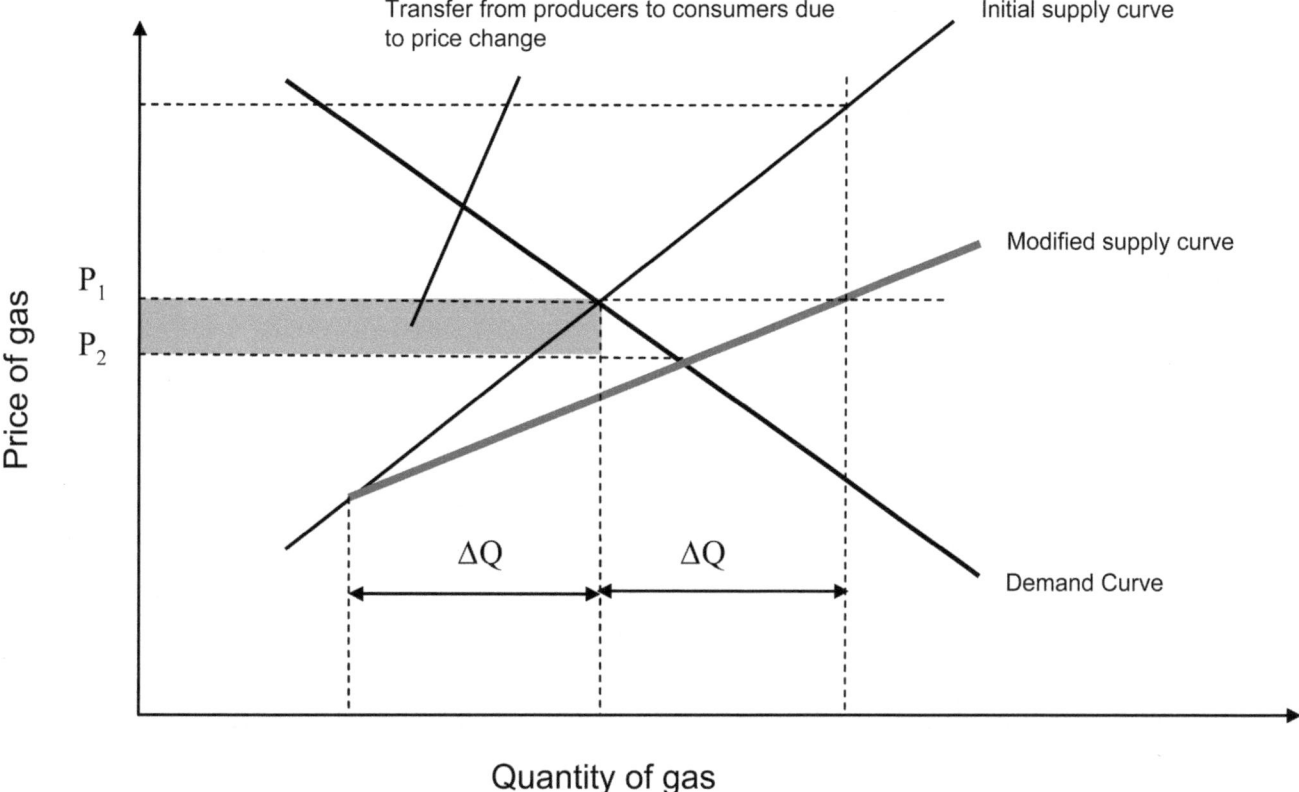

FIGURE J-7 Demand curve and two supply curves for natural gas.

$$E_t = 0.5 * \Delta Q_t * \Delta C_{pt}$$

where E_t are economic benefits in year t, ΔQt_t is the incremental gas produced in year t, and ΔC_{pt} is the change in the costs of production for the incremental gas resulting from the program. Total economic benefits are the discounted sum of annual benefits.

Implementation

Estimates of ΔQ_t can be derived from DOE's estimates of incremental production attributed to the program, scaled down to match the panel's estimate of the increases attributable to the program. These estimates are shown in Figure J-5.

Estimating ΔC_{pt} requires some assumptions. The section on the drilling, completion, and stimulation subprogram (decision tree shown in Figure J-2) provides a starting point. Each branch on the tree corresponds to a different benefit. Consider the top branch, or most optimistic result. The branch posits that a 5 percent decrease in drilling costs will result in a 13 Tcf increase in ERR in 2015. Using the scaling factors defined in that section on drilling, completion, and stimulation would lead to a cumulative increase in production of about 1.7 Tcf. Since none of this gas was produced absent the program, its initial cost of production must have been at least as high as the market price of gas; thus, with the program, the cost of this gas is at least 95 percent of the price of gas, or ΔC_p in 2015 is 0.05 multiplied by the market price of gas in 2015.

For branches on the tree with a smaller reduction in drilling costs, the ΔC_p would be smaller (specifically, 3 percent, 1 percent, or 0 times the market price of gas for the other branches).

While ERR estimates are given for each of the other three subprograms, the cost reductions due to the drilling, completion, and stimulation subprogram were not estimated directly. It can be noted that the magnitude of ERR in each of the programs is reasonably close to that of the drilling, completion, and stimulation subprogram. A reasonable estimate of the cost reduction thus may be the middle value provided for that subprogram, or 3 percent for all incremental production attributed to the other three subprograms.

Note on Domestic Versus Foreign Production of Gas

The savings to consumers may be considered an economic benefit to the U.S. economy to the extent that producers are

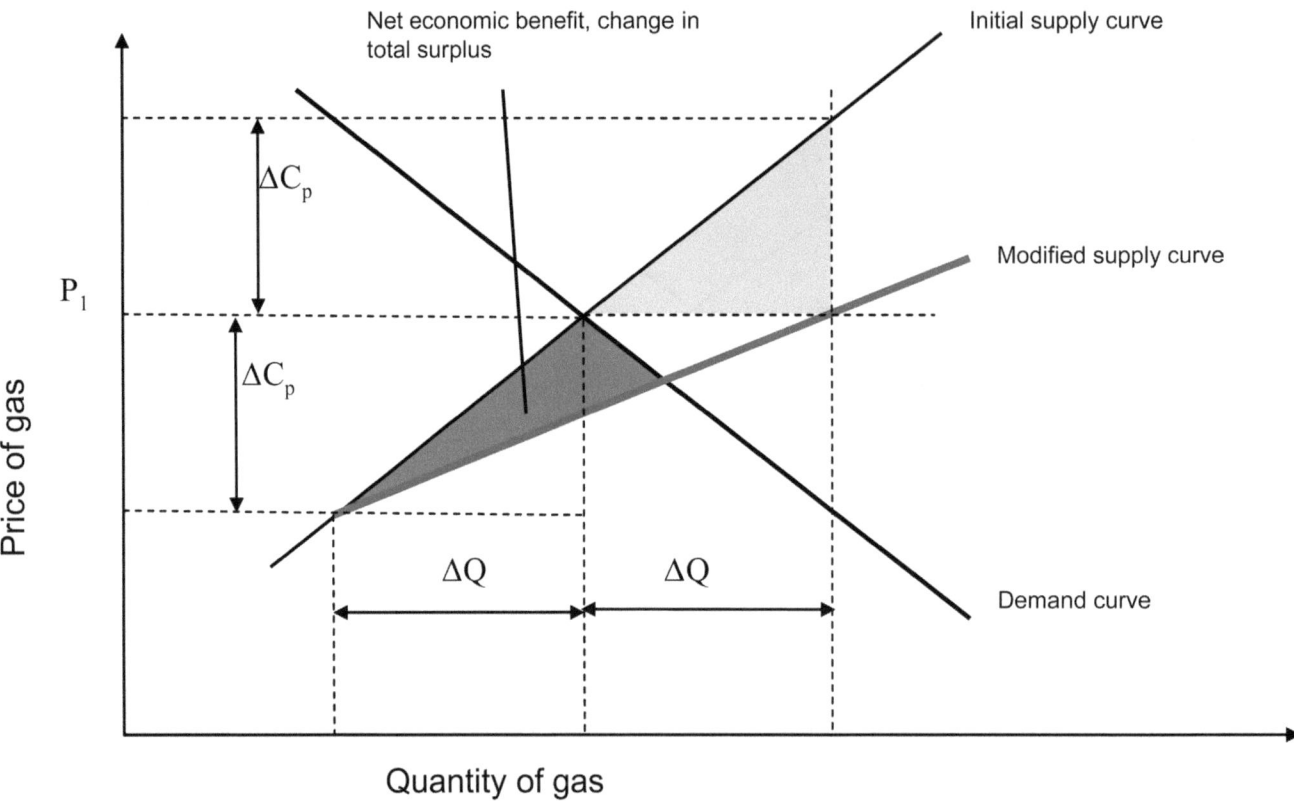

FIGURE J-8 Change in total surplus from the change in the supply curve. The DOE program is assumed to result in a decrease in the costs of production for ΔQ units of gas that were not economic to produce absent the DOE program. The cost of this marginal gas assuming no technical improvements from the program is shown by the initial supply curve to the right of the preprogram equilibrium price (P_1). The reduction in costs of production for gas that was previously uneconomic results in a change in the supply curve (the "modified supply curve"), and the net change in total consumer and producer surplus that results is equal to the area of the dark gray triangle. For the benefits calculations, this area is approximated by the area of the light gray triangle, with the change in production costs, ΔC_p, taken from the panel's assessment of the potential changes in costs attributable to DOE research.

foreign and hence their loss does not offset consumer gains. However, if a substantial share of gas consumed in the United States is imported (LNG), then the price elasticity assumption that underlies the DOE analysis is incorrect. According to the NEMS statistics, ERR (in the most optimistic case) adds 5 percent to the 2015 gas supply in the United States and results in a price decrease of at least 5 percent (a savings of $0.40 out of $7 to $8), implying a total price elasticity (supply plus demand) of, at most, 1. If LNG accounts for a substantial share of the market, then the price of gas is determined by world supply and demand. The contribution of the natural gas E&P program is then small relative to supply, and the market price will be much less sensitive to small changes in domestic supply. Thus, the savings to consumers, represented by a price change times the quantity of imported gas consumed, is, at best, trivial. As in the domestic case, the economic benefits of the program remain contingent on the value added by the additional gas produced by the program.

K

Report of the Panel on DOE's Distributed Energy Resources Program

INTRODUCTION

This report presents an estimate by the Panel on DOE's Distributed Energy Resources Program of the prospective benefits of the combined heat and power (CHP) component of DOE's distributed energy resources (DER) R&D program. The charge given to the panel by NRC's Committee on Prospective Benefits of DOE's Energy Efficiency and Fossil Energy R&D Programs, Phase Two, was to apply the methodology and approach of the Phase One committee to estimating benefits of the CHP program. The panel applied that methodology, taking into account technical and market (including regulatory issues) risks, both of which affect the program's ability to meet its goals. The methodology required the panel to make observations regarding program emphasis, benefits, and other factors that might assist it in its assessment. This report describes the panel's work and presents its observations about the program and suggestions for improvement as well.[1]

In summary, the CHP component of the DER program is designed to demonstrate four integrated CHP applications, each having greater than 70 percent combined electric and thermal efficiency, that could be manufactured and installed (assuming commercial-scale production) cheaply enough to result in a 4-year payback to customers by 2008. The CHP budget request for 2006 is $20.5 million. Key findings of the panel are (1) the 70 percent efficiency goal is achievable in the stated time frame in the end-use applications targeted by the program, but the payback goal is not, except in areas of the country where electricity is constrained or costs are high relative to natural gas; (2) the program produces a positive net present value (NPV) economic benefit under the four global scenarios assessed, ranging from $46 million to $83 million; and (3) the CHP program as a component of the larger DER program is too small to be modeled accurately using conventional models, and any further application of the committee's methodology should be to larger-impact programs. For many participating facilities, originally drawn to the CHP program for economic reasons, the protection against blackouts provides an important value-added service. For others, the economic benefits do not justify investment in CHP; instead, security is the deciding factor. In any case, security is recognized as an important benefit of CHP and one likely to become even more critical in the future. Understanding the value that customers assign to selected CHP attributes and the trade-offs they are willing to make and assigning a dollar value to those attributes would go a long way toward positioning the CHP program in the marketplace. Monetizing the value of attributes would add to the positive net economic benefit figures reported above in item 2.

DISTRIBUTED ENERGY RESOURCES PROGRAM: SUMMARY AND BUDGET

The DER R&D program is designed to improve the technology and encourage the adoption of distributed energy technologies. Many, but not all, of the technology applications in the DER program are configured as CHP or cogeneration projects in which the thermal energy recovered from the engine is used at a customer facility to provide hot water, space heating, or cooling. Of the $56.6 million DOE budget request for its DER program for 2006, $20.5 million, or 36 percent, is directed at the activities called "End-Use System Integration and Interface," which is the program area subject

---

[1] A committee member chaired the panel and two committee members served on the panel; however, the committee itself did not participate in the panel's work. The eight-member panel, represented academia, industry, and government and possessed the balance recommended by the committee responsible for the Phase One report. Panel members' biographies are provided in Attachment A. Throughout its work, the panel met twice for two days each, as recommended by the committee, and the chair had several interactions with DOE staff regarding panel information and data needs. The panel is grateful for the support provided by Patricia Hoffman, program manager of the CHP program at DOE, and her contractors for their support of the panel's work.

to this panel assessment.[2] The technologies and applications supported within this program area are predominately CHP projects. The DER program has the mission of improving the prime movers (microturbines, for instance) and their related technology development, while the CHP component of the DER program is focused on demonstrating system integration and application while reducing costs. DOE works with component equipment and technology manufacturers to develop and demonstrate CHP systems, that provide both electrical power and heating or cooling at customer sites with minimal site-specific engineering. The program strives to standardize the technical and engineering analysis required to install and operate CHP systems, making them applications-ready.

The stated goal of the CHP program is to demonstrate four integrated CHP applications, each having >70 percent combined electric and thermal efficiency, that could be manufactured and installed (assuming commercial-scale production) cheaply enough to give customers a 4-year payback by 2008.[3] The program is targeting four integrated systems applications—one each in an office building, a hospital, a college building, and a supermarket. On the one hand, the panel believes that the 70 percent efficiency goal is not difficult to achieve, as CHP systems are available today that meet or exceed the 70 percent efficiency goal. The attainment of the 70 percent efficiency goal, however, is highly dependent on the characteristics of the CHP system's application, including the electrical load and thermal load it is serving and the profiles of both. On the other hand, it believes that the 4-year payback criterion presents a significant challenge owing to the high capital costs of CHP systems and the need for sufficient thermal heating or cooling load. Moreover, the price differential between the CHP host facility's electricity prices and CHP fuel prices can dramatically affect the economics of CHP projects and, as a result, the payback. The panel applauds DOE's program focus on demonstrating CHP applications in the broader, more challenging markets targeted and, in particular, on demonstrating applications that the panel believes are unlikely to be developed without government support. If successful, the CHP program should result in more systems being designed and installed that meet the efficiency and payback goals, which would lead to more rapid deployment and greater customer confidence in system performance.

The panel heard presentations from DOE and its contractors and had the opportunity to ask questions and speak informally with program representatives at each of its meetings. DOE provided information and data about specific CHP system applications funded through the program, and it shared freely with the panel problems and difficulties with the technology and component integration as well as program successes. For this the panel is grateful. DOE and its contractors were also very responsive to the data requests of the panel.

BENEFITS ANALYSIS OF THE CHP PROGRAM

DOE Estimate

DOE estimates the benefits of the entire DER program using the U.S. Energy Information Administration's National Energy Modeling System (NEMS) and the MARKet ALlocation model (MARKAL). DOE estimated the contributions from its DER program through 2025, in terms of gigawatts (GW) generated, natural gas savings, oil savings, consumer energy savings, carbon emission reductions, and nonrenewable energy savings. The benefits estimates provided by DOE are generally consistent with its Government Performance Results Act (GPRA) reporting.[4] Estimates were provided for 2010, 2015, 2020, ending in 2025 (from NEMS) and for 2020, 2030, 2040, and 2050 (from MARKAL). By providing data for overlapping years, 2020 and 2025, the panel was able to consider the differences in estimates from the two models. The panel decided to base its probabilities and benefits estimate on the NEMS model, because panel members were more familiar with it and its horizon was shorter.

Since the NEMS analysis was conducted for the DER program as a whole, the panel requested that DOE and its contractors develop a simplified approach for parsing out the benefits that would likely be attributable to the CHP program component. DOE complied with this request by first scaling the total benefits of the DER program by the CHP proportion of the total DER budget. Following this, benefits were further adjusted to reflect CHP penetration in the commercial sector only and to account for the CHP program focus in office buildings, hospitals, college buildings, and supermarkets. Industrial and utility sector CHP applications are not included in this scaling. DOE estimated that the 2.2 GW of CHP capacity added through 2025 is attributed to the CHP program (of the total 64 GW of DER added nationally by 2025, as used in the GPRA analysis).[5] The panel accepted this as a reasonable first-order approximation for CHP penetration in the marketplace attributable to the DOE R&D program.[6]

DOE's estimates of the quantity of CHP added as a re-

---

[2]The remaining activities supported by the DER program and not considered by the panel in its review include development of industrial gas turbines, microturbines, advanced reciprocation engines, and related materials and sensors.

[3]For the payback goal to be met, CHP installations by 2008 will provide customers with a four-year simple return of the original investment (defined as revenue realized, in the form of savings, divided by total system first-cost).

[4]See Table 4.6 of DOE (2005b).

[5]Frances Wood, OnLocation, Inc.; Chip Friley, Brookhaven National Laboratory; and Chris Marnay and Kristina Hamachi LaCommare, Lawrence Berkeley National Laboratory, "Estimating benefits of EERE combined heat and power R&D," Presentation to the panel on October 24, 2005.

[6]The panel agreed with the simplified approach, viewing it as a reasonable approximation of the benefits likely to result from the CHP program component.

sult of its program are based on several key assumptions: principally, the expected costs and payback period for the CHP applications; the estimated maximum portion of the commercial buildings sector (new and retrofit) representing the total potential market; and market penetration of CHP as a function of system economics, defined by the payback period. In its benefits analysis, DOE evaluates a "program" case, where the technical goals of the program are assumed to be met and the total potential commercial CHP market is assumed to be 50 percent of new buildings and 5 percent of buildings being retrofitted. DOE compares this to a "without DOE" set of assumptions, where the technical and payback goals are reached 10 years later than with the program, and the total retrofit market is much smaller. DOE's estimate of the incremental capacity attributed to the DOE R&D program incorporates an assumption that without the DOE program, technical achievement would be delayed by 10 years.

Panel Estimate

The panel explicitly assessed the likelihood of meeting technical and market outcomes with and without the DOE program using the decision tree analysis prescribed by the committee, rather than assuming a 10-year delay without the program.[7] Because of this difference in assumptions, when the panel applied the scaling method suggested by DOE for estimating the benefits of the CHP program, it started with the incremental installed DER capacity in the "program case" over and above that of the NEMS reference case. The program case analysis versus the AEO Reference Case used by the panel begins with a DER contribution of 77 GW rather than the 64 GW used in the GPRA analysis. The estimated incremental contribution from the CHP program assuming DOE's goals are met then becomes at 2.65 GW by 2025.[8]

Modeling Observations

The panel found neither NEMS nor MARKAL to be well suited for determining the benefits of such a small niche set of technologies and applications.[9] Since the models are national in scope and CHP technology is likely to be valued differently in different regions of the country, the panel does not believe that the models are able to reflect market penetration accurately. At the panel's request, DOE developed a simplified approach to parse the likely benefits of the CHP program from the larger DER portfolio that was modeled in NEMS, which the panel appreciated and readily accepted. Since both models evaluate market penetration on a national level, they provide an overly optimistic projection of the reduction in environmental emissions that results from displacing electricity generated in coal-burning power plants. The models also have no reasonable mechanism for calculating the intrinsic benefits of this technology, such as security and electric system reliability. Use of these technologies will provide greater benefits to the end-user based on electricity reliability in the face of possible outages. Many commercial and institutional customers consider security and insurance-against-business-disruption factors as being important considerations when contemplating the use of DER—a factor that NEMS is not able to capture. The NEMS model's shortcomings were fully considered when the panel developed its probabilities of CHP market adoption.

Electricity System Constraints Scenario

The methodology prescribed by the committee requires the panel to evaluate the program under three different global scenarios: an *Annual Energy Outlook* (AEO) Reference Case scenario, a High Oil and Gas Prices scenario, and a Carbon Constrained scenario. As allowed by the methodology, the panel elected to define a fourth global scenario, where there are assumed to be severe constraints on the electricity system. The Electricity Constrained scenario captures benefits of the CHP program where its impacts might be of greatest value—that is, in regions of the country where electricity service is most constrained. The scenario assumes that the electric distribution system is unable to meet the peak electricity demand of its customers.[10] As defined by the panel, the scenario assumes that sufficient power is not available for some extended time (weeks or months) in one or more high-demand-load pockets in the affected regions of the country. The critical power shortage can be mitigated by taking several actions, including reducing voltage; imposing

---

[7]The panel had some difficulty initially differentiating between technical and market success in the DOE CHP program goal—meeting the 70 percent efficiency and 4-year payback goals. The two goals, although very different, represent a single program goal. Improving efficiency is in its own right a legitimate goal; the ability to meet this goal also has a bearing on the program's ability to meet the payback goal. The panel viewed the technical success of the program as achieving both the efficiency and payback goals. It viewed market success as meeting the CHP program's projected market penetrations, as defined and used by DOE in its NEMS modeling. The panel assigned its probabilities of technical and market success with this understanding in mind.

[8]The CHP program benefit reported by DOE to the panel is greater than the program benefit it reported in its GPRA analysis. There could be a number of reasons for this discrepancy, but the panel did not fully investigate the reasons for it, because the panel believed it to be beyond the scope of its charge.

[9]The time frames employed by both models are long and of limited use in helping DOE to determine success in meeting near-term program goals. The impacts of the CHP technologies themselves, as modeled on such a large scale, are essentially small enough to be marginal at best. Since the program goals focus on near-term acceleration of these technologies to commercialization, it is not clear how longer-term models, particularly MARKAL, can be useful for a program whose goal is to be met in 2008.

[10]Other reasons could be cited for severe reductions in power availability, such as an ongoing terrorist attack on several critical links in the grid, an accidental failure of one or more large generators in a high demand load pocket, a reduction in fuel availability, or lack of water for hydroelectric generation due to an extended drought.

selected rolling blackouts; terminating supply to selected high-demand customers with interruptible electricity service contracts; increasing real-time electricity prices; improving energy efficiency in the service area; and installing CHP systems in load-constrained areas.

At the panel's request, DOE worked with Lawrence Berkeley National Laboratory staff to estimate the gigawatt contribution of CHP nationally by 2025 under the Electricity Constrained scenario. While the market penetration of the CHP technologies improves under this scenario, it still does fully capture the reliability and security value of CHP technologies.

The Panel's Decision Tree Analysis

To estimate the expected benefits of the CHP program, the panel developed a decision tree representing various decision paths and technical and market outcomes, and assigned probabilities to each uncertainty. The decision tree developed by the panel is illustrated in Figure K-1. The first node represents a decision about DOE funding: the likelihood of technical success and market success depends on whether the DOE program is funded, and the benefit of the program will ultimately be estimated as the difference between the yes and no branches.

Since the payback period depends on local electricity prices relative to natural gas prices (assuming that CHP systems under study burn natural gas), two types of local market conditions were defined: locally high electricity prices and locally low electricity prices, as represented by the second node in the decision tree. The panel then estimated the likelihood of achieving specific payback periods and market penetration rates separately for the different local market conditions.

The third node represents three possible outcomes for technical success, defined by the panel as the payback period for a system with at least 70 percent efficiency (less than 4 years, 5 to 7 years, or 8 years or more). The final node represents three possible market success outcomes, defined by the portion of the market adopting CHP technologies assuming a 4-year or shorter payback period. High market adoption was defined by the panel as implementing CHP systems in 50 percent of new buildings and 5 percent of existing buildings; moderate market adoption was defined as 25 percent of new and 2.5 percent of existing; and low was defined as 10 percent of new and 1 percent of existing buildings. Market penetration is a function of the payback period to customers having installed the CHP technology. DOE gave the panel its assumed market penetration curve (function) for use in estimating probabilities of technical and market success.

Assessment of Technical and Market Success

Panel members estimated the probability of achieving each specified level of technical success as defined by the payback, both with and without DOE funding. This was done for the two local market conditions and each of the four global benefits scenarios described previously.

The panel discussed each member's probability of technical success and the reasons for the assignment of the probabilities, and developed consensus estimates. The panel then developed estimates of the probability of achieving each of the three specified levels of market success both with and without DOE funding, under each of the two local market conditions and for each of the four global scenarios.

The consensus estimates are illustrated in Figures K-2 though K-6, which show the cumulative probability for selected payback periods for a CHP system that is 70 percent or more efficient in a locally high-electricity-price market with DOE funding. These figures allow comparison of the estimated impact of the four global scenarios on technical success. The panel estimated the highest likelihood of technical success in the Electricity Constrained scenario and the lowest in the High Oil and Gas Prices scenario (because higher natural gas prices will make it more difficult for CHP technologies to achieve a shorter payback).

Figures K-4, K-5, and K-6 show the cumulative probability for various payback periods under each of the four global scenarios and illustrate the effect of DOE funding and of local market conditions. The panel estimated a higher likelihood of achieving a shorter payback with the DOE program than without it and higher likelihoods in the locally high electricity price market than in the locally low-electricity-price market.

Finally, the panel also estimated the likelihood of achieving different levels of market adoption for CHP technologies. The consensus probabilities are shown in Figure K-7. The panel estimated a higher likelihood for strong market adoption with DOE funding than without DOE funding. It estimated the highest likelihood of market adoption in the Electricity Constrained scenario.

Prospective Benefits Results

The panel estimated CHP program benefits for each of the four global scenarios by using its probabilities for technical and market success and DOE's estimate of incremental CHP capacity installed under conditions of high technical and market success to develop an estimate of incremental CHP capacity attributable to the DOE program. That value was then multiplied by the dollar value per gigawatt assessed by the panel to yield the expected economic benefits of the program.

The estimated increase in CHP installed, given the panel's assessment of the technical and market risks, is illustrated in Figure K-8. The figure shows the expected value of the CHP that DOE estimates will be added due to its program (the top solid line) and the panel's estimate of the CHP that will be added if both technical and market success are low (the bottom solid line). For each global scenario, and for each of

APPENDIX K

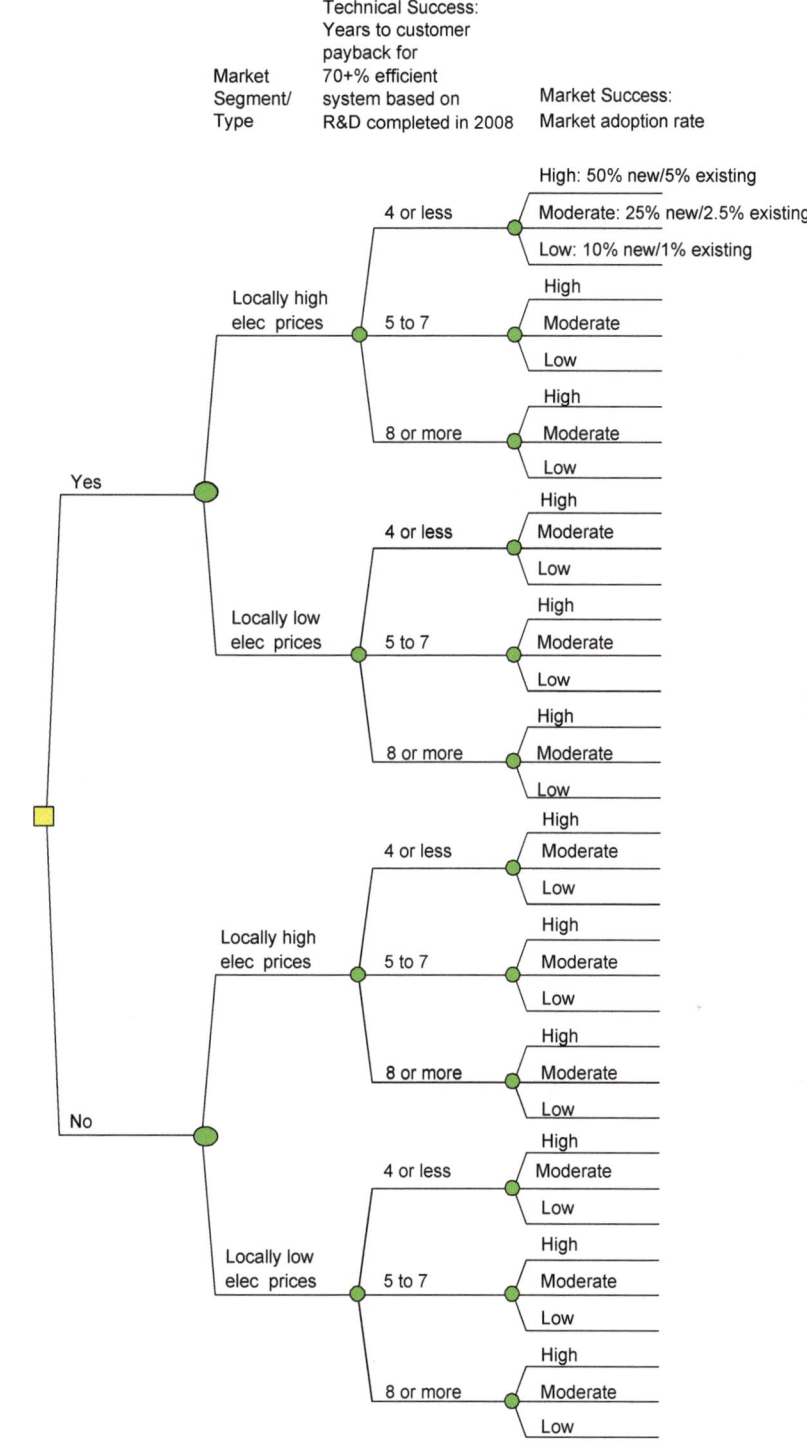

FIGURE K-1 Decision tree for combined heat and power program.

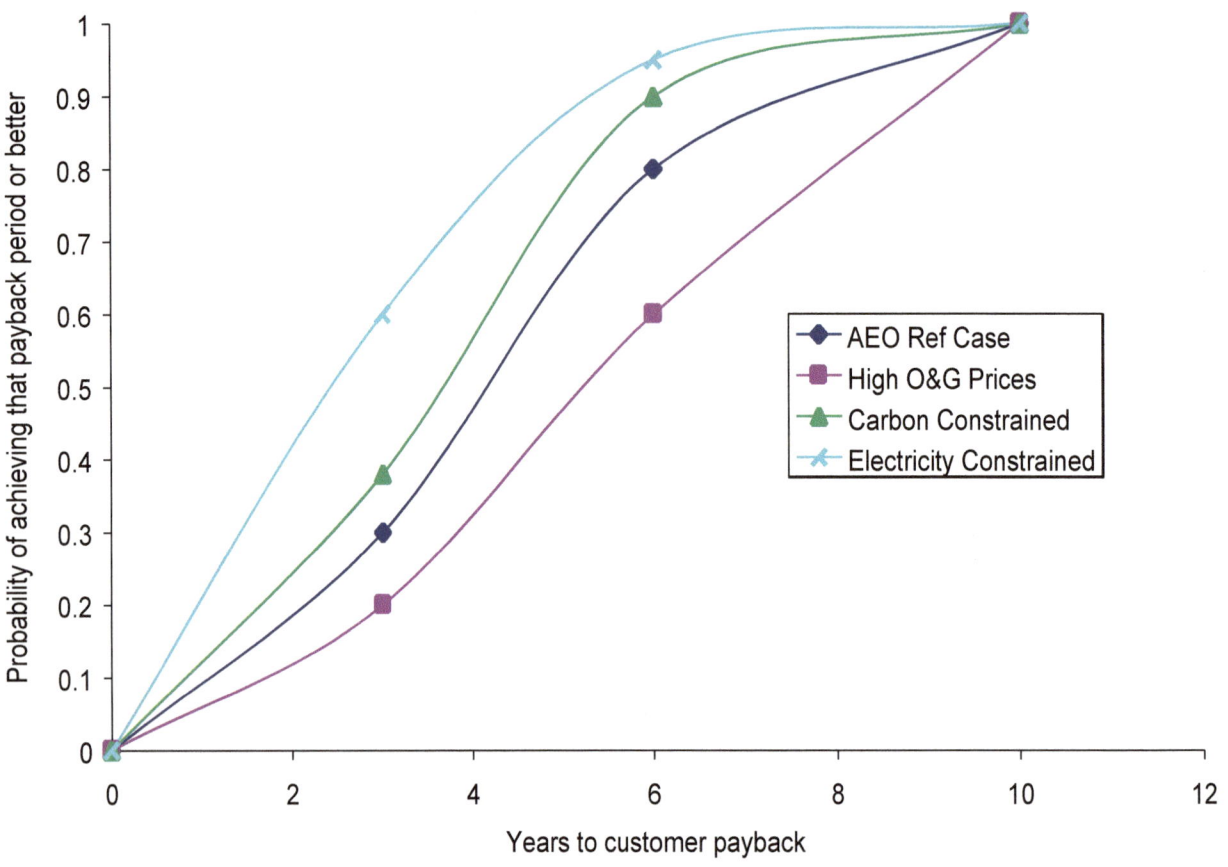

FIGURE K-2 Assessment of technical success with DOE funding and high electricity prices.

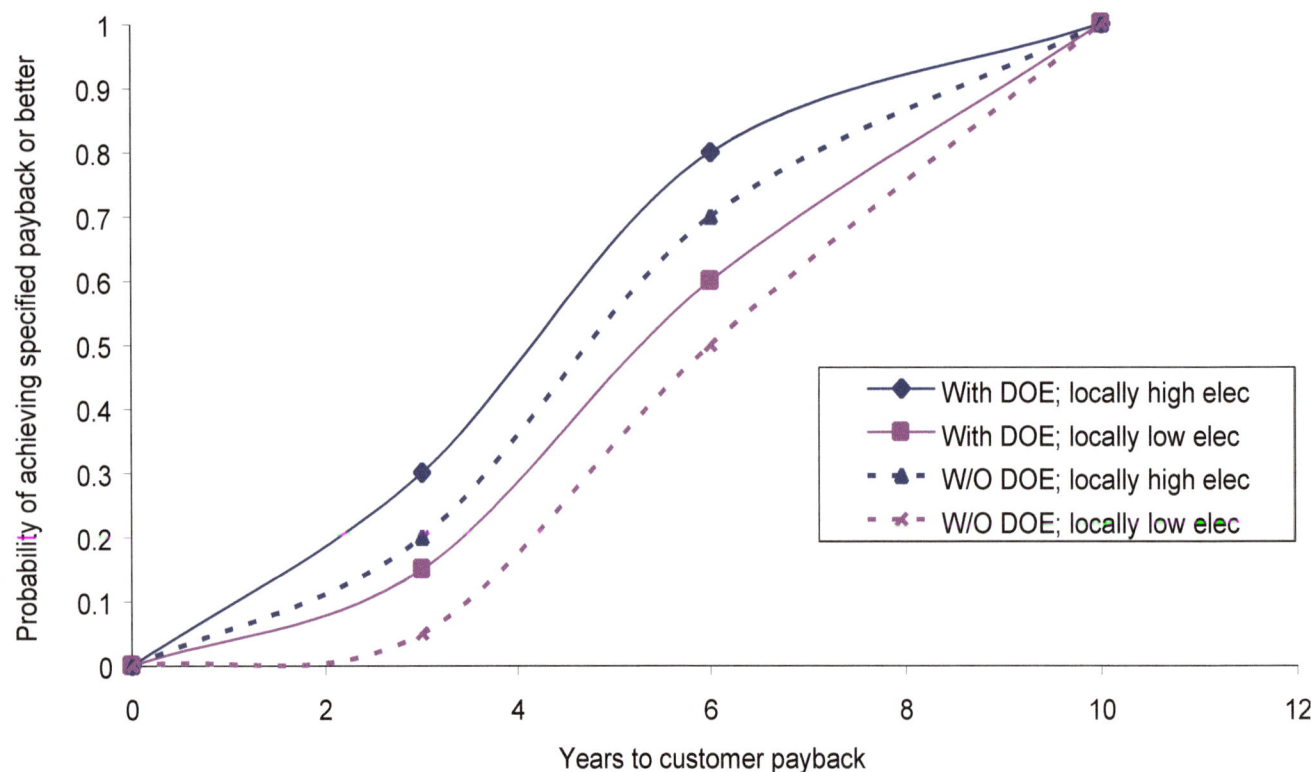

FIGURE K-3 Assessment of technical success for the AEO Reference Case scenario.

APPENDIX K

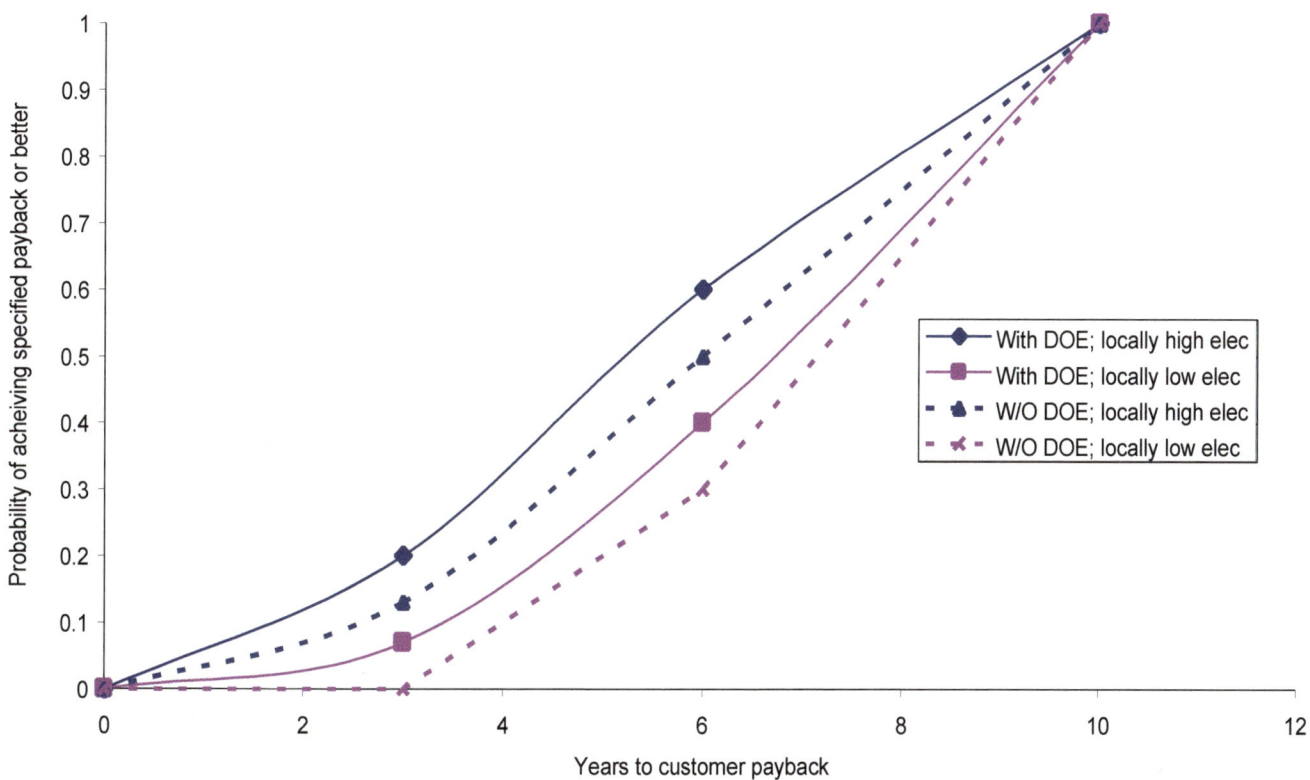

FIGURE K-4 Assessment of technical success for the High Oil and Gas Prices scenario.

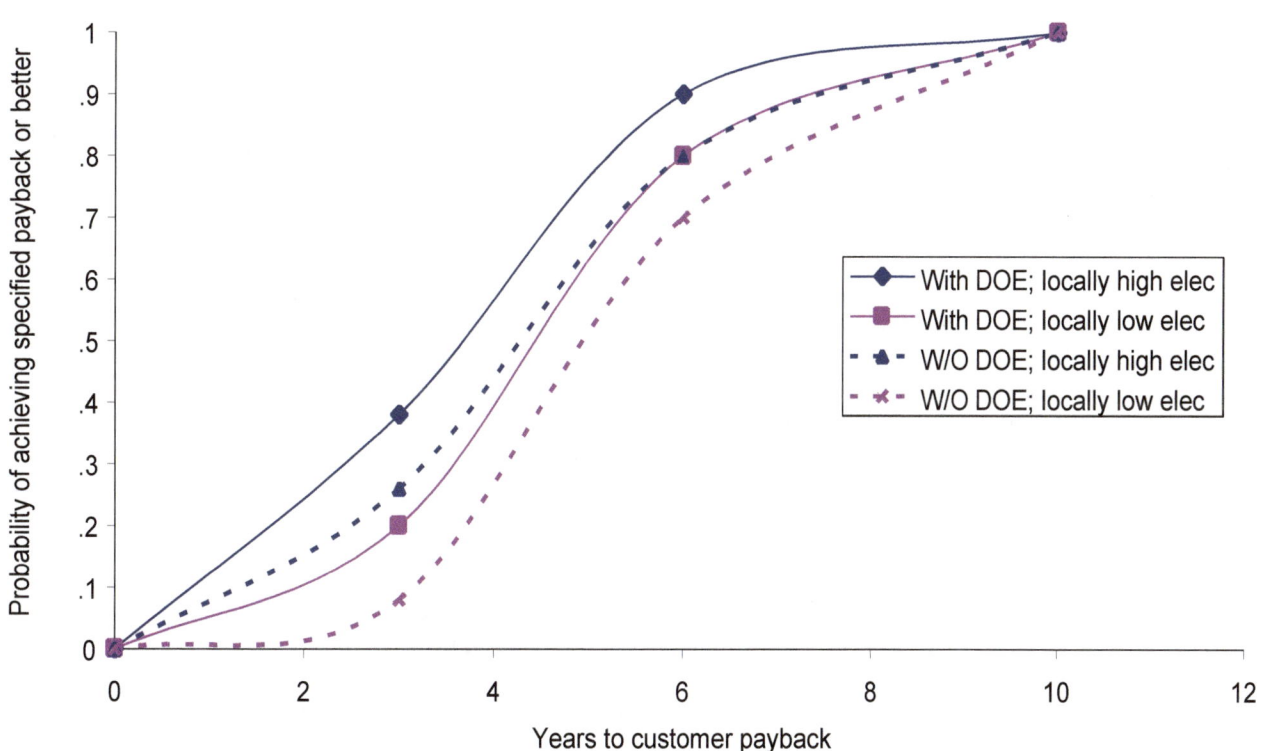

FIGURE K-5 Assessment of technical success for the Carbon Constrained scenario.

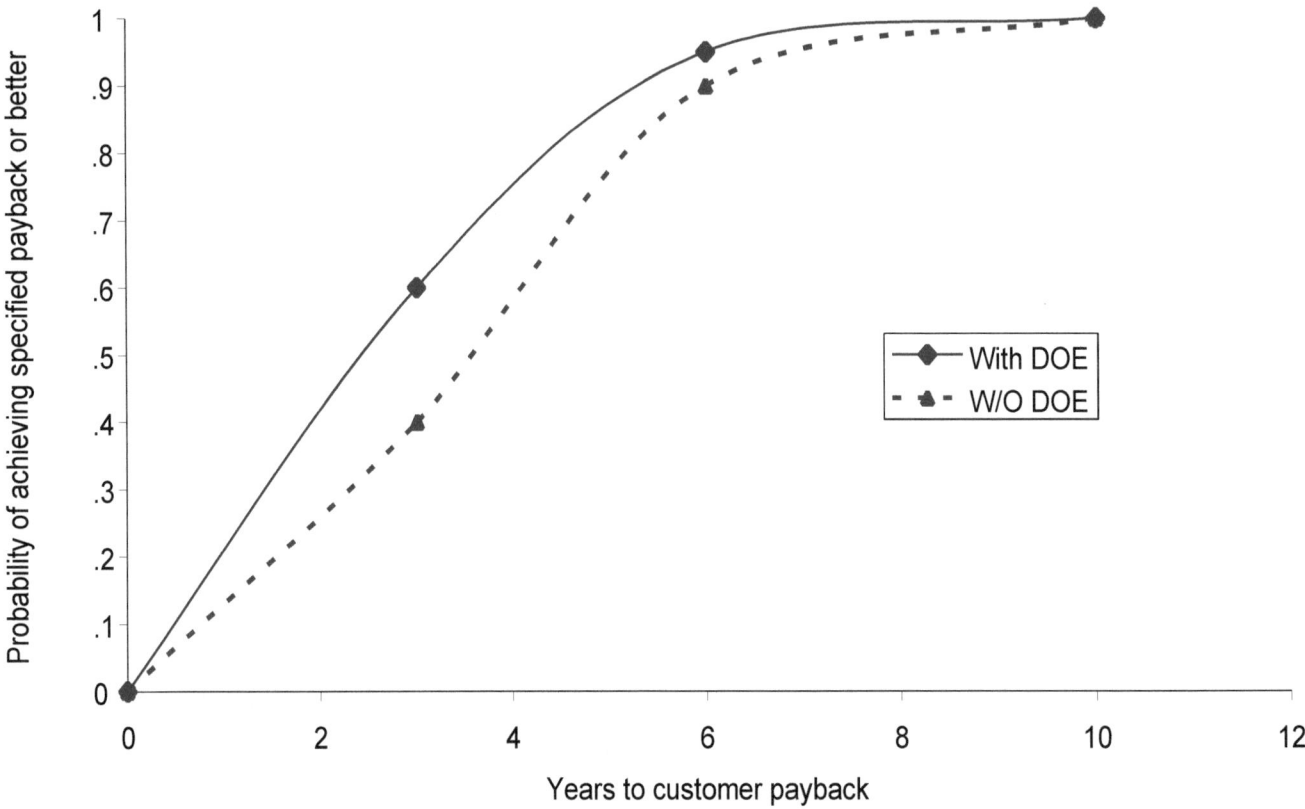

FIGURE K-6 Assessment of technical success for the Electricity Constrained scenario. The panel assessments of the likelihood of technical success in the locally low and locally high electricity prices market conditions were identical for this scenario, so only one set of curves is shown.

the two local market conditions, the expected value of the CHP addition is calculated as the probability-weighted average of all possible technical and market success outcomes specified in the decision tree (and shown with the diamond and cross markers). Separate estimates are shown for CHP additions with and without the DOE program. Figure K-9 shows the uncertainty around the expected Electricity Constrained additions for each scenario and each local market condition. This figure reproduces Figure K-8 but includes uncertainty bars representing the 10th to 90th percentiles of the estimated incremental CHP. This uncertainty is derived directly from the estimated incremental CHP associated with each of the technical and market success end points from the decision tree and the panel's estimates of the probability of each of those technical and market outcomes.

Economic Benefits. The panel used secondary research to assign a dollar value to CHP savings (it looked at CHP economic analyses studies conducted elsewhere in the country). Generally, in today's economic climate, CHP technologies are barely breaking even on a net economic resource basis—the present value of life-cycle benefits are roughly equal to the present value of life-cycle costs. The panel assumed that this situation is likely to improve in the near term as CHP technology improves and niche applications are being found that would provide CHP with an economic advantage over the next-best alternative, the purchase of electricity and thermal energy. The panel determined that CHP could easily provide a 10 percent net benefit above its costs over the study period, leading to a benefit-cost ratio of 1.1. While the net benefit could be higher or lower than 10 percent, the panel assumed the 10 percent was a reasonable approximation, given that the technology is expected to improve and the economics of CHP, given current and expected future energy costs, are also expected to improve. This assumption translates into a net economic benefit of approximately $230 per kilowatt of installed CHP. This is calculated on an electricity system avoided cost basis and takes into account for the initial CHP investment and life-cycle operating costs.

To calculate the NPV of the economic benefits, the panel considered the incremental capacity attributable to the program in each year from 2006 to 2025, estimated the economic value by year, and then calculated the NPV of that benefits stream using a 3 percent discount rate, as recommended by

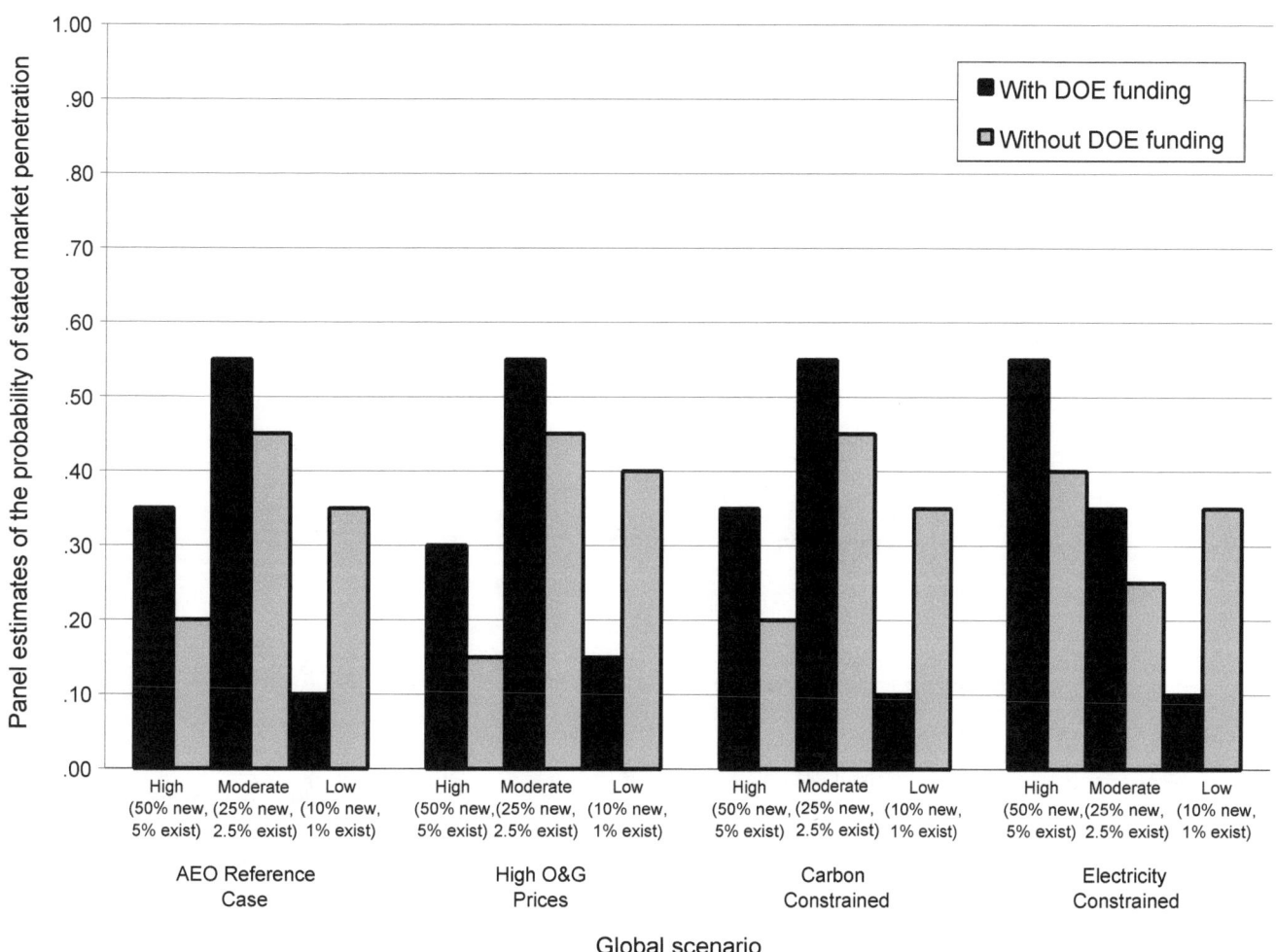

FIGURE K-7 Assessment of market success. Solid bars represent results for the "with DOE program" case. Shaded bars represent results for the "without DOE program" case. Numbers in parentheses are market penetration in new construction/existing buildings.

the full committee. The panel further assumed that installed CHP would deliver economic benefits for a minimum of 10 years.

The panel's estimate of expected economic benefit from DOE's CHP program is shown in Figure K-10 for each of the four global scenarios. Net economic benefits range from a low of $46 million in the High Oil and Gas Prices scenario to a high of $83 million in the Electricity Constrained scenario. This net benefit is assumed to be made available for a DOE CHP program investment of approximately $20 million per year. The panel estimates that the DOE CHP program provides a net economic benefit to the country under each of the four global scenarios.

Environmental Benefits. Clean and efficient CHP has the potential to reduce overall emissions since it uses clean-burning natural gas as a fuel and it displaces central station generation power and local combustion boilers and furnaces used to generate steam, heating, or hot water. If the thermal use of the CHP system is high, and especially if the steam, heat, or hot water was previously produced using a less clean burning fuel, a switch to CHP clearly confers air emissions benefits both locally and regionally. If the displaced heat was previously produced by burning natural gas or if the thermal output of the CHP system is used to displace electric space cooling, there might still be regional air emission benefits from CHP, since the overall high fuel utilization efficiency reduces emissions of CO_2 and criteria pollutants; the local effects are less clear. Even a very clean CHP unit still produces some NO_x and other gases close to loads. Whether these small emissions are significant compared to background emissions from cars, trucks, buses, boilers, furnaces, and food cooking is debatable. The benefit of using less central station power also varies depending on whether the source is a dirty old coal plant, a modern combined cycle facility, or a nuclear or hydro plant.

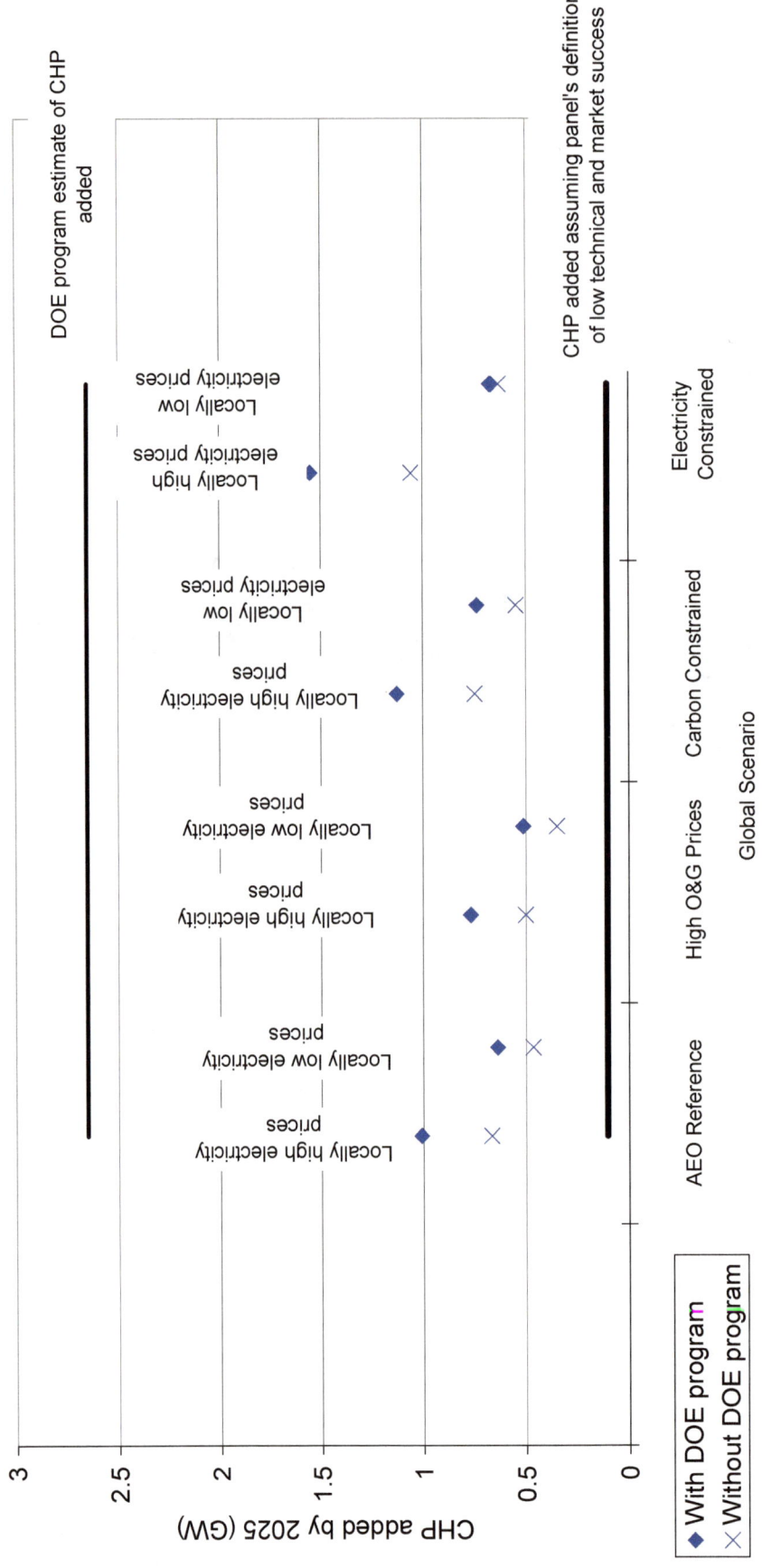

FIGURE K-8 Panel's estimate of the amount of CHP added with and without the DOE program for each scenario.

As a practical matter, local air permits, especially in non-attainment areas, act as barriers to CHP. Important issues related to CHP penetration include the size of systems and hence their need to comply with local laws and regulations, the availability of fuels and environmental permitting for the use of such fuels, and proximity to population. Unfortunately, the parts of the country in which electric and gas rates favor CHP also tend to be the same ones with high population densities and air pollution concerns. The panel believes that at modest levels of penetration, the cleaner burning CHP systems considered in this DOE program will not materially degrade anyone's environment nor will they help address major pollution concerns. At higher penetrations, CHP will help reduce global CO_2 emissions and SO_2, NO_x, particulates, and mercury to the degree that it displaces fuels like coal and oil, either at central power plants or locally, to produce steam, heat, or hot water. Even if a clean fuel such as natural gas is displaced by a CHP installation, either centrally or locally, there will still tend to be environmental benefits. The environmental benefits accrue from the decreased energy use attributable to CHP systems that are more efficient than the baseline of central power and local boilers or furnaces. High thermal utilization and fewer emissions tend to improve the overall environmental impact of CHP. DOE programs to further reduce the already low emissions from CHP and to enable even greater heat utilization with cost-effective technologies will enhance CHP's already overall favorable environmental performance.

Security Benefits. Security benefits arise from the reduction in primary energy use associated with the implementation of CHP technologies. The term "security" as defined by the panel for the CHP program has several meanings, including (1) invulnerability to terrorist attacks and natural disasters, (2) insensitivity to energy disruptions caused by reductions in oil imports, and (3) customer protection from the effects of disruptions to utility electric service. Like all energy efficiency measures, CHP reduces primary energy use and helps improve energy security of the type described in item (2) above. However, the unique benefit of CHP is to decentralize power production and locate it at or close to loads, providing benefits of the types described in (1) and (3) above. Considerable concern has been expressed in recent years about the need to reduce the threat and potential costs of acts of terrorism through greater use of distributed energy resources. However, CHP has the ability to reduce the risk and costs of many other adverse events, including utility outages from storm events, cars hitting power poles, squirrels chewing through wires, to more unusual but devastating outages such as the daylong loss of power to 50 million people in the Northeast on August 14, 2003. It must be noted, however, that reliance on natural-gas-fired distributed generation is also subject to terrorist attacks on pipelines.

When properly configured, CHP protects many different customers and loads from utility service disruptions. The least expensive type of clean, grid-connected CHP operating in parallel with the utility grid uses induction generators, which immediately stop generating when the utility power that supplies their excitation is removed. This automatic safety feature makes many utilities favor induction equipment. Unfortunately, the induction systems do not directly help customers during supply disruptions. To get the benefit

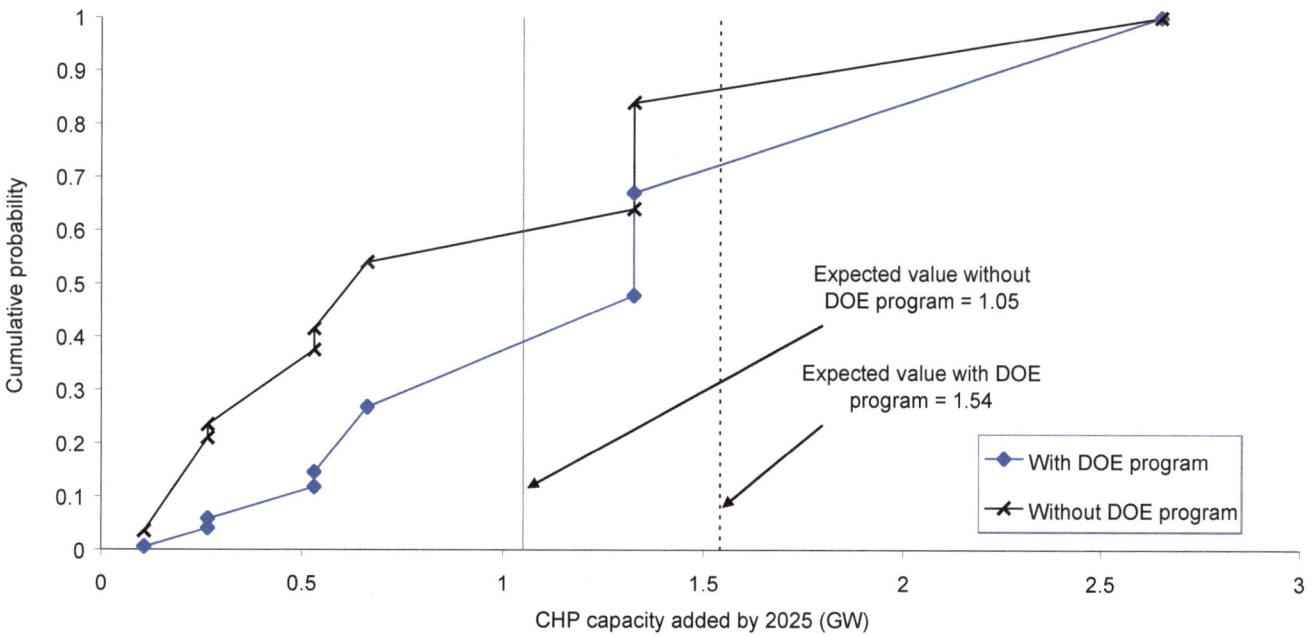

FIGURE K-9 Uncertainty around estimated CHP additions for the Electricity Constrained scenario.

		Global Scenario			
		AEO Reference Case	High Oil and Gas Prices	Carbon Constrained	Electricity Constrained[a]
Program Risks	Technical Risks	Technical risks identified as uncertainty in the ability to reach the 4-year-or-less payback target identified by DOE (for a 70+% efficient system). Panel specified two alternative payback periods (5-7 years, 8 years or more) as other possible technical outcomes. Estimated likelihood of achieving a 4-year payback ranged from a low of 20% (without the DOE program under the Reference Case scenario) to a high of 60% (with the DOE program in the Electricity Constrained scenario).			
Program Risks	Market Risks	Market risk characterized by the estimated fraction of the commercial buildings sector that would adopt CHP technologies, given a 4-year or less payback. High market penetration was defined as 50% of new construction and 5% of existing. Moderate market penetration was defined as 25% of new and 5% of existing, and low market penetration was defined as 10% of new and 1% of existing. Estimated likelihood of high market penetration ranged from a low of 15% (without the DOE program in the High Oil and Gas Prices scenario) to a high of 55% (with the DOE program in the Electricity Constrained scenario).			
Program Benefits	Expected Economic Benefits (see note)	Expected net economic benefits are presented as the present value of the annual economic benefit of incremental CHP capacity installed between 2006 and 2025 attributable directly to the DOE program.			
Program Benefits	Expected Economic Benefits (see note)	$57 million at 3%	$46 million at 3%	$64 million at 3%	$83 million at 3%
Program Benefits	Expected Economic Benefits (see note)	$40 million at 7%	$32 million at 7%	$45 million at 7%	$58 million at 7%
Program Benefits	Environmental Benefits	The net environmental impact of CHP is not known. While CHP may result in lower emissions at central generation facilities, the CHP systems themselves are natural gas powered and result in local environmental emissions. At the relatively modest levels of deployment anticipated, the environmental benefits associated with the program are negligible.			
Program Benefits	Security Benefits	Security benefits arise from the displacement of primary energy use by the incremental installed CHP capacity. This benefit is due to dispersed generation being less vulnerable to terrorist and natural disaster events. Reduction in total primary energy use from all incremental CHP capacity installed between 2006 and 2025 attributable directly to the DOE program is			
Program Benefits	Security Benefits	10 trillion Btu	8 trillion Btu	11 trillion Btu	15 trillion Btu

NOTE: Benefits are presented as the difference in the expected value with and without the DOE program. See Figure K-9 for a discussion of the uncertainty surrounding benefits. Economic benefits are shown as present values discounted at both 3% and 7% real; environmental and security benefits are discounted at 3%.

[a]Fourth scenario defined by the panel to represent a future state of the world where there are severe constraints on the electricity system. The scenario assumes that the capacity of the electric distribution system is seriously deficient with respect to meeting the electricity and peak demand needs of its customers in one or more high-demand-load pockets.

FIGURE K-10 Results matrix of the Panel on DOE's Distributed Energy Resources Program.

of grid parallel operation and stand-alone capability, a CHP system needs to use synchronous generators and more expensive automatic switchgear to provide emergency load isolation and utility system protection. The problem becomes more difficult and more expensive to solve if the generator is operating on a network system, such as in New York City and many other large urban areas, rather than on the more common radial distribution systems.

Many critical facilities, including hospitals, nursing homes, computer warehouse facilities, and some tall buildings dependent on elevators, are required to have backup generators to carry all or part of their load during a blackout. These are almost invariably cheap, dirty, diesel-powered generators, permitted to run a very limited number of hours in a year. Emergency backup diesels need to be regularly started and actively maintained if they are to be relied on during an emergency. Such generators also often have only very limited fuel storage on-site, making operation during an occasional longer outage problematic. For these reasons as well as in the interest of lower emissions, a clean, natural-gas-based CHP system might offer a better solution than an emergency diesel generator. Natural-gas-powered CHP would provide energy-efficiency-based savings during normal times and superior reliability during an emergency. In addition, there is usually no fuel limitation during prolonged blackouts.

For many facilities originally drawn to CHP for economic reasons, the added protection against blackouts is an important value-added service. For others for which the economics are not attractive enough to compel investment in CHP, security is the deciding factor and economic benefit is a by-product. In any case, security is recognized as an important benefit of CHP and one likely to become even more critical in the future. The DOE CHP program should fully recognize this fact and direct efforts at reducing the cost and improving the performance of interconnection technology to allow CHP systems to safely operate both grid-parallel and grid-isolated as well as on network systems. The current DOE systems integration program has not emphasized security enough, although all CHP R&D will tend to provide some security benefit. The panel feels security should be a much more important program goal.

PROGRAM OBSERVATIONS

In addition to applying the committee's methodology to DOE's CHP program for estimating prospective benefits, the panel also discussed at some length the focus of the program and the projects being funded. As a result of these discussions, the panel suggests that DOE consider addressing the following items as it selects CHP projects for funding that could improve both program payback and ultimate program success:

1. Electrical energy is worth approximately three times as much as thermal energy. Overall system efficiency is less important than knowing the relative amounts of electrical and thermal energy. A system with 38 percent electrical efficiency and 32 percent thermal efficiency creates a much higher value and shorter payback than a system with 26 percent electrical efficiency and 44 percent thermal efficiency even though both have an overall efficiency of 70 percent.

2. In calculating payback, it is important to know the electrical load factor. Clearly a system operating 24 hours per day will have a much shorter payback than a system operating 12 hours per day.

3. In calculating payback, it is important to know the thermal load factor and how this profile coincides with the electrical load profile. When the thermal load drops to zero, as can happen in comfort conditioning in mild weather, the fuel cost of electricity effectively doubles.

4. To optimize payback and load profiling (items 2 and 3 above), it might be advantageous to power only part of a facility's electrical load and leave the rest to be powered by the local utility company. With fewer or smaller generator sets online, less heat would be produced, thus more closely matching the minimal thermal load profile. Correspondingly, steady electrical loads with flat load factors maximize total kilowatt-hours produced and minimize payback. Thus, powering selective steady loads such as lighting will produce the greatest return on investment.

5. Capital cost can be reduced by eliminating redundant CHP units. If lighting is the load to be served by CHP, loss of one generator set may reduce the lighting level but the facility will continue to operate. The need for redundancy is reduced.

6. The heat rejected from most prime movers is at a lower temperature than the heat from a fired boiler or water heater. As the final exhaust temperature is typically 300°F to prevent condensation and acid corrosion, the percentage of heat that can be recovered is less than with a fired boiler or water heater. However, if the exhaust heat is used in a drying, baking, or preheating operation, the 300°F limitation no longer applies and the thermal efficiency is much higher. In addition, when the exhaust heat exchanger is eliminated, the capital cost is lower.

DOE should consider working more closely with utilities to encourage utility ownership of CHP systems. Paralleling CHP with the utility grid eliminates the need for redundant units, improves system efficiency by allowing the equipment to follow thermal load, and improves facility security by providing an independent source of electricity. If the utility owns and/or operates the CHP systems, it can dispatch them as needed to maximize overall efficiency and minimize pollutant emissions. It can also use them as peaking units when necessary. Utility ownership of distributed generation would be consistent with the obligation of utilities to serve the public. DOE should support replicable pilot applications where there are offsetting capital costs for, say, standby generators in hospitals or standby generators and uninterruptible power systems in data centers. DOE should address the problem of institutional barriers and work with regulatory agencies and utilities to eliminate or reduce them. The panel further observed that

- Funding for CHP from entities other than DOE was fairly easily identified, with many states investing more in CHP than DOE.
- NEMS and MARKAL are reasonable models for examining the integration of larger technologies and energy systems into the U.S. energy markets. However, to properly evaluate the potential of niche CHP technologies to penetrate the marketplace, more specific models must be used. These systems might do well to address regional issues associated with technology use and factors related to electrical system security and reliability.
- DOE staff and contractors made an effort to tease out the benefits of the CHP program that the panel was interested in (simple approach to estimating benefits), which made the panel's work more manageable.

The panel reached the following conclusions on technical risks:

- The best results of the DOE program will be realized where CHP exhaust heat is used for drying and baking food products, preheating combustion air for large conventional boilers, or inputting to spray dryers. The fuel cost per kilowatt-hour of CHP systems is lowest in these applications, and electrical efficiency is less critical. Equipment costs will be less also, as there is no downstream heat transfer.[11]
- The economics of a conventional CHP system with an exhaust-heat water heater is more difficult. The cost is significantly higher, from the addition of the downstream heat transfer to the need for more installation engineering. Also, either a utility interface or a redundant unit is generally required.[12]

OBSERVATIONS ON THE PROSPECTIVE BENEFITS METHODOLOGY

The panel found the prospective benefits methodology and matrix framework to be workable and generally easy to understand. It found the application of the methodology to be valuable in helping to quantify the expected benefits of the end-use system integration and interface aspects of DOE's CHP program. Nonetheless, the panel struggled with several key issues that required extensive discussion and additional information from DOE. While the overall efficiency and payback goals of the program are clearly stated, the CHP program strategies do not clearly align with the goals, making it more difficult to apply the methodology to the program. For example, the 70 percent system efficiency and the 4-year financial payback goals by 2008 are defined for different applications in different market segments. Goals might be met in one market segment sooner than in another. Also, without a clear cost goal (equivalent to a CHP system cost per unit of output), the payback goal is not well enough defined. The DOE articulation of goals made it difficult for the panel to consider assigning probabilities to technical achievement and market penetration. Also, because CHP penetration will vary depending on the relative costs of electricity and natural gas, as well as on electricity system constraints, probabilities were not easily assigned across the national market. The panel suggests that DOE consider setting a system cost goal (cost per kilowatt or per million Btu) rather than payback. The panel also suggests that the NRC consider applying the prospective benefits methodology to larger DOE R&D programs, not to subprogram components.

The panel decided to expand the benefits matrix to include a fourth scenario, Constrained Electricity, which it believes would more accurately capture the full benefits of DER technology generally and CHP in particular. Since the benefits of CHP are more regional than national, perhaps CHP program goals could be specified regionally, with applications focused on particular market segments in different parts of the country. This regional specification of the CHP goal would have greatly simplified the probability estimates of the expert panel members. In addition, the committee might consider selecting for review using its methodology DOE R&D programs that are national in scope and whose success is not highly dependent on where they are.

ATTACHMENT A
PANEL MEMBERS' BIOGRAPHIES

Paul A. DeCotis, *Chair*, is director of energy analysis at the New York State Energy Research and Development Authority (NYSERDA), where he oversees statewide energy planning and policy analysis, corporate strategic planning, program evaluation, and energy emergency planning and response. Prior to joining NYSERDA, Mr. DeCotis was chief of policy analysis at the New York State Energy Office. He is the record access officer to the State Energy Planning Board and chair of the Interagency Energy Coordinating Working Group, made up of staffs of the New York state departments of public service, environmental conservation, transportation, and economic development, which is charged with preparing New York's energy plan. He is also a member of the New York Independent System Operator (NYISO) Management Committee, the Business Issues Committee, and the Energy Working Group of the Coalition of Northeastern Governors (CONEG). Mr. DeCotis is president of Innovative Management Solutions, a management consulting practice, specializing in strategic planning and policy development, mediation, and organizational and executive management training and development. He is an adjunct professor in the MBA program at the Sage Graduate School and in the Public Policy Department at Rochester Institute of Technology, and was formerly at the School of Industrial and Labor Relations at Cornell University. He is currently on the board of directors of the Association of Energy Service Professionals, serving as executive vice president and U.S. Department of Energy experts review panel chair for the Weatherization Program evaluation. Mr. DeCotis was past

---

[11] Further, the recuperator, which is expensive, might be unnecessary, and the processes to which CHP is applied generally have high load factors. A major advantage is the ability to provide enough power to keep processes operating if the grid fails. With air bearings, there are no lubricants in the gas turbine to contaminate food products. A 25 percent efficient CHP system has an operating cost of less than six cents per kWh if it is assumed that CHP electricity generation costs 5.42¢/kWh for fuel and 0.5¢/kWh for maintenance. If CHP displaces 16 cent electricity, the saving is 10¢/kWh, or $800/kW per year based on 8,000 operating hours per year. This is highly scaleable.

[12] The way around this is to use CHP to serve selective loads such as lighting; connecting just enough lights to the generator set to exactly match the maximum continuous rating of the generator set. As a result, a CHP system failure simply reduces the amount of lighting available rather than having the building go dark. No paralleling, load sharing, load shedding, redundant unit, or utility interface is needed. Lighting is a very large market with high load factors and no step loads. Enough CHP can be installed to match the thermal load for maximum efficiency.

peer review panel chair of the U.S. DOE Federal Energy Management Program, and was also a member of the Committee on Prospective Benefits of DOE's Energy Efficiency and Fossil Energy R&D Programs. He has a B.S. in international business management from the State University of New York College at Brockport, an M.A. in economics from the State University at Albany, and an M.B.A. in finance and management studies from Russell Sage College.

James W. Dally (NAE) has had a distinguished career in industry, government, and academia and is the former dean of the College of Engineering at the University of Rhode Island. Dr. Dally is Glenn L. Martin Institute Professor of Engineering (emeritus) at the University of Maryland at College Park. His former positions include senior research engineer, Armour Research Foundation; assistant director of research, Illinois Institute of Technology Research Institute; and senior engineer, IBM. Currently, he is also an independent consultant. Dr. Dally is a mechanical engineer and the author or coauthor of six books, including engineering textbooks on experimental stress analysis, engineering design, instrumentation, and the packaging of electronic systems, and has published approximately 200 research papers. He has served on a number of NRC committees and is currently on the Panel on Air and Ground Vehicle Technology for the Army Research Laboratory Technical Assessment Board and on the Committee on Review of Federal Motor Carrier Safety Administration's Truck Crash Causation Study. He has a B.S. and an M.S. from the Carnegie Institute of Technology and a Ph.D. from the Illinois Institute of Technology.

Marija Ilić holds a joint appointment at Carnegie Mellon as professor of electrical and computer engineering and engineering and public policy, where she has been a tenured faculty member since October 2002. Her principal fields of interest include electric power systems modeling; design of monitoring, control, and pricing algorithms for electric power systems; normal and emergency control of electric power systems; control of large-scale dynamic systems; nonlinear network and systems theory; and modeling and control of economic and technical interactions in dynamical systems with applications to competitive energy systems. She is an IEEE fellow and an IEEE distinguished lecturer, as well as a recipient of the First Presidential Young Investigator Award for Power Systems. In addition to her academic work, Dr. Ilić is a consultant for the electric power industry and the founder of New Electricity Transmission Software Solution, Inc. (NETSS, Inc.). From September 1999 until March 2001, Dr. Ilić was a program director for control, networks and computational intelligence at the National Science Foundation. Prior to her arrival at Carnegie Mellon, Dr. Ilić held the positions of visiting associate professor and senior research scientist at the Massachusetts Institute of Technology. From 1986 to 1989, she was a tenured faculty member at the University of Illinois at Urbana-Champaign, where she taught since 1984. She has also taught at Cornell and Drexel. She has worked as a visiting researcher at General Electric and as a principal research engineer in Belgrade. Dr. Ilić has coauthored several books on large-scale electric power systems. Dr. Ilić received her M.Sc. and D.Sc. degrees in systems science and mathematics from Washington University in St. Louis and earned her M.E.E. and engineering diploma from the University of Belgrade.

Lester B. Lave (IOM) is the Harry B. and James H. Higgins Professor of Economics and University Professor; director, Carnegie Mellon Green Design Initiative; and codirector, Carnegie Mellon Electricity Industry Center. His teaching and research interests include applied economics, political economy, quantitative risk assessment, safety standards, modeling the effects of global climate change, public policy on greenhouse gas emissions, and understanding the issues surrounding the electric transmission and distribution system. He is a member of the Institute of Medicine and a recipient of the Distinguished Achievement Award of the Society for Risk Analysis. He has a B.S. in economics from Reed College and a Ph.D. in economics from Harvard University.

Robin Mackay was the founder of Capstone Turbine Corporation in 1988 and served as vice president of marketing until he retired in 1996. He also served on the board of directors from the company's inception until 2000, when it went public. Prior to that Mr. Mackay spent 24 years with the Garrett Corporation (later AlliedSignal Aerospace, now Honeywell), where he was director of industrial market development. He was responsible for the sale of several hundred gas turbine generator sets into cogeneration applications, as well as booking research contracts for advanced concepts such as microturbines, gas-turbine-driven air conditioners, closed-cycle gas turbines, subatmospheric gas turbines, and gas turbines mounted on air bearings. Prior to that he spent 6 years with Boeing, where he sold, installed, and brought on line the first all-turbine cogeneration system using two 140-kW Boeing gas turbines in 1962. Mr. Mackay graduated from McGill University with a degree in mathematics and economics. He holds eight patents and has two pending. He has authored numerous papers for the SAE, ASME, AEE, and other organizations. The most recent is a paper entitled "High Efficiency Vehicular Gas Turbines," presented at SAE's Future Transportation Technology Conference in September. Mr. Mackay has a small company, Agile Turbine Technology, LLC, in Manhattan Beach, California. Agile cooperates with other companies who wish to use Mr. Mackay's patents.

Ali Nourai received his doctorate in engineering from the Rensselaer Polytechnic Institute in 1978. He is a strategic technology consultant in American Electric Power and is responsible for distributed generation and energy storage programs. During his 26 years of activities in the utility industry, Dr. Nourai has developed and applied many techniques to improve the energy efficiency and performance of power systems. He holds seven patents and has published a number of technical papers. He works closely with the energy storage program in DOE's Office of Electricity Delivery and Energy Reliability and serves as a regular peer reviewer for the energy storage projects of DOE. He is a registered professional engineer in the state of Ohio and received the Walter Fee Award from IEEE's Power Engineering Society for professional contributions and technical competence through significant engineering achievements.

Terry Surles is currently director for the Pacific International Center for High Technology Research. PICHTR's activities focus on the demonstration and deployment of renewable energy technologies in Pacific island nations (PINs). These activities also include technical training and capacity building for PIN nationals for the operation and maintenance of renewable energy systems. Previously, he was vice president at the Electric Power Research Institute (EPRI) and its subsidiary, the Electricity Innovations Institute. Before joining EPRI, Dr. Surles was program manager at Public Interest Energy Research (PIER) and assistant director for science and technology of the California Energy Commission. Dr. Surles was the associate laboratory director for energy programs at Lawrence Livermore National Laboratory, following his time at the California Environmental Protection Agency as deputy secretary for science and technology. Dr. Surles was at Argonne National Laboratory for a number of years, holding a number of positions in the energy and environmental systems area, with his last position being general manager for environmental programs. Dr. Surles holds a B.S. in chemistry from St. Lawrence University and a Ph.D. in chemistry from Michigan State University. He was a member of the Phase One committee.

Gunnar Walmet has been with the New York State Energy Research and Development Authority (NYSERDA) for over 20 years. He has been the director of NYSERDA's industry and buildings R&D programs since 1989. These two R&D programs helped companies implement dozens of innovative, energy-efficient, environmentally beneficial technologies and processes. Many of NYSERDA's industry and building projects have resulted in patents and awards, including "The Best of What's New" award by *Popular Science*, DOE's National Award for Energy Innovation, and the Governor's Pollution Prevention Award, as well as awards from such diverse groups as R&D 100, the National Center for Appropriate Technology, the American Society of Mechanical Engineers, Renew America, and the U.S. Environmental Protection Agency. The work has resulted in several new businesses, thousands of jobs created or saved, and tens of millions of dollars in product sales each year. Successful buildings R&D projects include creating the internationally renowned Lighting Research Center at Rensselaer Polytechnic Institute; developing the nation's first non-ozone-depleting supermarket refrigeration and air-conditioning system; commercializing a pulse combustion boiler and a gas-fired hydronic boiler; developing several district heating systems; and innovative programs in demand reduction and real-time pricing. The industry program has been instrumental in developing a radio-frequency induction heating system, commercializing an adaptive controller for resistance welding, constructing the state's first full-scale indoor fish-production facility, an academic center for remanufacturing technology, developing and demonstrating an environment-friendly paint booth for solvent-based coatings, and developing an optical lens system to monitor wafer contamination in semiconductor production. Under Mr. Walmet, NYSERDA has also implemented the nation's most aggressive program to promote CHP, or cogeneration, and to demonstrate superconducting power systems. Prior to joining NYSERDA, he was an engineer at GE's R&D Center and Medical Systems Division in Schenectady, where he helped develop a new membrane blood oxygenator for use in open-heart surgery, helped design the gas cleanup train for a pilot-scale coal gasifier, and developed biogas purifying systems. His research at GE resulted in 14 issued patents. In 2003 Mr. Walmet won national recognition from the American Council for an Energy-Efficient Economy, which named him a Champion for Energy Efficiency. He has an M.S.M.E. from Union College and a B.S.M.E. from Trinity College.

L

Report of the Panel on DOE's Light-Duty Vehicle Hybrid Technology R&D Program

Basic hybrid electric vehicle (HEV) technology is relatively well established today and is reasonably well recognized by consumers owing to publicity surrounding the Toyota Prius and other hybrid light-duty vehicles now on the market or soon to be introduced. Current market penetration of HEVs in the United States is less than 2 percent of new vehicles, but it is increasing rapidly, assisted by significant federal and state incentives and to some extent by industry subsidies such as repair/warranty cost absorption. Extensive R&D on technologies for hybrid vehicles is being conducted around the world by automotive companies and their suppliers, as well as by government programs such as DOE's FreedomCAR and Vehicle Technologies (FCVT) program.

INTRODUCTION AND OBJECTIVE OF STUDY

The Panel on Prospective Benefits of DOE's Light-Duty Vehicle Hybrid R&D Technology Program is one of six expert panels convened under the auspices of the Committee on Prospective Benefits of DOE's Energy Efficiency and Fossil Energy R&D Programs (Phase Two) to conduct a beta test of the committee's methodology for assessing prospective benefits of R&D programs. The NRC-appointed panel, composed of nine members with a mix of industrial and academic experience (see Attachment A), was asked to assess the potential benefits of DOE's R&D activities that are focused on HEV technologies using more efficient internal combustion engine (ICE) power trains for light-duty vehicles.

The panel met in Washington, D.C., on October 3 and 4 and November 7 and 8, 2005. Both meetings included open, information-gathering sessions attended by DOE headquarters staff and contractors. The DOE representatives briefed the panel on light-duty vehicle R&D programs within the Office of Energy Efficiency and Renewable Energy (EERE) and on the approach used by DOE to estimate the prospective benefits of these programs in accordance with the requirements of the Government Performance and Results Act (GPRA).

The panel received guidance from its parent committee in developing its meeting agendas and applying the prospective benefits methodology. Its work was facilitated by a consultant who assisted all six panels and was, therefore, able to ensure consistency in the application of the methodology. In assessing the likelihood of technical success for DOE's R&D programs, the panel drew on its own expert judgment and on the findings of the recent review of the research program of the FreedomCAR and Fuel Partnership (NRC, 2005b). That review included more detailed assessments of the current status of and future prospects for DOE's R&D on light-duty hybrid vehicle technologies than were possible within the time and resource constraints of the current effort.

The panel's detailed comments on the prospective benefits methodology are included later, but some observations are made here to set the context for the remainder of this report. The panel found that, in general, the methodology provided a good framework for establishing upper and lower bounds on the ultimate impact of the DOE program and for focused conversations among panel members and DOE representatives. However, the panel emphasizes that there is significant uncertainty in the estimate of prospective benefits of DOE's program, measured in terms of reductions in fuel consumption. These benefits depend not only on the success of the research itself—its timely achievement of technical and cost goals—but also on factors beyond DOE's control, including the outcomes of research being conducted by other organizations around the world and the timing and rate of commercialization of new vehicle technologies. The benefits will also depend on the future market penetration of light-duty hybrid vehicles, which is likely to be affected by factors such as oil prices, emissions regulations, and fuel economy standards. Estimates of anticipated benefits of research often depend on expert judgment regarding numerous technical and market factors about which there is considerable uncertainty. In the present case, the number of vehicles potentially affected is very large, because 16 million to 17 million new light-duty vehicles are sold in the United States each year.

Thus, relatively small differences in the probability of research success or in the market penetration rate—both of which are difficult to estimate with certainty—result in large differences in estimates of overall benefits. Consequently, the panel cautions that the quantitative results presented in its report should be considered in the context of a range of potential outcomes, not as a precise prediction of the results and benefits of the research program.

SUMMARY OF DOE PROGRAM ON LIGHT-DUTY VEHICLE HYBRID TECHNOLOGY

DOE's R&D on technologies for light-duty hybrid vehicles with ICE power trains is conducted under the auspices of the FCVT program in EERE. The mission of the FCVT program is to "develop more energy-efficient and environmentally friendly highway transportation technologies that enable America to use less petroleum" (DOE, 2005c, pp. 1-13). Within the broad context of the FCVT program, light-duty hybrids with gasoline or diesel-fueled ICE power trains are seen as a step on the transition pathway toward the ultimate goal of fuel-cell-powered vehicles running on hydrogen.

Because the FCVT program comprises much more than R&D for light-duty ICE hybrids, the panel's first step in analyzing the materials received from DOE was to decide which parts of the program to include in its assessment. The FCVT budget for R&D related to passenger vehicles covers work on energy storage (high power energy storage, advanced battery development, and exploratory technology research), advanced power electronics and electric motors, materials (automotive lightweight materials and automotive propulsion materials), advanced combustion and fuels, and systems.[1] The panel selected three of these areas on which to base its assessment: high power energy storage, automotive lightweight materials, and advanced combustion and fuels.

The reasons for selecting these three technical areas are twofold. First, in the panel's judgment, these areas are concerned with critical technologies for more fuel-efficient light-duty vehicles. Second, they consistently received the largest share of the FCVT funding for passenger vehicles in FY02, when the FCVT program was initiated, through FY05.[2] Over that 4-year period, high power energy storage received 20 to 22 percent of each year's funding (a total of $69 million), automotive lightweight materials received 18 to 21 percent of annual funding (a total of $62 million), and advanced combustion and fuels received 25 to 30 percent of annual funding (a total of $88.5 million). DOE's FY06 budget request indicates a continuing emphasis on these three technology areas.[3]

At the second panel meeting, DOE representatives questioned the panel's decision to focus on three technology areas rather than on the entirety of EERE's R&D on light-duty vehicles with ICE power trains. The panel considers its approach appropriate for the purposes of the present assessment, in which reduced fuel consumption is the metric of success. The technology areas selected are critical if light-duty hybrid vehicles are to achieve greater commercial success. DOE's research in these three areas could result in faster and/or broader market penetration by hybrid vehicles, and important fuel savings could result.[4] The panel sees DOE's role as focused on high-risk R&D on critical technologies, while leaving to others relatively low-risk technology development and the integration of vehicle subsystems into a marketable product.

DOE PERFORMANCE GOALS AND PANEL ASSESSMENTS OF THE TECHNICAL AND MARKET RISKS

DOE and its industry partners have developed performance goals for activities under the FCVT program. These goals comprise target dates, technical characteristics, and cost. The panel identified this last factor as particularly important, because the biggest challenge to market acceptability of hybrid vehicles is likely to be the incremental vehicle cost of achieving adequate vehicle performance, safety, and durability. As noted in the recent review of the research program of the FreedomCAR and Fuel Partnership (NRC, 2005b), the cost savings projected to be attributable to the higher fuel mileage of HEVs will probably not offset the higher initial cost of the vehicle at foreseeable fuel prices. Thus, further cost reductions may be necessary for hybrid vehicles to gain widespread acceptance and have a significant impact on fleet fuel mileage. The following sections identify the relevant DOE performance goals and discuss the panel's assessments of technical and market risk for each of the three technical areas identified earlier—namely, high power energy storage, automotive lightweight materials, and advanced combustion and fuels.[5]

---

[1] Ed Wall, director DOE Office of FreedomCAR and Vehicle Technologies, "Prospective Benefits of DOE's Energy Efficiency and Fossil Energy R&D Programs (Phase Two)," Presentation to the panel, October 3, 2005, Washington, D.C.

[2] Ed Wall, director DOE Office of FreedomCAR and Vehicle Technologies, "Prospective Benefits of DOE's Energy Efficiency and Fossil Energy R&D Programs (Phase Two)," Presentation to the panel, October 3, 2005, Washington, D.C.

[3] The next largest budget category from FY02 through FY05 was advanced power electronics and electric motors, which received 16 to 17 percent of the budget each year (a total of $54 million). Other categories each received less than 10 percent of annual funding.

[4] As discussed below, improvements in energy efficiency resulting from DOE's program may be manifested in the marketplace as vehicle attributes that are even more attractive to the consumer than greater fuel economy.

[5] The benefits of DOE's R&D may extend to vehicles with conventional power trains.

High Power Energy Storage

Hybrid vehicle drive trains use three main technologies: energy storage (batteries), electric motors, and power electronics. Of the three, the panel chose to focus on energy storage (batteries) as a surrogate for electric hybrid drive technology because, in the panel's view, battery technology presented the largest technical and cost hurdles. Nonetheless, significant cost and technical concerns also exist with electric motors and power electronics.

DOE Performance Goals

DOE's performance goal for hybrid energy storage is to develop, by 2010, electric drive train energy storage with 15-year life at 300 Wh with discharge power of 25 kilowatt (kW) for 10 seconds and at a cost of $20 per kilowatt (/kW) (DOE, 2005c). DOE activities on light duty vehicle hybrid technology aims to demonstrate technical achievements and to use cost modeling to determine whether cost targets could be achieved if the technology were implemented on a large scale. The panel notes that achievement of DOE's performance goal would not eliminate technical risk or assure market availability of the technology at the stated target, and that several additional years (from 3 to 10) would be necessary for scale-up to high-volume vehicle production.

Technical Risks

Chemical batteries and ultracapacitors are the primary alternative devices for energy storage in HEVs. In addition to the energy storage requirements, technical challenges include cost, durability (number of charge cycles before performance deteriorates), low-temperature operation, and safety. Adequate durability (for example, 15-year calendar life or 150,000 miles) is essential, because the cost of battery replacement—approximately the same as replacing today's conventional engine—would be a serious market deterrent to hybrid vehicles. Ultracapacitors have excellent durability, power capability, low-temperature performance, and safety, but are considered unlikely to achieve the cost objective in the foreseeable future. Nickel metal hydride (NiMH) batteries are relatively well developed today and commercially available, but they are projected by DOE to have little chance of meeting the long-term cost objective and 15-year durability. DOE anticipates that lithium ion (Li ion) or equivalent technology is necessary to meet the cost target of $20/kW. Other emerging energy storage technologies, such as lithium-sulfur batteries or a combination of batteries and capacitors, may be necessary to achieve the cost and performance targets.

The recent review of the FreedomCAR and Fuel Partnership notes that efforts directed to the development of new materials and electrochemical couples in DOE's electrical energy storage program present "the best chance to remove the major barriers of abuse tolerance, cost, and calendar life for high-power batteries" (NRC, 2005b, p. 77). The panel agrees that the fundamental research being supported by DOE may play a significant role in identifying potential breakthrough storage technologies.

In the panel's judgment, DOE's technical and cost targets are unlikely to be achieved by 2010 because proven NiMH technology is unlikely to achieve the cost targets and 15-year life. The next-generation battery technology (Li ion) still has significant limitations, including safety and low-temperature performance, that require further development before volume commercialization. Li ion battery technology is more likely to be developed in an additional 10 years (i.e., by 2020), but even with the extended time frame, achieving the cost targets will still be difficult. For example, achieving adequate low-temperature performance and safety may entail added costs for enhancing the battery system.

To reflect these important technical risks associated with meeting DOE's performance goal, the panel estimated the probabilities of three alternative outcomes (see later discussion of decision tree):

- Meeting the performance goal by 2010,
- Meeting the performance goal by 2020, and
- Meeting the technical targets but a less aggressive cost target ($28/kW instead of $20/kW) in two time frames (present through 2010 and through 2020).

Market Risks

Adequate vehicle performance is essential for market acceptance of hybrid vehicles. The performance of well-engineered hybrids is generally considered acceptable and sometimes superior to that of CVs, but under some driving conditions (e.g., up long hills) the limited battery energy storage may be unable to maintain adequate performance.

Other market risks associated with hybrid vehicle drive trains include unknown durability because of the greater complexity associated with the battery, electric motor, and power electronics. Adequate battery durability (15-year calendar life) is essential, because the high replacement cost—approximately that of replacing today's conventional engine—would be another serious market deterrent to hybrid vehicles.

Automotive Lightweight Materials

The Automotive Lightweighting Materials activity covers a broad range of structural materials for body, chassis, and power train (engine and transmission) applications. The general objective is to reduce the weight of passenger vehicles without sacrificing performance, safety, or the recyclability. Another very challenging goal is to achieve significant weight reduction at little or no added cost. Weight reduction can be an important enabler for reducing a vehicle's fuel consumption, but, in general, vehicle manufacturers have been reluctant to use advanced lightweight materials because they are more expensive than conventional mild steel. A 5 percent reduction in vehicle mass can yield fuel savings of between 3 and 4 percent (Sovran, 2003). But premium materials add significant cost. For example, aluminum typically costs 1.3 to 2.0 times as much as steel, magnesium costs 1.5 to 2.5 times as much, and carbon-fiber-reinforced composites cost 2.0 to 10.0 times as much (NRC, 2005b, Table 3-7). Therefore progress in reducing the costs of lightweight materials, coupled with progress in reducing their fabrication and assembly costs, will expand their application.

Materials activities include R&D in high-strength steel, aluminum, and both glass- and carbon-fiber-reinforced composites for body and chassis applications. Additional research is focused on cast aluminum and magnesium, aluminum metal matrix composites, titanium, and other advanced materials for potential drive-train and power-plant applications.

DOE Performance Goals

The specific goals for body, chassis, propulsion, and fuel system applications have been outlined by DOE (2005c). The embodiment of individual research objectives into systems applications to automotive vehicles will yield reductions in total vehicle weight. The system goals evaluation by the panel are as follows: By 2012, develop and validate advanced material technologies that will do the following:

- Enable reductions in the weight of body and chassis components of at least 60 percent and overall vehicle weight of 50 percent (relative to comparable 1997 vehicles);
- Exhibit performance, reliability, and safety characteristics comparable to those of conventional vehicle materials; and
- Enable commercially available aluminum, lightweight metal alloys, high-strength steels, and glass- and carbon-fiber composite materials with life-cycle costs equivalent to that of conventional steel.

These performance improvements are referenced to a typical 1997 vehicle, for which the weight of the three primary components breaks down approximately as follows: (1) the complete body, 35 percent (the body-in-white—the body shell without doors, glass, or other closures such as hoods, trunk lids, and tail gates—is about 20 percent); (2) the chassis, 34 percent; and (3) the power train, 27 percent (NRC, 2000, p. 47).

Technical Risk

Clearly, to achieve a 50 percent weight reduction of the complete vehicle, it is necessary to make significant reductions in all three main components. If no weight reduction can be achieved in the power train (as might be the case for a hybrid vehicle, for example), the body and chassis weight must be reduced by substantially more than 50 percent to achieve an overall 50 percent weight reduction; this goal is extremely aggressive. For that reason, the panel elected to consider three levels of vehicle weight reduction: 10 percent, 25 percent, and a stretch goal of 50 percent. The panel assumed in each case that cost parity would be achieved and that vehicle structural performance requirements and recyclability goals would be met. It is logical to assume that manufacturers will apply premium materials in increasing order of cost/benefit. For example, it should be possible to achieve a weight reduction of 10 percent through the application of high-strength steels in the body and chassis. A more aggressive application of high strength steel and the use of aluminum for closures, as well as reductions in power-train weight, might be necessary to achieve a 25 percent weight reduction. Previous studies showed that extensive application of carbon-reinforced composites in the body and chassis would be required to achieve the very aggressive goal of 50 percent vehicle weight reduction (NRC, 2000, 2005b, Table 3-7).

In body applications, high-strength steel has been shown to meet mechanical performance requirements (e.g., stiffness, crashworthiness). Indeed, in some production vehicles today, the body-in-white contains as much as 50 percent high-strength steel. Aluminum has seen extensive light-duty vehicle application, although in more restricted volumes due to its higher cost. Certain manufacturing issues (such as joining) also mean that aluminum has been used primarily for closures in high-volume applications. However, some lower volume vehicles have been aluminum intensive, proving the viability of an all-aluminum body. Carbon-reinforced composites have not seen extensive application in high-volume automotive products for several reasons. The material costs remain very high, and there are no high-volume fabrication and assembly systems for composite-intensive vehicles (this issue is discussed in more detail under the next subsection on market risks). While the panel does not believe that the technical goal of a 50 percent reduction in vehicle weight can be achieved by 2012, it assigned nonzero probabilities for success in achieving intermediate goals of 10 percent and 25 percent by 2012.

DOE research in materials has contributed to the application of high-strength steels, aluminum, and glass-reinforced composites in vehicles and is expected to eventually contribute to vehicle weight reduction. Because DOE research in weight reduction of power-train components and other smaller components can be used across a broad range of vehicle applications, not just in light-duty or hybrid vehicles, the panel believes it is more likely to lead to success and reflected this opinion in its assignment of probabilities on the decision tree branch with DOE funding. However, the panel also believes that applications of high-strength steels and aluminum in the body or chassis require little additional research—as noted above, these materials are already in production. The panel expects, moreover, that material suppliers and automobile manufacturers will continue to focus on further reductions in manufacturing costs, even without DOE participation.

Market Risks

Automobile manufacturers have been slow to utilize lightweight materials in body and chassis applications for several reasons. Certainly, cost has been an impediment. For body and chassis applications, perhaps the most important impediment is that the introduction of new materials impacts the existing fabrication and assembly processes. Market fragmentation has forced manufacturers to increase their ability to fabricate and assemble multiple body styles on a single assembly line. This is done using advanced assembly systems and technologies that have been refined for application to a steel body. This manufacturing footprint is amenable to the introduction of high-strength steels with little disruption. Furthermore, it can accommodate aluminum closures (doors, hoods, deck lids) with minimal disruption. On the other hand, to convert fabrication and assembly systems to accept carbon- or glass-reinforced composites would require a major development activity to prove the feasibility of manufacturing composite-intensive bodies in high volume. Commercialization would require an enormous capital investment for converting assembly plants. The panel is not aware of any significant development activity that would enable such a transformation in vehicle body manufacturing. For this reason, it believes that extensive application of advanced composites in vehicle bodies is virtually impossible by 2012. This would limit vehicle weight reduction opportunities to well below the 50 percent goal. This conclusion is consistent with the findings of the Committee on Review of the FreedomCAR and Fuel Research Program (NRC, 2005b).

In addition to capital costs, ease of implementation is important for the introduction of new technology into high-volume production. This explains why manufacturers have relied primarily on continuous improvement of the power train for efficiency improvements (NRC, 2001). For new material systems, first applications will occur in subsystems and components that can be easily integrated into final assembly. Early applications will probably occur in systems that are not safety critical until field experience can assure that the material systems will not degrade overall product safety. Applications to specialty vehicles can provide valuable field experience, but low-volume production may not easily carry over to high-volume applications unless the manufacturing technology is scalable.

The price of fuel can influence the rate at which new lightweight systems are commercialized. The panel noted that light-duty vehicle weight has remained nearly constant over the past several years (EPA, 2005). In fact, U.S. vehicle fleet weight has increased with the shift from cars to trucks and sport utility vehicles, although this trend may begin to reverse itself with the current relatively high cost of gasoline. To improve overall vehicle performance, automobile manufacturers have focused more on improving the efficiency of the power train than on material substitution (NRC, 2001). Indeed, in the recent past, they have used the improvements in engine efficiency to enhance performance rather than improve fuel economy (EPA, 2005). Therefore, a significant market risk is that weight savings through material substitution may not be applied in high volume applications if fuel costs remain similar to those that prevailed in the early 2000s. On the other hand, improvements in fuel efficiency attributable only to power-plant and drive-train improvements will eventually reach technological limits. At that point, weight reduction technologies can provide very important options for additional efficiency.

Finally, the panel notes that weight reduction will be an important enabler for the market success of hybrid vehicles. Purpose-built hybrid vehicles that optimize the entire vehicle system for fuel efficiency will very likely achieve the greatest market acceptance.

Advanced Combustion and Fuels

The focus of the engine, emissions, and fuels research activities in FCVT is to support the development of improved internal combustion engines that have the potential for high efficiency while achieving near-zero emissions. Recognizing that the engine and its fuel and emission control systems are interdependent, the research also looks at potential advancements in emission control technologies and considers a new generation of transportation fuels, both petroleum- and nonpetroleum-based.

The ultimate goal of the FCVT program is to reduce the U.S. dependence on petroleum-based fuels. Therefore, particular attention is being paid to high-efficiency, compression ignition (diesel-like) engines and the trade-offs between maximum thermal efficiency and engine-out exhaust emissions, including a fundamental understanding of in-cylinder combustion processes. Depending on the resulting level of criteria pollutants and the chemistry of the exhaust gas, which varies over a range of operating conditions, different exhaust aftertreatment technologies, their conversion efficiencies, and durability issues must be considered.

However, in most cases, the fuel preparation (injection) systems, intake air boosting systems, exhaust gas recirculation (EGR) system (if needed), engine control systems (fuel, boost, EGR, etc.), and aftertreatment systems (diesel particulate filter, oxidation catalyst, NO_x-trap, selective catalytic reduction, etc.) being researched represent significant cost penalties compared to the port-injected, naturally aspirated, three-way-catalyst power trains that dominate the U.S. light-duty vehicle market, justifying the combustion, aftertreatment, and fuels research being funded under the FCVT program. Furthermore, the panel believes that a cooperative effort between government, the automotive industry, and the transportation fuels industry is necessary to expedite advancements that can also be applied to more conventional vehicles and future generations of hybrid vehicles (see also NRC, 2005b).

DOE Performance Goals

The technical goals of the combustion, emission control, and fuels activities are to demonstrate significant improvements in engine peak- and part-load brake thermal efficiencies (compared to current production, EPA-compliant, gasoline-fueled engines) that meet Tier 2, Bin 5, emission standards and 150,000-mile durability requirements and can be produced at reasonable cost in high volume. Specific targets include increasing peak brake thermal efficiency from the levels found in production gasoline (about 33 or 34 percent) to 42 percent in 2007, 44 percent in 2009, and 45 percent in 2010. In addition, part-load—2 bar brake mean effective pressure (bmep) at 1,500 rpm-brake thermal efficiency is targeted to improve from about 20 to 22 percent (for current, throttled, lambda-1-controlled power trains) to 27, 29, and 31 percent in 2007, 2009, and 2011.

Technical Risks

Conversion to unthrottled, direct-injection diesel engines has already demonstrated 40-42 percent peak brake thermal efficiencies in light-duty passenger vehicles sold in Europe. The technical risk is to accomplish this while controlling full-load, lean, stratified combustion-generated particulate and NO_x emissions to stringent Tier 2, Bin 5, standards for 150,000 miles at a total power train cost of $30/kW. Further advancements, approaching 45 percent peak brake efficiency while minimizing the efficiency penalties associated with existing lean-combustion aftertreatment systems, will be extremely challenging. The technical target is to keep the efficiency penalties associated with emissions control to <3 percent in 2007 and <1 percent in 2009 and beyond.

The goal of achieving 45 percent peak brake efficiency would require significant advancements in many areas, including high-pressure, direct-injection fuel injection systems; more efficient boosting (turbocharging/supercharging) systems; reduced friction components; reduction in parasitic and accessory loads; higher strength/lighter weight power-train materials to accommodate combustion pressures, which could approach 220 bar or more; advanced fuel injection and combustion process controls to support low-temperature combustion (LTC) processes; and new methods for noise control. LTC methods, including homogeneous charge compression ignition (HCCI), are interesting because they could significantly reduce engine-out NO_x emissions, which would facilitate less complex and lower cost aftertreatments. However, some LTC processes also produce high combustion pressures, require extremely high boosting levels, and still require aftertreatment for hydrocarbon and particulate emissions. Since LTC has seen only limited applicability under relatively low engine load conditions where combustion stability can be maintained, the need for transition to more conventional combustion processes under high-load conditions may remain, which would still necessitate NO_x aftertreatment.

Additional R&D is required to define the petroleum and nonpetroleum fuels that will facilitate the combustion performance and decrease exhaust emissions needed to achieve future fuel efficiency targets and the tightening of exhaust emission standards. Programs must specify these fuels and involve the fuels suppliers early on to avoid delays in commercial introduction of more efficient engine systems.

For advanced conventional spark-ignited engines, higher octane gasoline could be needed for turbocharged systems. This could require gasoline with higher aromatics or ethanol content, raising the question of cold-start performance in winter. In addition, lower sulfur levels could be needed to improve the performance of catalysts, depending on the oxygen content of the exhaust gas and the potential for catalyst poisoning. Sulfur content would be even more important if spark-ignited LTC is employed, to minimize the potential for particulate formation.

For HCCI-based LTC, a more volatile fuel with a higher cetane number could be needed to enhance combustion. Greater volatility would compensate for lower fuel injection pressure (e.g., 100 bar), which would otherwise produce larger droplets that inhibit mixing. This suggests the need for a new diesel fuel with enhanced cetane number, higher volatility, and lower sulfur content.

The desire to expand the use of non-petroleum-based alternative fuels adds to the complexity of the plethora of fuel mixtures and combinations that might be considered. Most alternative fuels currently being considered, such as Fischer-Tropsch-derived diesel fuel, so-called biodiesel (methyl ester-based), and alcohol-based fuels, contain little or no sulfur but exhibit other properties that could significantly influence combustion and emissions, such as very low vapor pressure (alcohol fuels) or low lubricity (biodiesel).

New and modified engines could also require modifications to the lubricating oils used so that lubricant requirements will need to be defined further as the program moves forward. While it is directionally easier to modify commercial lubricants than fuels, in part because of the much smaller quantities involved, changes to commercial facilities could be needed and must be anticipated.

Based on these technical risks, the panel thought it extremely unlikely that the 45 percent target could be achieved by 2010 in a production-feasible configuration that would also meet Tier 2, Bin 5, emission standards at 150,000 miles and for a cost of $30/kW. However, the panel believes that advancements beyond the state of the art will be facilitated by the planned research activities and has assessed the likelihood of reaching some intermediate performance levels, including efficiency, emissions, and cost parameters.

In addition, the panel believes that the research being done on advanced combustion, emissions control, and fuels will produce advancements that can also be applied to the power trains of conventional light-duty vehicles and produce significant fuel consumption savings in parallel with the planned application to hybrid vehicle platforms.

Market Risks

The market risk associated with the advanced combustion, emissions control, and fuels activity is primarily related to the uncertainty of achieving the target performance parameters at the right cost and to questions about the ability of the advanced concepts to satisfy the durability and reliability requirements for high-volume automotive production.

In addition, there are customer acceptance issues associated with several of the technologies currently being developed. Diesel combustion and other compression-ignition-type processes achieve very high combustion pressures, and the associated rate of pressure rise produces noise levels that have historically been unacceptable for light-duty passenger vehicle applications. Techniques to control power-train noise without sacrificing engine efficiency, power-train weight, and exhaust emissions must therefore be developed, as is being done in Europe.

The durability of increasingly complex aftertreatment systems is also a market risk that must be overcome. In addition to the requirement to achieve regulated emission levels for 150,000 miles of consumer use, many of the techniques currently being researched suffer from premature failures due to wear, thermal cycling, excessive temperatures, and variable fuel composition. These must all be overcome in order to achieve reliability levels that are acceptable for high-volume automotive markets.

Furthermore, variations in fuel and feedstock composition or in handling procedures would call for significant changes in fuel refining, formulation, and distribution. Perhaps the most difficult change would be the introduction of an additional fuel, especially a gaseous fuel. This would entail either the replacement of an existing fuel or new facilities to provide sufficient fuel distribution. The nationwide introduction of lead-free gasoline in the 1970s exemplifies not only the ability of the fuel industry to respond but also the time and investment required. The minimal penetration of alcohol fuels, which were introduced in the late 1980s, is an example of market failure (to date) of an alternative fuel strategy.

Reducing sulfur levels in any existing fuel also has important implications. Refineries would require additional hydrotreating facilities to remove sulfur and to manufacture the requisite hydrogen.

Higher octane gasoline would also require additional refining facilities to boost aromatics and isoparaffins contents. Higher cetane diesel fuel would require changes to reduce aromatics content while maintaining adequate cold flow properties in winter. A significant change in volatility could change the need for conversion (cracking) in refining. New facilities would be needed to produce alternative fuels such as alcohols or biodiesel.

All of these potential changes would involve added investment by the fuels industry and would increase the cost of fuel to the consumer. As the trade-offs between engine performance/cost and fuel properties are defined, refiners will need to define preferred routes to providing alternative fuels. From these studies, a plan and schedule for commercialization could be set up. The studies would also provide guidance on increased fuel cost, allowing the overall engine system to be optimized from a consumer perspective.

RESULTS AND DISCUSSION

Decision Trees and Estimated Probabilities of Technical Success

The panel used the decision analysis methodology developed by the parent committee to generate quantitative estimates of the likelihood of achieving DOE's performance goals in the three key technology areas (high power energy storage, automotive lightweight materials, and advanced combustion and fuels) both with and without the DOE program. The estimated probabilities of success with the DOE program assume that sufficient funding will be available for DOE to continue its work at current levels. Because the research projects in the three key areas are essentially independent and nonoverlapping, the probabilities of technical success under each scenario were evaluated separately for each area.

High Power Energy Storage

The panel identified two key areas of uncertainty related to achieving DOE's goals for the hybrid and electric propulsion subprogram. The first is uncertainty about the ability to reach the battery performance and cost targets specified by DOE: "electric drive train energy storage with a 15-year life at 300 watt-hours (Wh) with a discharge power of 25 kW for 10 seconds and $20/kW cost." The second is uncertainty about whether those technical goals can be achieved by 2010, DOE's target year for such improvements.

Figure L-1 illustrates a decision tree for the technical uncertainties in battery performance. The first node represents a decision about DOE funding of the research program. The second node represents uncertainty about the level of technical improvements in batteries that will be enabled by research completed by 2010. Three levels of technical success were defined as follows:

• High success is the achievement of DOE's goal for both battery performance and cost: 15-year life at 300 Wh with

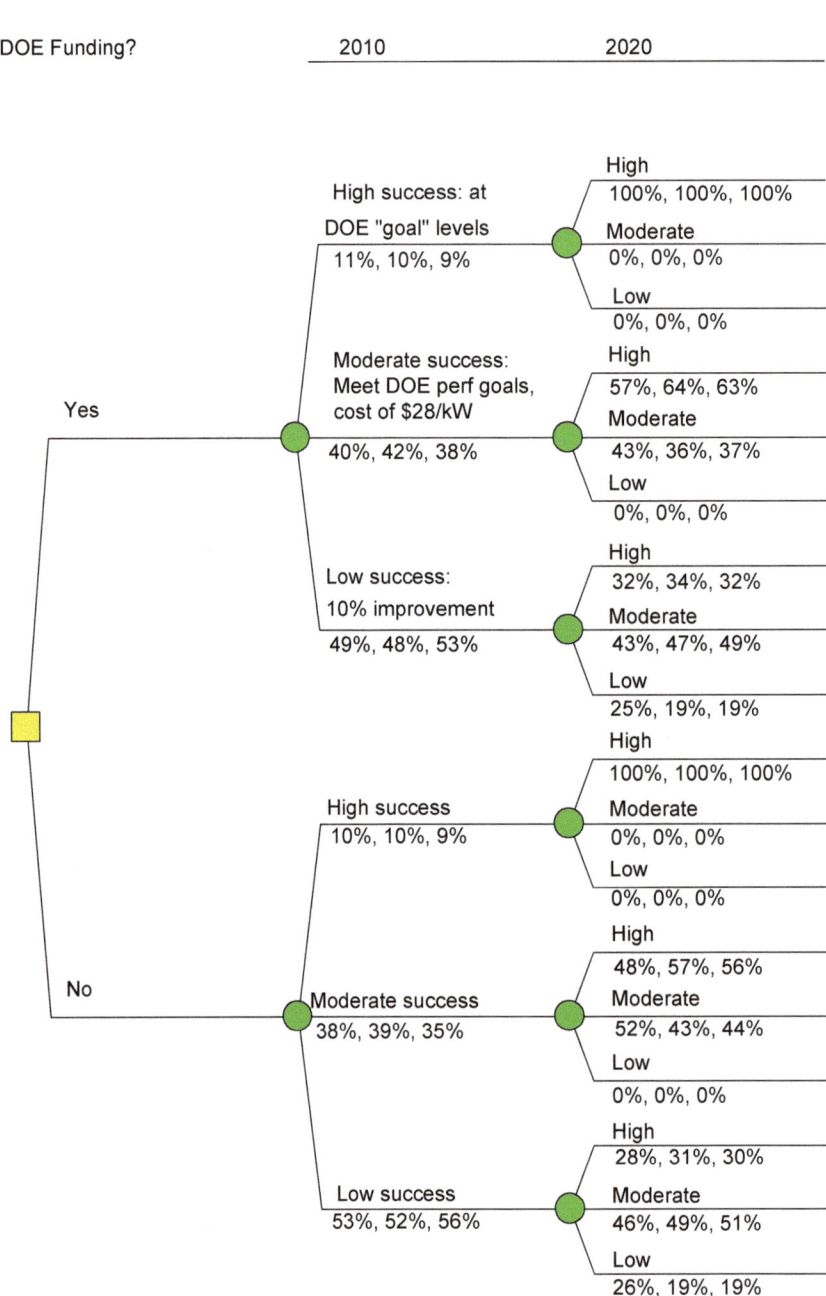

FIGURE L-1 Decision tree representing the panel's evaluation of the batteries program.

a discharge power of 25 kW for 10 seconds and $20/kW cost.

• Moderate success is the achievement of DOE's goals for battery performance, but at a cost of $28/kW.

• Low success is making incremental improvements over current levels of battery performance and costs (10 percent incremental improvement in 2010).

The third node in Figure L-1 represents uncertainty about the level of technical improvements in batteries that will be enabled by research completed in 2020. The same three levels of technical success were used to characterize this uncertainty except that "low success" was defined as a 30 percent incremental improvement over current performance.

Each panelist estimated the likelihood of achieving each

APPENDIX L

of these levels of technical success in each time period, both with and without the DOE research program, under each of the three global scenarios. The probabilities on each branch in Figure L-1 show the average of the panelist's individual assessments of the likelihood of each outcome under the three global scenarios: AEO Reference Case, High Oil and Gas Prices, and Carbon Constrained. In discussions, the panelists focused first on whether and to what degree the DOE program would increase the probability of achieving the higher levels of technical success and second on the absolute probabilities of being able to achieve those levels of success. In interpreting Figure L-1 (and similar Figures L-2 and L-3), it is important to note that the method used requires that the probability estimates for alternatives on each branch sum to unity. Thus, if DOE funding increases the probability of achieving more challenging goals, the probability of achieving a lesser goal can appear to be lower with DOE funding than without DOE funding. This is an

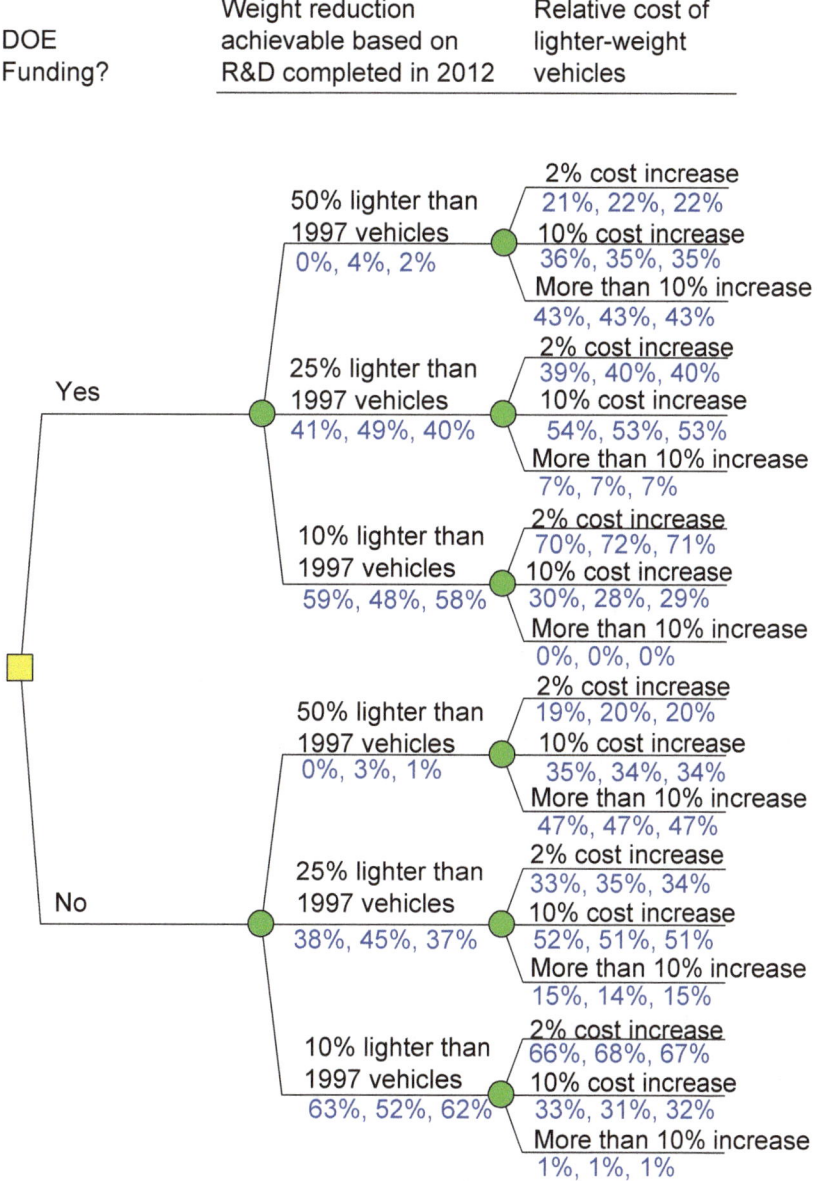

FIGURE L-2 Decision tree representing the panel's evaluation of the lightweighting research program.

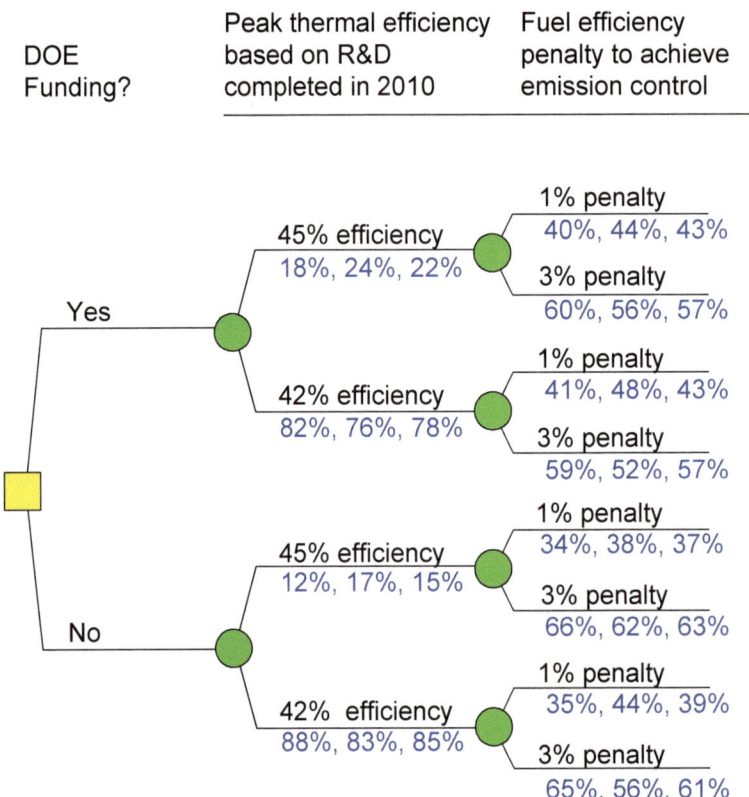

FIGURE L-3 Decision tree representing the panel's evaluation of the advanced combustion engines program.

artifact of the method, implying that all the probabilities for a given scenario need to be considered simultaneously when interpreting the figures.

Automotive Lightweight Materials

The panel identified two key technical uncertainties relating to DOE's goals for the projects in the automotive lightweight materials area. The first uncertainty is the weight reduction that is technically achievable by 2012, and the second uncertainty is the increased cost of such a lighter vehicle. Figure L-2 illustrates the decision tree representing the panel's assessment of these uncertainties. Uncertainty about weight reduction was characterized by the vehicle weight reduction enabled by technology demonstrated by 2012 (compared to 1997 vehicle weight). Uncertainty about the cost of lightweighting was characterized by the cost of the lighter vehicles relative to the cost of heavier conventional vehicles. Again, each panelist estimated the likelihood of achieving each level of technical success in each time period, both with and without the DOE research program, under each of the three global scenarios.

Advanced Combustion and Fuels

The panel also identified two areas of uncertainty related to DOE's goals for the advanced combustion engines and emissions control projects. The first uncertainty was identified as the peak brake thermal efficiency that is technically achievable based on technologies demonstrated by 2010, and the second was the fuel efficiency penalty to meet EPA emissions guidelines. Figure L-3 represents this structure in the decision-tree format.

Effect of Technology Improvements on Vehicle Fuel Economy and Cost

In the model used by the panel, benefits from DOE's hybrid vehicle R&D program are assumed to accrue when more fuel-efficient vehicles are adopted in the marketplace (see later discussion). These vehicles need not necessarily be hybrids because the benefits of DOE's R&D may extend to vehicles with conventional power trains. Thus, both HEVs and CVs need to be considered when estimating benefits.

Each of the three technology areas evaluated—high-power energy storage, automotive lightweight materials, and advanced combustion and fuels—would affect fuel efficiency and vehicle cost if implemented in new vehicles. Thus, to estimate the prospective benefits, the technical success scenarios had to be translated into fuel economy estimates for vehicles (conventional and hybrid electric) implementing those technologies. In addition, the effect on vehicle costs of implementing the technologies had to be derived using the estimated costs for the improved technologies.

Fuel Economy

Technical advances in the lightweight materials and advanced combustion engines areas could lead to fuel economy improvements in both CVs and HEVs as a result of reduced vehicle weight and improved engine efficiency. Improved high-power battery performance could further improve the fuel economy of hybrids by enabling increased hybridization benefits, such as increased brake energy capture and more electric assist during the drive cycle. However, improvement in hybrid fuel economy due to improvement in battery performance is likely to be relatively small (perhaps 5 percent), and better battery technology would primarily benefit HEVs by reducing vehicle costs (see later).

The panel's estimates of the improvements in fuel economy that could result from specific technical improvements are summarized in Table L-1.

On the assumption that the fuel economy improvements from each of the technical improvements are additive, the estimates in Table L-1 were translated into estimated fuel economy improvements for vehicles with combinations of technical improvements, as shown in Table L-2. Table L-2 also shows the estimated probability of achieving the technical improvements associated with each combination of technical improvements. The table shows the probabilities

TABLE L-1 Panel Estimates of Fuel Economy Improvements Relative to Conventional Vehicles

R&D Activity	Weight Reduction[a] (%)	Engine Efficiency[b] (%)	Fuel Economy Penalty Due to Emissions Control (%)	Estimated Fuel Economy Improvement[a]
Advanced combustion	n/a	45	1	1.25
		45	3	1.22
		42	1	1.15
		42	3	1.12
Batteries and energy storage	n/a	n/a	n/a	1.3
Automotive lightweighting	25	n/a	n/a	1.12
	10	n/a	n/a	1.05

NOTE: n/a, not applicable.
[a]Relative to a conventional vehicle.
[b]Brake thermal efficiency.

TABLE L-2 Panel Estimates of Fuel Economy Improvement for Vehicles with Specified Technical Improvements

Technical Improvements (%)			Estimated Fuel Economy Improvement Relative to 2006 Conventional Vehicles		Probability of Achieving Improvements	
Vehicle Weight Reduction	Engine Efficiency	Emissions Penalty	Conventional Vehicles	Hybrid Electric Vehicles	With DOE Program	Without DOE Program
25	45	1	1.40	1.82	.03	.02
25	45	3	1.37	1.78	.04	.03
25	42	1	1.29	1.67	.14	.11
25	42	3	1.25	1.63	.19	.21
10	45	1	1.31	1.71	.05	.03
10	45	3	1.28	1.67	.06	.05
10	42	1	1.21	1.57	.20	.19
10	42	3	1.18	1.53	.29	.36

with and without the DOE program based on panel averages for the Reference Case.

DOE uses the Powertrain Systems Analysis Toolkit (PSAT) to estimate fuel economy and high-level, market-relevant characteristics based on specific vehicle technical characteristics. DOE staff provided the panel with a summary of the PSAT personal vehicle assumptions and results from the FY2006 GPRA analysis (DOE, 2005b), which the panel compared with its own estimates. The panel's estimated fuel economy improvements for specific technical advances mapped fairly closely with improvements from the PSAT analysis.

Vehicle Costs

The panel's assessments of technical success for batteries and for the lightweight materials research areas included explicit consideration of the manufacturing costs required to achieve the technical improvement. Thus, for every set of technical outcomes shown in Table L-2, we can derive an approximate assessment of the costs for a vehicle with those characteristics, including an estimate of the uncertainty surrounding the costs.

For estimating benefits, the relevant metrics are (1) the incremental cost for HEVs over conventional vehicles and (2) the incremental cost of lighter weight, more efficient conventional vehicles over today's vehicles (2006). To estimate the incremental cost of HEVs associated with the batteries research, the following assumptions were made:

- Current (2006) incremental retail costs for hybridization are about $2,500 for a midsize vehicle[6] (incremental manufacturing costs are about $1,984).
- The battery costs account for about 64 percent of the increased manufacturing costs for hybrids.[7]
- Current (2006) battery costs are about $35/kW.

Based on these assumptions, the impact on vehicle costs of the batteries research is to reduce the incremental cost of hybrids by $690 if battery costs are reduced to $20/kW, by $320 if battery costs are reduced to $28/kW, and by $160 if battery costs are reduced by 10 percent.

To estimate the incremental costs of lighter, more efficient conventional vehicles, the panel used the incremental costs defined as part of the assessment of technical success for automotive lightweighting: a 2 percent increase, a 10 percent increase, and an increase of >10 percent. The panel noted that if the vehicle costs for lightweighting increase more than 10 percent, the technologies would not be implemented because they would not be viable in the market.

Additional Unquantified Benefits

Work on electric-hybrid drive technology—batteries, power electronics, and electric motors—may also yield fuel economy benefits for conventional and, in the future, fuel cell vehicles. For example, this work may help to reduce costs and improve the performance of integrated starter-alternator and 42-volt systems, which provide fuel economy benefits and are in the process of being commercialized today. Also the work supports the development of fuel cell vehicles that will require similar motor and power electronics and (probably) battery technology. For the purposes of the approximations in Tables L-1 and L-2, the estimated fuel economy improvement associated with battery technologies reflects only the greater fuel economy of hybrids compared to conventional vehicles. The improvements resulting from increased hybridization benefits and the implementation of better batteries, electric motors, and power electronics in conventional vehicles were not quantified for this assessment.

The estimated vehicle cost reductions associated with the success of work in the batteries and energy storage area are based solely on the estimated reductions in battery costs and therefore do not include any cost reductions that might be associated with other advances in power electronics and electric motor technologies resulting from DOE's electric hybrid drive technology program.

Effect of Technology Improvements on Market Risk

Fuel economy is not the only real or perceived benefit of hybrids. However, it is generally believed that fuel cost savings must pay back the incremental vehicle cost to the consumer within a few (3-6) years for hybrids to achieve substantial market penetration, and the panel believes, accordingly, that the greatest challenge to substantial market penetration of hybrid vehicles is their incremental cost.

HEVs have penetrated less than 2 percent of the U.S. market for new vehicles, but this penetration is increasing rapidly, assisted by significant federal and state incentives and somewhat by industry subsidies (such as repair/warranty cost absorption). Large financial subsidies by government or industry are unlikely to be viable in the long term, however, and alternatives to hybrids for similar fuel savings—such as more fuel-efficient conventional engines—are under development and may be preferred if the price of hybrid vehicles does not drop sufficiently.

In determining the benefits of R&D on hybrid technologies, it is necessary to project when and to what extent vehicles with these new technologies will enter the vehicle fleet. The DOE target years represent the time by which DOE expects to demonstrate technical success incorporated into cost modeling that predicts that cost targets could be achieved if the technology is implemented in high volume. The panel recognizes that this definition of success does not

---

[6] Based on EPA (2005b).
[7] Based on EPA (2005b).

eliminate technical risk or assure market availability of the technology at the stated target, and that several additional years (from 3 to 10) would be necessary for scale-up to volume vehicle production.

The panel does not presently have adequate information to determine if DOE's cost and performance targets will allow HEVs to penetrate the market or if DOE-projected fuel economy benefits are reasonable. Recent vehicle fuel economy modeling by DOE (2005b) suggests that the fuel economy benefit of hybrid vehicles versus advanced conventional vehicles may have been overstated in DOE's previous NEMS and MARKAL models. This would increase the market risk and cost challenge for hybrid vehicles.

Quantifying Benefits

The economic, environmental, and security benefits of DOE's research in light-duty vehicle hybrid technology depend on the degree to which the technical goals are reached and to which the technologies penetrate the marketplace. Specifically, the research, if successful, will lead to more fuel efficient vehicles in the market and on the road, which will result in reduced gasoline consumption. The reduced gasoline consumption leads directly to other benefits: economic benefits from reduced consumer expenditures for gasoline, environmental benefits from reduced carbon dioxide and other emissions, and security benefits from reduced demand for oil.

To quantify these benefits, the panel needed to estimate the reduction in gasoline consumption over time that could be attributed to the light-duty vehicle hybrid technology program. The model used by the panel is adapted from a vintage stock model developed by the Committee on Alternatives and Strategies for Hydrogen Production and Use (NRC, 2004b). It produces estimates of the total vehicle miles driven by year and by vehicle type based on assumptions about the vehicle sales and average vehicle lifetime (14 years and about 142,000 miles). Combined with estimates of the number of HEVs in the fleet, the panel used this model to derive estimates of the total number of vehicle miles per year for conventional vehicles and HEVs.

The panel considered two different "market success" scenarios for HEVs: one where sales of HEVs were estimated to grow relatively quickly ("High HEV") and one where that market growth is significantly slower ("Low HEV"). Figure L-4 shows the fraction of new vehicles of each type (conventional and HEV) sold in each year under the two HEV market success scenarios. Figure L-5 shows the fraction of total vehicle miles driven by each type of vehicle by year.

The panel's three decision trees (Figures L-1 through L-3) identified many different possible outcomes of research on light-duty hybrids, all of which could be translated into estimated changes in fuel economy and estimated changes in vehicle costs. Overall, the trees specify 145 different possible

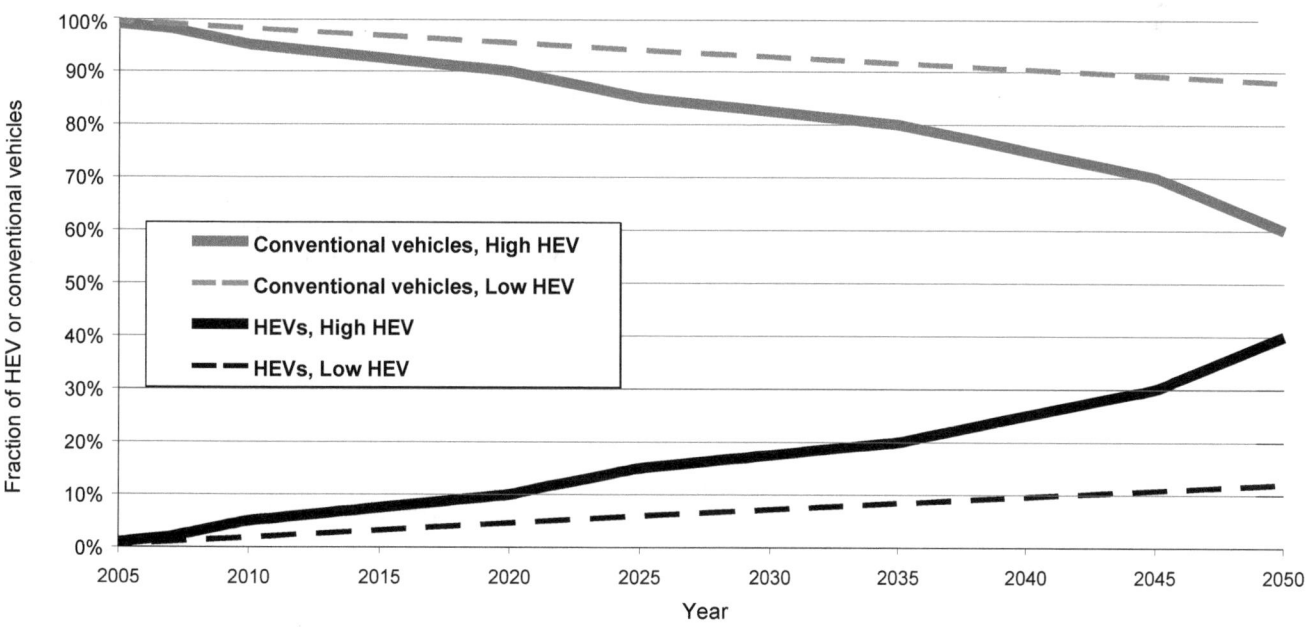

FIGURE L-4 Fraction of new vehicles purchased that are conventional and hybrid electric, for two HEV market scenarios.

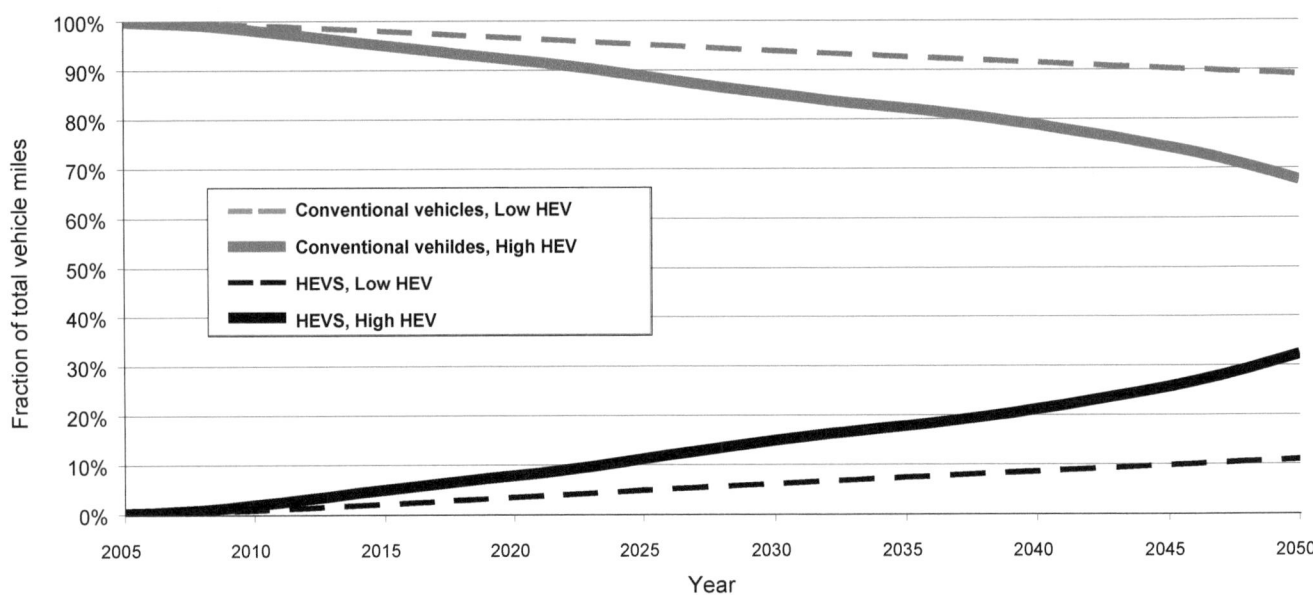

FIGURE L-5 Fraction of total vehicle miles driven by year by vehicle type, for two HEV market scenarios.

outcomes (hereinafter called "cases") for fuel economy and costs for conventional and HEVs.[8]

For each of these cases, the benefits model calculates the total fuel consumption, emissions, and consumer expenditures on vehicles and gasoline by year for each year through 2050. Attachment B describes the model and these calculations in more detail.

The panel's assessments result in two probabilities that can be assigned to each of the 145 cases: the probability of the outcome with DOE's research, and the probability of the outcome without DOE's research.

Economic Benefits

The expected economic benefits of DOE's research are calculated as the difference in expected value of total consumer expenditures with and without the program. For example, Figure L-6 shows the expected value (the probability-weighted average) of consumer expenditures for vehicles and fuel in the Low HEV market scenario assuming Reference Case prices, with and without the program,

as well as the uncertainty about those expenditures.[9] The expected economic benefit of the program is the difference in the expected value with and without the program: about $5.9 billion. Results of these benefits calculation for the two different HEV market success cases and for the three global scenarios are shown in the results matrix in Figure L-7.

Environmental Benefits

The primary environmental benefits anticipated from this research are a reduction in total carbon emissions as a result of having more fuel-efficient conventional vehicles and HEVs on the road. The benefits model estimates the total carbon emissions associated with each case by multiplying the total fuel use by 3.04 kg carbon per gallon of gasoline burned.[10] As with the economic benefits, the model produces an estimate by year of the total carbon emissions from automobiles for each case, and the expected environmental benefit of DOE's program is the difference in the expected value of total emissions with and without the program. Using a 3 percent annual discount rate, the total carbon emissions reduction attributable to the program in the Reference Case

---

[8]Twelve possible outcomes for fuel economy improvements based on the results of lightweighting and the engine efficiency research, multiplied by two possible outcomes for the costs of lightweighting, multiplied by six possible outcomes for the costs of batteries, which impact the incremental cost for hybridization, results in 144 possible outcomes. The 145th outcome is associated with incremental vehicle costs of more than 10 percent, judged by the panel not to be viable in the market. In that outcome it is assumed that the technologies are not implemented and that the benefits are not realized in the market.

[9]Annual expenditures from 2006 through 2050, discounted at 3 percent real, as recommended by the full committee. Note that the summary matrix includes economic benefits calculated at two different discount rates.

[10]It is assumed that a gallon of gasoline, when used in an internal combustion engine, would release 2.42 kg of carbon (or 8.87 kg of carbon dioxide). The supply chain (reservoir to pump) for gasoline is about 79.5 percent efficient. Therefore, about 3.04 kg of carbon is released into the atmosphere per gallon of gasoline consumed (3.04 is calculated as the ratio of 2.42 to 0.795).

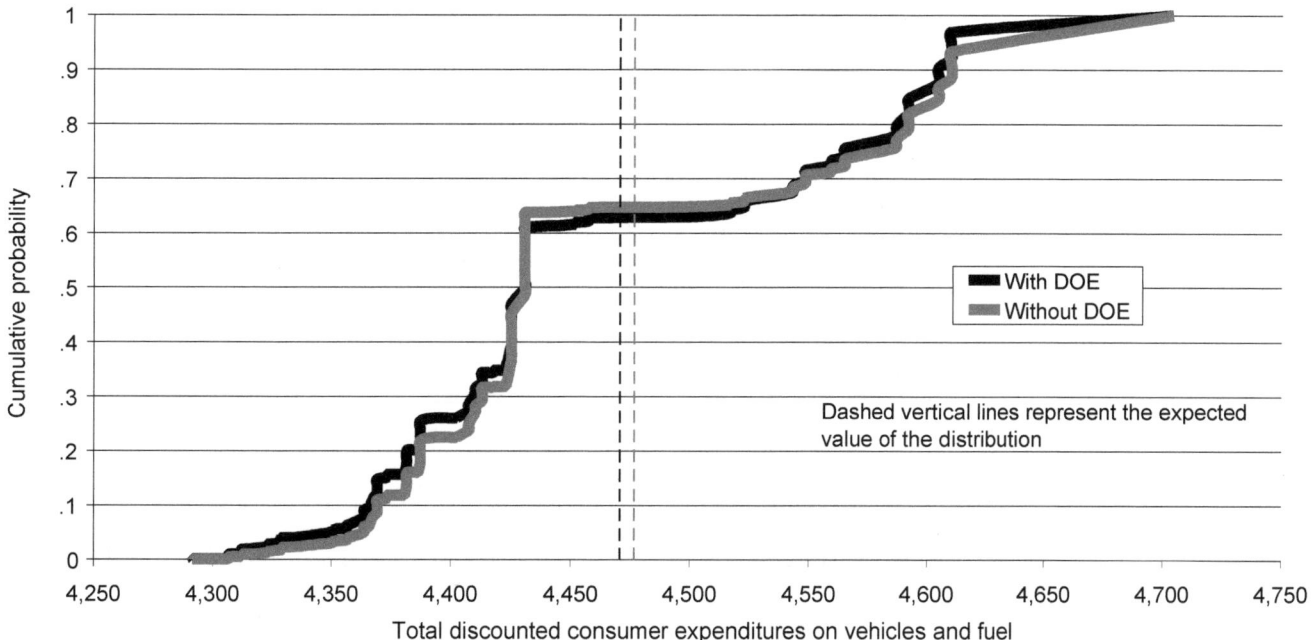

FIGURE L-6 Range and likelihood of discounted total consumer expenditures for vehicles and gas between 2006 and 2050, assuming the Low HEV market scenario and Reference Case prices. The distributions of outcomes with and without the DOE program are shown. The vertical lines represent the expected value of each distribution (the probability-weighted average), and the expected economic benefit of the program is the reduction in the expected total consumer expenditures.

is about 28 million metric tons. Results of these benefits calculations for the two different HEV market success cases and for the three global scenarios are shown in the results matrix (Figure L-7).

Security Benefits

The primary security benefits anticipated from this research derive from a reduction in gasoline use and, as a result, a lower demand for imported oil. As described above, the benefits model produces and estimate of the total amount of gasoline consumed in each case evaluated. The security benefit of DOE's program can be estimated by the reduction in the expected total gasoline demand with and without the program. Using a 3 percent annual discount rate, the expected value of the total reduction in gasoline consumption from 2006 to 2050 attributable to the program is about 9.8 billion gallons in the Reference Case. Results of these benefits calculation for the two different HEV market success cases and for the three global scenarios are shown in the results matrix (Figure L-7).

FINDINGS

Benefits of DOE's Light-Duty Hybrid Vehicle R&D

Finding 1: DOE's light-duty hybrid vehicle R&D program is likely to yield important technology advances that could improve the fuel economy for light-duty vehicles in the United States.

The panel notes that the technology advances resulting from DOE's program will not necessarily be used to improve the fuel economy of new vehicles. Depending on regulatory requirements and market drivers, automobile manufacturers may choose to use new and improved technologies to enhance vehicle performance and other attributes that are more attractive to consumers than improved fuel economy. Accordingly, while DOE's program will probably make it possible to achieve higher fuel economy, there is no guarantee, under existing regulations that the desired reduction in petroleum use will result. Demand-side policies to complement supply-side technology development are likely to be critical in achieving fuel economy benefits.

Finding 2: The methods currently used by DOE to assess the potential fuel economy benefits of its light-duty hybrid vehicles R&D tend to be overly optimistic in estimating the impact and timing of technology advances.

When DOE estimates the potential fuel economy benefits of its light-duty hybrid vehicles R&D efforts, it assumes that the program's very ambitious performance and cost goals will be met by the relevant target date(s). In contrast, the panel considered it unlikely that any of DOE's R&D efforts in electric hybrid technology, lightweight materials,

		Global Scenario		
		AEO Reference Case	High Oil and Gas Prices	Carbon Constrained[a]
Program Risks	Technical risks	Evaluated for four subsets of projects in the program. Technical risks identified as uncertainty in the level of technical advancement, uncertainty in the vehicle cost impact of the new technologies, and, for batteries, uncertainty in the time that the new technologies would be market-ready. See decision trees for the probabilities of technical success in each area.		
	Market risks	The panel addressed two types of market risks: market acceptance of more fuel-efficient conventional vehicles and market acceptance of HEVs. The market acceptance of more fuel-efficient conventional vehicles was assumed to be strictly a function of the trade-off between increased capital costs and decreased lifetime fuel costs. To address the market acceptance of HEVs, two HEV market conditions were defined. In the Low HEV condition, HEV sales increase linearly from 2003 market share to about 12% of new vehicle sales in 2050. In the High HEV condition, HEV sales increase exponentially from 2003 and account for about 40% of new vehicle sales in 2050.		
Program Benefits	Expected economic benefits	Economic benefits are calculated as the reduction in the expected consumer expenditures for vehicles and fuel from 2006 to 2050 attributable to the DOE program.		

		In the Low HEV market condition:
		  $5.9 billion at 3%
		  $3.7 billion at 7%
		In the High HEV market condition:
		  $7.2 billion at 3%
		  $4.2 billion at 7%
			In the Low HEV market condition:
			  $27.5 billion at 3%
			  $15.7 billion at 7%
			In the High HEV market condition:
			  $28.2 billion at 3%
			  $15.9 billion at 7%
				In the Low HEV market condition:
				  $7.3 billion at 3%
				  $4.7 billion at 7%
				In the High HEV market condition:
				  $8.5 billion at 3%
				  $5 billion at 7%

	Environmental benefits	Environmental benefits are calculated as the reduction in total carbon emissions from vehicles from 2006 to 2050 that can be attributed to the DOE program. Difference between Low HEV and High HEV market conditions is less than 2%; only one value is shown.		
		28 million metric tons	51 million metric tons	32 million metric tons
	Security benefits	Security benefits arise from reduced gasoline consumption and associated reduction in oil imports. The estimated reduction in gasoline use by vehicles from 2006 to 2050 that can be attributed to the program for Low HEV and High HEV market conditions are shown.		
		9.2 to 9.4 billion gallons	16.7 to 17 billion gallons	10.4 to 10.6 billion gallons

NOTE: Benefits are presented as expected values. Economic benefits are shown as present values discounted at both 3% and 7% real; environmental and security benefits are discounted at 3%.

[a]For the Carbon Constrained global scenario, the panel assumed Reference Case prices for oil and gas. The differences in expected benefits come from differences in the probability of technical and market risks.

FIGURE L-7 Results matrix of the Panel on DOE's Light-Duty Vehicle Hybrid Technology R&D Program.

and combustion would achieve the ambitious (stretch) goals within the specified time frames. As discussed later, partial success could certainly result in important benefits, albeit more modest than indicated by DOE's estimates.

In the panel's judgment, DOE underestimates both the lead time required for new technologies to be implemented in production vehicles and the time associated with commercialization ramp-up. A lead time of several years is required to introduce a new technology into a production vehicle. For technologies that are radically different from those in use, the lead time can be far longer, particularly if major changes in manufacturing and assembly processes are needed, as would be the case with the use of carbon-fiber-reinforced composites for body and chassis applications. Such radical changes are likely to require large capital investments by the automobile manufacturers. Thus, market penetration of new technologies is likely to be slow unless there are clear opportunities for manufacturers to amortize their investments over a relatively short period or unless new regulatory requirements (e.g., more demanding emissions standards) drive the implementation process.

The panel experienced some difficulty in establishing

what baseline DOE used in its benefits assessment. For such an assessment to be useful, the baseline for the comparison needs to be clearly defined. In the present case, selecting an appropriate baseline presents some challenges. If a late-1990s vehicle is the starting point, the anticipated incremental improvements in such a vehicle over time need to be taken into account. If no allowance is made for these improvements, then the benefits of DOE's program will be overestimated. However, some parts of DOE's program, notably in the advanced combustion and fuels area, are likely to contribute to the incremental improvements in the efficiency of conventional (nonhybrid) vehicles and should, therefore, be included when assessing the overall benefits of the program.

Finally, the panel noted that DOE does not distinguish the R&D it funds from R&D funded by others, including private industry and research organizations within the United States and overseas. All outcomes "associated with" the types of technologies DOE is investigating are used in the agency's benefits calculations. Because no attempt is made to extract the benefits directly attributable to DOE's program, the benefits of the program are overestimated.

Finding 3: Important fuel economy benefits could accrue even if DOE's R&D on light-duty hybrid vehicles fails to achieve its ambitious cost and performance goals.

Advances in the three areas of DOE's hybrid vehicle R&D program examined by the panel—electric hybrid technology, lightweight materials, and combustion and fuels—could result in improved fuel economy for conventional ICE vehicles even if there are only relatively modest improvements in performance and cost. Incremental technological improvements may be more readily incorporated into production vehicles than more ambitious advanced technologies and could have important benefits because the number of vehicles potentially affected is larger and the time required for significant implementation is shorter. Incremental innovation rather than radical innovation is a special skill of industry, and DOE's industry partners will play a key role in guiding the R&D.

As noted in the recent review of the FreedomCAR and Fuel Partnership (NRC, 2005b), R&D in the advanced combustion and fuels area appears particularly likely to yield commercial benefits. Improvements in engine and aftertreatment technologies could be incorporated into new vehicles quite rapidly, and even relatively small incremental improvements could have an important impact on the nation's fuel consumption when implemented in the 16 to 17 million new light-duty vehicles sold in the United States every year.

Finding 4: DOE's R&D on light-duty hybrid vehicles has benefits above and beyond the potential for improved fuel economy.

Although the potential for improved fuel economy was the only metric used by the panel for quantifying the prospective benefits of DOE's R&D, other benefits have resulted, or are likely to result, from the program. For example, improvements in vehicle reliability, recyclability, and performance can be anticipated. Also, hybrid vehicles with ICE power trains are one step on what might be a transition pathway to hydrogen-fueled fuel cell vehicles, and some of DOE's R&D may find application in those more futuristic vehicles.

Other benefits are harder to measure but nonetheless important. The program provides educational and training opportunities for researchers and contributes to their professional development. There is also a widely held view that one of the most important benefits of DOE's advanced vehicle technology programs in general has been the leverage provided by joint government-industry research.

DOE programs are a small fraction of the worldwide effort being applied toward hybrid vehicle technologies. In the panel's view, DOE efforts are likely to have been a catalyst for some non-DOE development, although it is difficult to substantiate this assertion or to assess the benefits derived from such a catalyst function.

Methodology Used by the Panel to Assess Prospective Benefits

Finding 5: The prospective benefits assessment methodology used by the panel to assess DOE's light-duty hybrid vehicle R&D offers value for managing this and similar research programs and for reviewing progress. In particular, the methodology

- Provides a framework for efficient and focused conversations between the reviewers of a research program and its proponents;
- Makes the decision process more transparent; and
- Helps to focus the attention of managers and reviewers on the most valuable program elements.

Applying the prospective benefits methodology to DOE's light-duty vehicle hybrid technology R&D program required the panel to specify key items that were not always apparent from the documents and information provided by DOE. In particular, some of the program goals were not described explicitly and completely. For example, setting a cost target of $28/kW for a battery by the year 2010 does not describe the objective adequately for assessment purposes. Does the cost target mean a customer could actually buy a battery at that cost? Does it mean that the technology exists that in principle would allow a commercial firm to make such a product? Does it mean the 500,000th production unit or the first? All these conditions must be specified for the assessment method to succeed, and both reviewers and proponents are forced to state their goals quite explicitly.

Equally important, the method focuses much of the conversation between reviewers and DOE staff around the goals. The template for presenting evaluation results and the decision trees used for technical and market risks provide a framework for structured conversation between program managers and their reviewers. This framework highlights areas of disagreement and so helps to focus the discussions on the key issues. The same holds true when program managers use the evaluation method as a management tool.

The structured conversations do not dispense with the need for informed, subjective judgment in evaluating the research. The method requires the judgment of knowledgable professionals at every step. Indeed, the program being evaluated could not be considered research if all elements of judgment were replaced with fact.

LESSONS FOR FUTURE APPLICATIONS OF THE METHODOLOGY

The panel offers six practical considerations, which may be useful for future applications of the methodology:

1. A review group unfamiliar with this approach will probably require a skilled facilitator to guide the members in applying it.
2. The use of conditional probabilities requires continued mental discipline on the part of the panel members.
3. Panel members who were more knowledgeable about a particular program element tended to be less optimistic about its probability of success, perhaps because they are more familiar with its problems.
4. The method requires that probability estimates for alternative options sum to one. This can make it difficult to interpret the decision trees. For example, if DOE funding increases the probability of achieving more challenging goals, the probability of achieving a lesser goal can appear to be lower with DOE funding than without DOE funding. This is an artifact of the method and means that reviewers must look at all the probabilities simultaneously to understand what the trees really mean. DOE also commented on this feature when completing the decision trees.
5. A research program can offer value if it achieves only a part of its goals. The method does not capture this well because the analyst must specify the partial achievement in advance and include that probability as an extension of the decision tree analysis.
6. Successful research can pay for itself by providing a range of benefits. This method, however, focuses on fuel economy as the single desideratum and does not quantify ancillary benefits—greater safety, superior vehicle performance, and so forth. However, these benefits can help the product embodying the technology to gain market share.

ATTACHMENT A
PANEL MEMBERS' BIOGRAPHIES

Wesley L. Harris (NAE), *Chair,* is Charles Stark Draper Professor of Aeronautics and Astronautics at the Massachusetts Institute of Technology and head of the Department of Aeronautics and Astronautics. Previously he was associate administrator of aeronautics at NASA and vice president and chief academic officer of the Space Institute at the University of Tennessee. His research interests include theoretical and experimental unsteady aerodynamics and aeroacoustics; computational fluid dynamics; and government policy impact on procurement of high technology systems. He is a member of the National Academy of Engineering, a fellow of the AIAA and of the AHS. He has been awarded several honorary doctoral degrees and has held several endowed professorships for visiting professors at U.S. universities.

David L. Bodde serves as a professor and senior fellow at Clemson University. There, he directs innovation and policy at the International Center for Automotive Research. Prior to joining Clemson University, Dr. Bodde held the Charles N. Kimball Chair in Technology and Innovation at the University of Missouri in Kansas City. Dr. Bodde serves on the boards of directors of several energy and technology companies, including Great Plains Energy, the Commerce Funds, and EPRI Solutions. His executive experience includes vice president, Midwest Research Institute; assistant director of the Congressional Budget Office; and deputy assistant secretary in the U.S. Department of Energy. He was once a soldier and served in the Army in Vietnam. He has extensive experience of energy policy and technology assessment, and his current work is directed at the role of entrepreneurs in the innovation and commercialization of energy technologies. He has served on a number of NRC committees, is a member of the Board on Energy and Environmental Systems, and a member of the NRC Committee on Review of the Research Program of the FreedomCAR and Fuel Partnership. He has a doctorate in business administration, Harvard University, M.S. degrees in nuclear engineering (1972) and management (1973), and a B.S. from the United States Military Academy.

Robert Epperly is president of Epperly Associates, Inc., a consulting firm. From 1994 to 1997, he was president of Catalytica Advanced Technologies, Inc., a company that develops new catalytic technologies for the petroleum and chemical industries. Prior to joining Catalytica, he was general manager of Exxon Corporate Research and earlier was director of the Exxon Fuels Research Laboratory. After leaving Exxon, he was chief executive officer of Fuel Tech, N.V., a company developing new combustion and air pollution control technology. Mr. Epperly has authored or coauthored over 50 publications on technical and managerial topics, including two books, and has 38 U.S. patents. He has extensive experience in fuels, engines, catalysis, air pollution control, and R&D management. He served as a member of the NRC's Standing Committee on Review of the Research Program of the Partnership for a New Generation of Vehicles (PNGV), which was a government-industry R&D program from 1993 to 2001 developing advanced vehicle technologies. He received an M.S. in chemical engineering from Virginia Tech.

David Friedman is the research director, Clean Vehicles Program, Union of Concerned Scientists (UCS), in Washington, D.C. He is the author or coauthor of more than 30 technical papers and reports on advancements in conventional, fuel cell, and HEVs and alternative energy sources, with an emphasis on clean and efficient technologies. Before joining UCS in 2001, he worked for the University of California at Davis in the Fuel Cell Vehicle Modeling Program, developing simulation tools to evaluate fuel cell technology for automotive applications. He worked on the UC Davis FutureCar team to build a hybrid electric family car that doubled its fuel economy. He previously worked at Arthur D. Little researching fuel cell, battery electric, and hybrid electric vehicle technologies, as well as photovoltaics. He served as a member of the NRC Panel on the Benefits of Fuel Cell R&D of the Committee on Prospective Benefits of DOE's Energy Efficiency and Fossil Energy R&D Programs (Phase One) and is currently a member of the NRC Committee on National Tire Efficiency. He earned a bachelor's degree in mechanical engineering from Worcester Polytechnic Institute and in 2005 was a doctoral candidate in transportation technology and policy at UC Davis.

Larry J. Howell is a consultant to industry and government, specializing in the management of R&D for business innovation, automotive technology, telematics, and vehicle structures and materials. Previous positions include executive director, science, General Motors Research and Development Center, in which capacity he served as chief scientist for GM, overseeing the GM R&D Center's six science labs: Thermal and Energy Systems; Electrical and Controls Integration; Materials and Processes; Enterprise Systems; Chemical and Environmental Sciences; and Vehicle Analysis and Dynamics. Dr. Howell had global responsibility for joint research with universities, government agencies, and GM's alliance partners. He also served as secretary to GM's Corporate Science Advisory Committee, which reports on technology issues to the General Motors board of directors. Other positions at GM included director of body and vehicle integration at GM Research; member of the General Motors Research Laboratories; and head of the Engineering Mechanics Department at GM Research. Prior to joining GM, he worked for General Dynamics Corporation as a principal investigator of research related to the structural dynamics of the space shuttle. In 1984, he completed the Executive Program at Dartmouth's Amos Tuck School of Business Administration. He is a member of the American Institute of Aeronautics and Astronautics (AIAA), the American Society of Mechanical Engineers (ASME), the Society of Automotive Engineers (SAE), and Sigma Xi. He served on an NRC study on the use of lightweight materials in 21st century army trucks, and has served on the College of Engineering advisory boards of the University of Illinois and Western Michigan University. He represented GM as a member of the Industrial Research Institute (IRI) and has served on the IRI board of directors. He is now an emeritus member of IRI. Dr. Howell earned his bachelor's, master's and doctoratal degrees in aeronautical and astronautical engineering at the University of Illinois, Urbana.

Allan D. Murray is president of Ecoplexus Inc., an automotive technology services company. Previously, he spent most of his career and has held a number of positions at Ford Motor Company, including technology director for the Partnership for a New Generation of Vehicles (PNGV) program, a government-industry partnership to develop advanced, affordable fuel-efficient vehicles; and manager, technology strategy, Plastic and Trim Products Division. As technology director of the PNGV program, he led government-industry research and development teams pursuing advanced vehicle construction, power trains, fuel cells, batteries, and power electronics. In his other positions he also led leading-edge automotive plastic and composite products, processes, and methodologies. He has extensive experience in bringing advanced automotive technologies and products from concept through production and has a broad-based knowledge of automotive systems and economics. He served as chairman and president of the nonprofit Michigan Materials and Processes Institute, the first automotive engineer elected a fellow of the Society of Plastics Engineers, and is a member of the Society of Automotive Engineers. He served as a member of the Panel on Benefits of Fuel Cell R&D for the Committee on Prospective Benefits of DOE's Energy Efficiency and Fossil Energy R&D Programs (Phase One). He has a Ph.D. and an M.S. in metallurgical engineering and materials science, Carnegie Mellon University; a B.S. in metallurgical engineering, University of British Columbia; and an M.B.A., Wayne State University.

William F. Powers (NAE) is retired vice president, research, Ford Motor Company. In his approximately 20 years at Ford he served as director, Vehicle, Powertrain and Systems Research; director, Product and Manufacturing Systems; program manager, Specialty Car Programs; and executive director, Ford Research Laboratory and Information Technology. Prior positions include professor, Department of Aerospace Engineering, University of Michigan, during which time he consulted with NASA, Northrop, Caterpillar, and Ford; research engineer, University of Texas; and mathematician and aerospace engineer, NASA Marshall Space Flight Center. He is a fellow, Institute of Electrical and Electronics Engineers; member, National Academy of Engineering; foreign member,

Royal Swedish Academy of Engineering Sciences. He has extensive expertise in advanced research and development of automotive technology. He is a member of the NRC's Board on Energy and Environmental Systems and recently served on the Committee on Alternatives and Strategies for Future Hydrogen Production and Use. He has a B.S. in aerospace engineering, University of Florida, and a Ph.D. in engineering mechanics, University of Texas, Austin.

Gary W. Rogers is president, chief executive officer and sole director, FEV Engine Technology, Inc. He is also president, FEV Test Systems, Inc. His previous positions included director, Power Plant Engineering Services Division, and senior analytical engineer, Failure Analysis Associates, Inc.; design development engineer, Garrett Turbine Engine Company; and exploration geophysicist, Shell Oil Company. He has extensive experience in research, design, and development of advanced engine and power train systems, including homogeneous and direct-injected gasoline engines, high-speed direction injection (HSDI) passenger car diesel engines, heavy-duty diesel engines, hybrid vehicle systems, gas turbines, pumps, and compressors. He provides corporate leadership for a multinational research, design, and development organization specializing in engines and energy systems. He is a member of the Society of Automotive Engineers, is an advisor to the Defense Advanced Research Projects Agency on heavy-fuel engines, and sits on the advisory board to the College of Engineering and Computer Science, Oakland University, Rochester, Michigan. He served as a member of the NRC Committee on Review of DOE's Office of Heavy Vehicle Technologies Program, and served on the NRC Committee on the Effectiveness and Impact of Corporate Average Fuel Economy (CAFE) Standards. He also recently supported the Department of Transportation's National Highway Traffic Safety Administration by conducting a peer review of the NHTSA CAFE model. He as a B.S.M.E. from Northern Arizona University.

James Lee Sweeney is professor and former chairman, Department of Engineering-Economic Systems and Operations Research, Stanford University. He has been a consultant, a director of the Office of Energy Systems, a director of the Office of Quantitative Methods, and a director of the Office of Energy Systems Modeling and Forecasting, Federal Energy Administration. At Stanford University, he has been chairman, Institute of Energy Studies; director, Center for Economic Policy Research; and director, Energy Modeling Forum. He has served on several NRC committees, including the Committee on the National Energy Modeling System, the Committee on Impact and Effectiveness of Corporate Average Fuel Economy (CAFE) Standards, and the Committee on the Human Dimensions of Global Change. He served on the Committee on Prospective Benefits of DOE's R&D on Energy Efficiency and Fossil Energy (Phase One), helping to develop the framework and methodology that committee applied to evaluating benefits. His research and writings address economic and policy issues important for natural resource production and use, energy markets including oil, natural gas and electricity, environmental protection, and the use of mathematical models to analyze energy markets. He has a B.S. degree from the Massachusetts Institute of Technology and a Ph.D. in engineering-economic systems from Stanford University.

ATTACHMENT B
DESCRIPTION OF THE BENEFITS MODEL

Model Description

To quantify economic, environmental, and security benefits, the panel used a simple model that tracks fuel use and costs, incremental vehicle costs, and carbon emissions from the light-duty vehicle fleet. The model used by the panel is adapted from a vintage stock model developed by the Committee on Alternatives and Strategies for Hydrogen Production and Use (NRC, 2004b). This model includes assumptions about the total vehicle miles driven by year as a vehicle ages (14 years and about 142,000 miles over that life, with more miles on newer vehicles), about total vehicle sales per year (assumed to grow at about 2 percent per year starting in 2004), and about current and future trends in vehicle fuel economy. Based on these assumptions, the model produces estimates, by year, of the total vehicle miles driven, the average fuel economy of vehicles on the road, the total gasoline consumed, and the total carbon emissions. The model covers the period from 1987 through 2050; the panel was concerned only with the projections from 2006 through 2050.

The panel modified the model by including three types of light-duty vehicles that are assumed to compete for the total volume of light-duty vehicles sold in each year. Base case conventional vehicles (CVs) are defined as having fuel economy performance that increases by about 10 percent between 2005 and 2015 and then increases slowly over time at about 1 percent per year. New CVs are defined as CVs that incorporate new technologies aimed at improving fuel efficiency, typically at some increased cost over the base case conventional vehicles. The fuel economy associated with new CVs is based on the results of the panel's decision tree assessments: Overall, 145 different fuel economy improvements and incremental vehicle costs "cases" were identified. Finally, the model also includes hybrid electric vehicles (HEVs). HEVs are assumed to have 30 percent better fuel economy than the CVs on which they are based. There are also incremental costs associated with HEVs: In 2006 that incremental cost is assumed to be $2,500; in future years, the incremental cost depends on the results of R&D on batteries. The total fuel consumption depends on the type and quantity of light-duty vehicles that are on the road in any given year.

Rather than explicitly model the competition between HEVs and CVs, the panel chose to define two HEV market conditions and evaluate the benefit of the program under both conditions. In the Low HEV market condition, HEV sales increase linearly from today's market share to about 12 percent of new vehicle sales in 2050. In the High HEV condition, HEV sales increase exponentially from today and ultimately account for about 40 percent of new vehicle sales in 2050.

The market acceptance of new CVs was assumed to be strictly a function of the trade-off between increased capital costs and decreased lifetime fuel costs. To determine which type of CV would be purchased in any given year, a simple cost comparison is made of the total capital and fuel costs assuming a 14-year vehicle life, annual vehicle mileage over the life of the vehicle, and a consumer discount rate of 7 percent. Whichever vehicle has the lower discounted total cost is assumed to capture the entire market for CVs in that year.

Calculating Fuel Usage, Fuel Expenditures, Incremental Vehicle Cost, and Carbon Emissions for a Given Set of Vehicle Characteristics

Based on the market assumptions (total vehicles sold, fraction of total vehicles that are HEVs, and fraction of CVs that implement new technologies) and fuel economy estimates for the three vehicle types, the model produces an estimate of annual gasoline usage by light-duty vehicles annually from 2006 through 2050. The estimated gasoline usage is then translated into economic expenditures and carbon emissions.

Economic expenditures on gasoline are estimated based on the projected price of gasoline (excluding taxes) by year and the total volume of gasoline used. The price of gasoline is estimated based on the price of oil: Refining and distribution are assumed to add about 42 cents per gallon to the crude oil price. The price of oil is defined by the global scenarios being considered: the 2005 Reference Case prices and twice those prices for the High Oil and Gas Prices scenario. The price of oil in the Carbon Constrained scenario is assumed to be the same as in the Reference Case. Prices were assumed to be constant after 2025 through 2050.

Annual carbon emissions are calculated based on the total gasoline usage and an estimated 3.04 kg carbon emitted per gallon of gasoline consumed.

Finally, HEVs and new CVs will cost the consumer more than base case CVs, and those incremental costs must be accounted for in the estimate of the total economic impact of the new technologies. In 2005, HEVs cost approximately $2,500 more than comparable CVs. As with the fuel economy estimates, the incremental costs of HEVs and of new CVs are defined for the specific case being evaluated. The incremental per-vehicle costs are multiplied by the number of vehicles of each type that are sold to produce an estimate of the incremental vehicle costs by year.

Estimating the Benefit of the DOE R&D Program

As described above, the panel's discussion and assessment of the technical risks associated with DOE's R&D activities resulted in the identification of 145 "cases," or different possible outcomes for fuel economy and incremental vehicle cost for new CVs and HEVs. The assessment also results in two probabilities for each case: the probability of that outcome with DOE's research program and the probability of the outcome without DOE's research program.

The expected economic benefit of the DOE program is the difference in the expected value of total consumer expenditures on vehicles and fuel with the program and those expenditures without the program. The expected value of total consumer expenditures in each case is calculated as the probability-weighted average of the expenditures in each of the 145 cases. The total consumer expenditures is calculated as the discounted net present value of fuel costs and incremental vehicle costs between 2006 and 2050, discounted at 3 percent or 7 percent real.

Similar calculations for the difference in the expected value of carbon emissions and gasoline consumption yield values for the expected environmental and security benefits of the DOE program.

M

Report of the Panel on DOE's Chemical Industrial Technologies Program

INTRODUCTION

The Committee on Prospective Benefits of DOE's Energy Efficiency and Fossil Energy R&D Programs, Phase Two, is applying a methodology for benefits evaluation that it developed in its Phase One study. The purpose of Phase Two is to give decision makers useful information on selected programs. Phase Two is also intended to refine the methodology developed in Phase One. The committee selected six programs for this purpose, one of which is the Chemical Industrial Technologies Program, the subject of this report.

The panel's assignment was to estimate the benefits of the Chemical Industrial Technologies Program, taking into account technical and market risks. The panel also was asked to make observations regarding program risks, benefits, and other factors that could help decision makers in evaluating the program. This report describes the work of the panel and presents its conclusions and recommendations on both the methodology and the program.

The panel members included persons from academia, industry, and government having extensive experience in chemicals processing. In addition, two members of the full committee served on the panel, one as chair. A consultant experienced in decision analysis assisted the panel. Biographies of panel members are presented in Attachment A.

The panel met twice for 2 days each time. On October 10-11, 2005, the panel received information on the program from DOE program representatives and the contractor who assists DOE in estimating benefits for the Chemical Industrial Technologies Program. At that meeting, the panel examined several of the program's individual projects in detail and from that examination determined how to apply the recommended methodology to this specific situation. Before the next meeting, which occurred on November 3-4, 2005, each panel member independently developed technical and market risk estimates for each of the 22 projects that currently make up the overall Chemical Industrial Technologies Program. At the second meeting, the panel reviewed its risk estimates and calculated the gross risk-adjusted benefits of the program. Following the second meeting, the panel developed a procedure for calculating net benefits. Members of the panel drafted sections of this report, circulated them electronically, and on January 9, 2006, conferred by telephone on the draft report. Based on the results of this conference call, the report was redrafted and circulated for further comment.

The panel is especially indebted to Dickson Ozokwelu, DOE's manager of the Chemical Industrial Technologies Program, and to Energetics, Inc., the program's technical assistance contractor, for outstanding support of the panel's work.

SUMMARY OF THE DOE PROGRAM

The Chemical Industrial Technologies Program, a subprogram of the DOE Industrial Technologies Program, has as its objective to reduce energy use in the chemicals industry. The chemical industry is a major energy consumer, accounting for 23.5 percent of the energy[1] among those industries included in DOE's Industrial Technologies Program.[2] The goal of the Chemical Industrial Technologies Program is to implement a successful strategy for DOE research that helps the chemical industry to achieve a 20 percent reduction by year 2020 in energy usage relative to its 2001 energy consumption of 6.6 quadrillion British thermal units (quads). This translates into a reduction in the chemical industry's energy use by 1.3 quads per year in 2020 and a proportional reduction in emissions.

To achieve its goal, the Chemical Industrial Technologies

---

[1]Dickson Ozokwelu, U.S. Deparment of Energy, "Overview of the chemicals subprogram," Presentation to the panel on October 11, 2005.

[2]ITP is focused on energy-intensive materials and process industries that account for over 55 percent of U.S. industrial sector energy consumption. These industries include aluminum, chemicals, forest products, glass, metal casting, mining, and steel (DOE, 2005e).

Program has a portfolio of relatively small projects, all of them competitively awarded and all of them involving 30-50 percent cost sharing. The amount of cost sharing increases as the projects get closer to commercial readiness. Industry partners are involved in planning and funding the research program. DOE has worked with a committee of 15 large chemical producers, the Chemical Industry Vision 2020 Technology Partnership, to select the projects to be funded in the Chemical Industrial Technologies Program. The partnership focused on 10 individual areas of future promise, such as computational fluid dynamics and ionic liquids. An exergy analysis[3] identified the major opportunities for energy savings in the chemical industry, focusing DOE project selection on those areas with the largest potential savings. This joint industry-government partnership is ongoing in helping DOE to select and evaluate projects. The primary selection criterion is energy savings in the United States.[4]

Twenty-two projects are currently being funded at $7 million per year. These projects, if they all were funded to completion and all were successful, are estimated by DOE to achieve a saving of 0.303 quads per year, or 23 percent of the overall program goal for the Chemical Industrial Technologies Program. Previous successful projects and new projects yet to be started will increase the savings.

The current Chemical Industrial Technologies Program as administered by DOE funds projects in three subprogram areas:

- *Reactions (13 projects).* This area includes oxidation catalysis, microreactors, new process chemistry, and biocatalysis. The total energy saving goal in the reactions area is 650 trillion British thermal units (TBtu) by 2020, 242 TBtu of which is accounted for by current projects. In several of the funded projects, reaction and separation are carried out in the same step, thus attempting to gain substantial saving in both energy consumption and capital equipment. This work is part of the current worldwide trend toward process intensification.
- *Separations (4 projects).* This area includes distillation hybrids, crystallization (not funded at the present time), and membrane separations. The total energy saving goal in separations is 420 TBtu, of which 40 TBtu is represented by current projects. This area also includes projects that combine separation and reaction (e.g., catalytic distillation) in the same equipment.
- *Enabling technologies (5 projects).* This area includes materials of construction, with an emphasis on corrosion, computations to improve the efficiency of dense fluidized beds, industrial energy systems such as an energy-conserving burner design, and sensors and control equipment. The total energy saving goal in this area is 260 TBtu, 21 TBtu of which is in current projects.

The budget of the Chemical Industrial Technologies Program has shrunk drastically. From $13 million in FY03, the budget decreased to $9 million in FY05 and $7 million in FY06. There is a clearly apparent contradiction between the ambitious goals of the program and the dwindling resources available to pursue them. The DOE management team seeks to cope with this situation by continuing its portfolio review, described below.

There have been a number of accomplishments from completed projects in FY03-05 in the Chemical Industrial Technologies Program. Examples include in situ analysis, distillation column flooding prediction, dimpled heat exchanger tubes, catalytic hydrogen retrofit reactors, and new alloys for ethylene cracker tubes.[5] All of these, with the possible exception of the retrofit reactor, have relatively specialized applications that will produce positive but modest benefits.

However, the program is also pursuing sweeping changes in process design in the hope that they can yield big energy savings. To focus their shrinking budget on the highest payoff projects, the DOE management team is trying to stimulate industry research and to fund projects that would not be possible without federal assistance, and is accelerating progress on projects that might be funded by industry sometime in the future. The portfolio is under continuing review, with the objective of ending projects that are not progressing or that promise only small energy savings.

A major new element complicating the management problem in this program is the surge in energy prices in 2005. This increase worsens the trend that few if any commodity chemical plants consuming large amounts of energy will be built in the United States in the foreseeable future. DOE intends to respond to this trend by emphasizing technologies that can be retrofitted to existing plants to increase their efficiency and lower their costs.

TECHNICAL AND MARKET RISK ASSESSMENT

In assessing the probabilities of technical success for the overall program, the panel needed to decide on whether to proceed from the top down or to rely on the project evaluations to build the program evaluation from the bottom up. Because of their number and heterogeneity, all 22 existing projects were assessed individually. The panel estimated the probability of technical and market success and estimated the benefit for the program as a whole by rolling up the project assessments. The panel was reasonably confident of its technical risk estimates but found market acceptance harder

---

[3]The exergy content of a system indicates its distance from thermodynamic equilibrium.

[4]For a more complete description and evaluation of the strategic planning process of the Chemical Industrial Technologies Program, see NRC (2004a). The report presents a positive evaluation of the strategic planning process for the Chemical Industrial Technologies Program.

[5]Dickson Ozokwelu, U.S. Department of Energy, "Overview of the Chemicals Subprogram," Presentation to the panel, October 11, 2005.

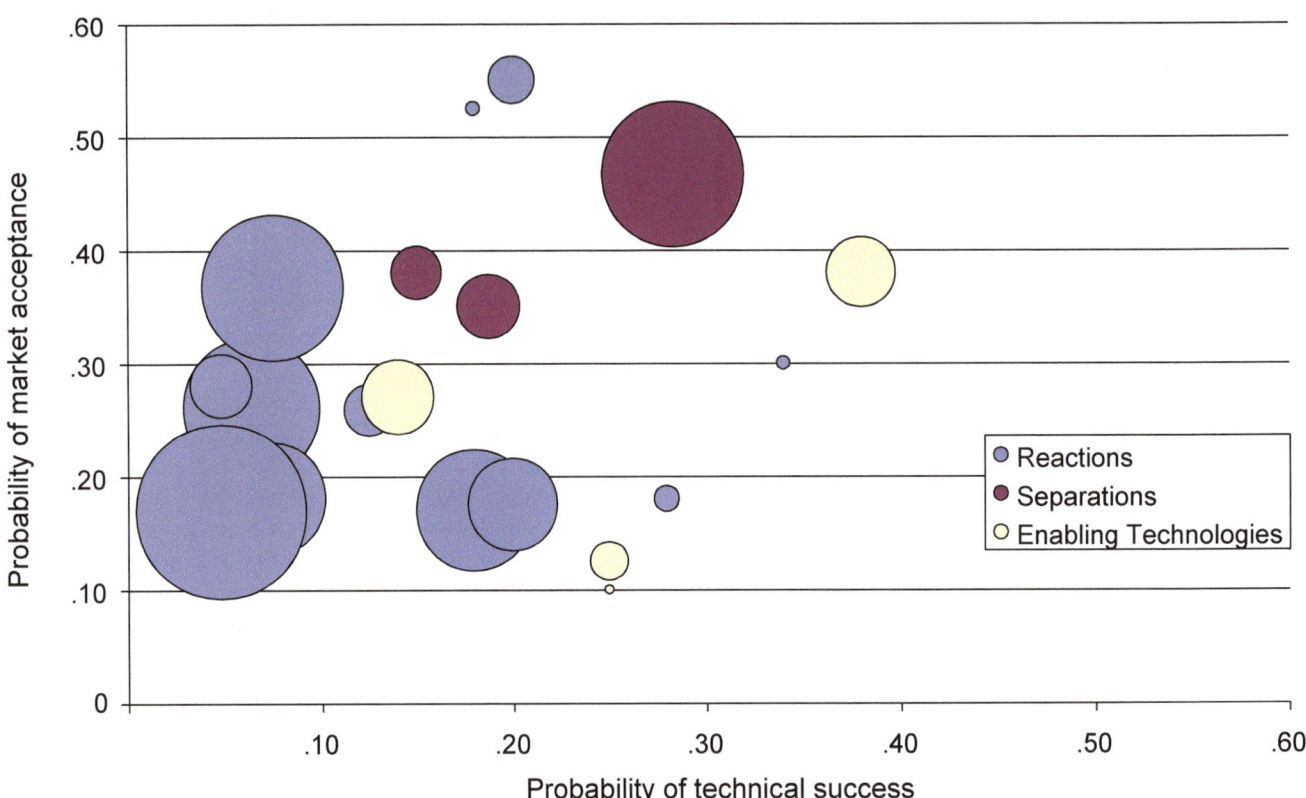

FIGURE M-1 Probability of technical success, probability of market success, and value of benefits if the project is successful. The area of a bubble is directly proportional to the estimated benefit for its associated project.

to evaluate because a number of the factors are difficult to predict at an early stage of research.

Although energy savings is the driver for DOE investment in these research projects, market acceptance of the technologies is determined by many additional factors. For example, industry also considers the total capital involved and its cost, the status of the industrial partners' manufacturing capacity at the time the technology is ready for consideration as a retrofit or possibly for use in a new plant, future raw material costs and availability, and principal market areas to be served. If all these factors are favorable, company managers will seek to convince themselves that an acceptable rate of return several years after a plant has started up can be achieved (usually about 20 percent return on investment for a new plant and often much more for a retrofit).

Notwithstanding these complexities, the panel believes it arrived at reasonable, and reasonably consistent, estimates of technical and market risk for all projects. The initial estimates from all the members were relatively close,[6] and the few outliers were quickly reconciled during the discussions of each project. This was aided by the fact that most projects were in areas where one or more committee members had familiarity or experience.

Overall, the panel assessed the projects in each category to be high risk and thus consistent with DOE's program strategy. Probability of technical success ranged from 5 to 50 percent, with market success from 10 to 70 percent. Combining these assessments the range for success for the entire portfolio is between 1 and 35 percent with an average of 7.5 percent.[7] Figure M-1 is a bubble chart that graphically summarizes the risks and potential benefits of each project. Each project is represented by a bubble on the chart: the size of the bubble is proportional to DOE's estimated gross economic benefit of the project if it is successful, and the location of the bubble indicates the technical and market risks assessed by the panel for that project.

RESULTS AND DISCUSSION

As noted in the Introduction, above, the Phase Two methodology suggests that an evaluation of the benefits of a specific program should explicitly consider (1) the role of DOE funding, (2) the technical and market risks that can affect the outcome and value of the program's activities, and

---

[6]The initial estimates were made between the first and second panel meetings and discussed at the second meeting.

[7]This result is consistent with research into the success rate of projects of this type. For example, in its presentations DOE compared the Chemical Industrial Technologies Program to studies that found an overall success rate of 16 percent for projects at a similar stage in the research chain.

(3) the net economic, environmental, and security benefits that will result from the portfolio. The methodology also requires that benefits be estimated under three global scenarios, representing three different possible future states of the world. The panel considered each of these considerations in their evaluation.

Technical and Market Success

As noted above, the panel estimated a probability of technical success and a probability of market success for each project in the Chemical Industrial Technologies Program. In the panel's judgment, the probabilities that might be assigned to the technical and market success for the projects would be the same for each of the three global scenarios—that is, in the AEO Reference Case, High Oil and Gas Prices scenario, and the Carbon Constrained scenario.

Benefits Estimation

The economic, environmental, and security benefits of the Chemical Industrial Technologies Program derive directly from the energy savings resulting from the projects. To the degree that energy usage in the chemical industry is reduced, there are net economic benefits to the nation in the form of lower energy demand and lower energy prices, environmental benefits in the form of reduced environmental emissions, security benefits in the form of less use of oil and natural gas and the associated diminished reliance on imports.

Benefits estimation thus begins with an estimate of the energy saving that would result from successful technology development and adoption. DOE's contractor, Energetics, uses a formal procedure to estimate the energy savings of a project, from which the gross economic benefit can be calculated. The procedure also estimates the reduction in the emissions of criteria pollutants. The panel reviewed this procedure and concluded that it was applied consistently for all projects and that it produced a reasonable estimate of the future energy savings for projects that are in the early stage of research.

Since the DOE estimates assume 100 percent success of each project, the panel calculated the expected total benefit of the portfolio by applying its probabilities of technical and market success to DOE's estimates of energy savings for each project in the portfolio and rolled up the individual project estimates into an expected value of gross benefits for the overall program.

To get some insight into the uncertainty surrounding this expected value, a probability distribution of the economic benefits of the portfolio was generated using the risk estimates for individual projects. Each project was assumed to succeed with the net probability estimated by the panel (technical and market success) and to yield the full benefits calculated by DOE if successful and no benefits if not successful. By simulating many cases using these probabilities it was possible to create a distribution of expected benefits. The panel recognizes this is a simplification that does not match actual outcomes—each research project could partly succeed and simply yield lower benefits than estimated. It believes, however, that the simplification is a reasonable first-order approximation.

Finally, the panel considered the benefits estimates for each of the global scenarios prescribed by the committee. Although the economic benefits of the research depend on the global scenarios, which have different energy prices, the environmental and security benefits are considered to be constant in all three global scenarios.[8] The panel thus calculated economic benefits using the procedure described above but plugged in energy prices appropriate to each scenario.

Role of DOE Funding

In developing their benefits estimates for each project, DOE assumed that the research being funded will accelerate the development and implementation of the identified technology, but that the technology would eventually have been developed by industry if DOE were not supporting the research. The benefit of DOE support is then defined as the energy and other savings that are realized by having the technology available earlier. The acceleration period was estimated technology by technology in the analysis performed by Energetics. For the projects evaluated in this report, they estimated the acceleration usually at about 3 to 5 years; one project claimed a 10-year acceleration, and one a 20-year acceleration.

The panel agreed that in almost all cases DOE support would be a significant accelerating factor, mainly because there are high risk projects. Indeed, in today's worldwide competitive chemical industry, companies are less likely to spend money on the early exploration of high-risk technology with a low expectation of a large benefit than on safer R&D to gain incremental advances. For this reason, some members of the panel believed that industry is reluctant to make investments in the kind of projects funded by the Chemical Industrial Technologies Program. If so, the acceleration of energy savings due to DOE investment would be much greater. Additionally, some projects in the DOE portfolio, such as developing an industry-wide database on corrosion management, are unlikely to be sponsored by a single company. On balance, the panel concluded that DOE's acceleration estimates were reasonable and in some cases probably conservative. Accordingly, the panel adopted these estimates as a basis for calculating benefits.

---

[8]Arguably, the market penetration of energy efficient technologies would increase in the High Oil and Gas Prices scenario. On the other hand, higher prices would have the effect of driving domestic chemical production offshore. The panel did not believe the net change in quantities would be small enough to ignore for the purpose of this analysis.

Decision Tree

In developing the decision tree required by the Phase Two methodology, the panel considered two issues. The first was to estimate the benefits associated with the "no DOE funding" branch of the tree. Because the benefits of the "DOE funding" branch were calculated as the incremental benefit arising from technology acceleration, the "no DOE funding" branch was assigned a value of zero.

The other issue was whether to develop a decision tree and assess probabilities of technical and market success for the overall program as a whole (from the top down) or to rely on the project evaluations to build the program evaluation from the bottom up. As described above, the panel implemented the bottom-up approach, evaluating each project individually and calculating the portfolio-level benefits directly from the project evaluations. To characterize the uncertainty in the benefits of the portfolio as a whole, the panel developed a probability distribution of expected total economic benefits of the portfolio based on the simulation described above. Figure M-2 illustrates the results of this evaluation of uncertainty in portfolio benefits. From this simulation and the assumptions regarding the role of DOE funding, an overall decision tree representing the benefits analysis was derived (Figure M-3). Both figures use gross economic benefits estimates, since the panel determined that net benefits can best be adjusted at the program level.

In the evaluation performed between the two meetings, panelists did estimate the likelihood of each subprogram achieving various levels of technical success. This top-down analysis was compared to the results of the bottom-up analysis described above. For the two smaller subprograms—separations (four projects) and enabling technologies (five projects)—the expected benefits estimated at the portfolio level were quite close to the benefits for the subprogram derived from the individual project evaluations (within 5 percent for the separations projects and within 15 percent for the enabling technologies projects). For the Reactions subprogram, however, with 13 projects, the estimate of benefits at the subprogram level was 10 times greater than the estimate derived from the project-level evaluation. The discrepancy is believed to be partly an artifact of the assessment questions asked, where very low subprogram benefits were not an option, and partly a result of the cognitive difficulties in aggregating a large number of unrelated project evaluations. Based on this result, the panel concluded that bottom-up methodology was more useful for the analysis of a portfolio of diverse and unrelated projects.

Net Economic Benefits Calculation

The economic benefits estimated above, like those estimated by DOE, represent the gross economic benefits of the program–that is, the estimate includes all the benefits potentially flowing from the research but not the full costs of achieving those benefits. Net economic benefits could be estimated for each project by reducing the gross benefits by the R&D costs, the additional development costs of bringing the technology to a point where it could be commercially implemented (weighted by the likelihood the R&D would lead to further development), and the costs of implementing the technologies should a decision be made to do so (weighted by the likelihood a commercialization decision would be made). However, because these projects are at such an early stage of R&D, the panel did not estimate the future development and commercialization costs, nor did it believe it had the information necessary to estimate those costs for each project.

Therefore, a simplified method was employed to estimate net economic benefits from the gross economic benefits calculated by DOE. First, it is assumed that for any technology that reaches market success, a decision will have been made by private industry to implement it. Such a decision is typically made on an economic basis, with the project passing some internal hurdle rate for financial return. A typical hurdle for a cost-reducing project would be a 3-year payback period. Therefore, the panel simply assumed that for any project that "succeeds," investment costs no more than the net present value of the economic benefits for first 3 years were required.

The net result of these assumptions is to reduce the gross benefits of each project by an amount equal to that associated with a 3-year delay in the gross benefits stream. Accordingly, the net economic benefits are taken as 91.5 percent of the gross economic benefits, corresponding to a decrease in benefits associated with a 3-year delay.[9]

Results Matrix

Figure M-4 summarizes the results of the panel's estimation of the benefits of DOE's Chemical Industrial Technologies program. Benefits are calculated as described in the report, for all three scenarios. Only the net economic benefits change in the three scenarios, as they are partially dependent on energy prices. The figure presents both the probability-weighted average of the net economic benefits (the statistical expectation) and the uncertainty around that estimate. Figure M-4 also presents the panel's observations about the program.

SUMMARY AND CONCLUSIONS

The panel believes that the benefits estimation methodology developed for Phase Two has successfully applied to the

---

[9]This reduction in estimated benefits is largely modest, because the payback assumption embodies a relatively high private sector hurdle rate, while public benefits are discounted at a lower rate (between 3 percent and 7 percent). The delay of 3 years at the lower public discount rate changes the benefits estimate only slightly.

APPENDIX M

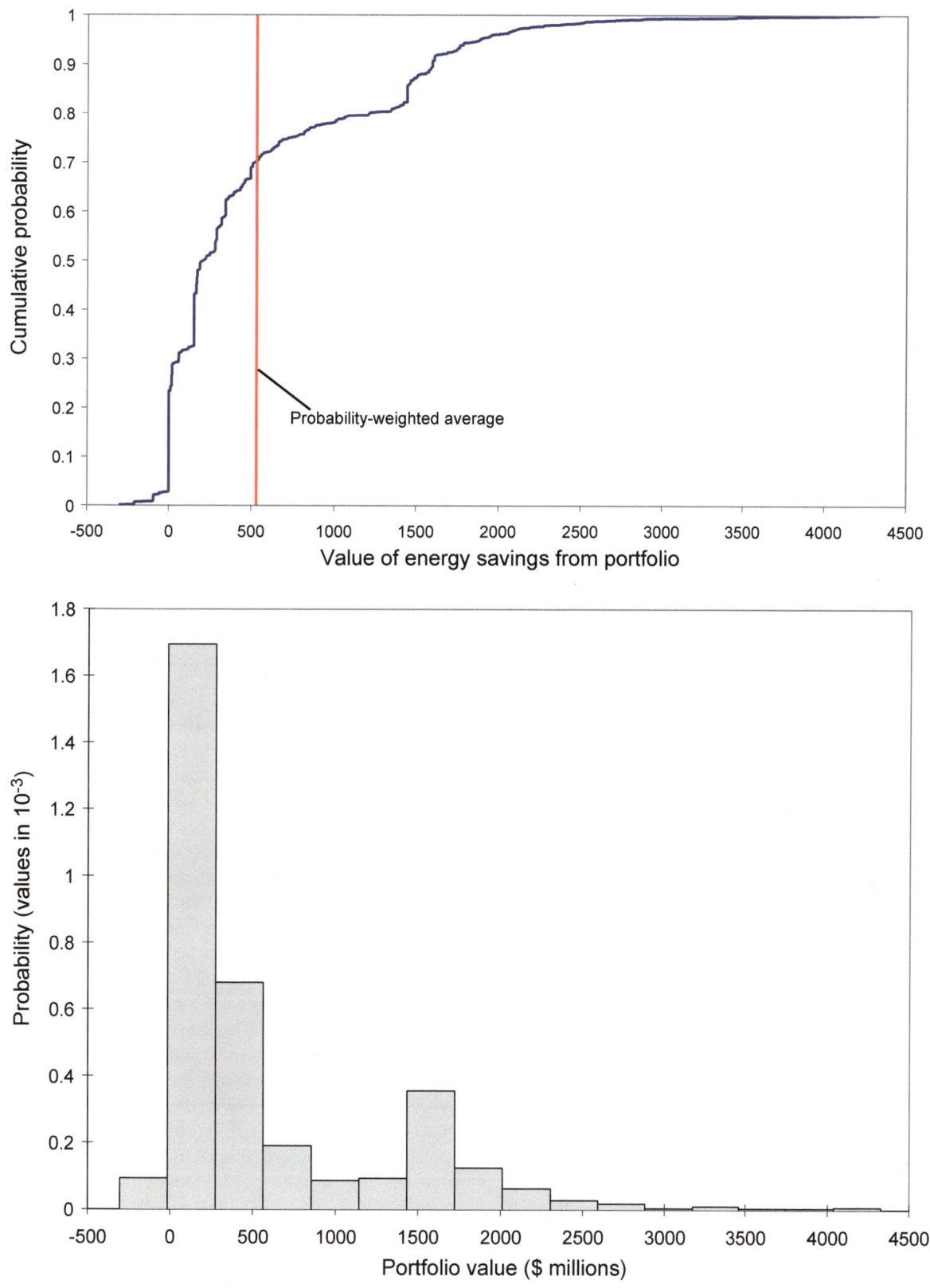

FIGURE M-2 Uncertainty surrounding estimates of program benefits.

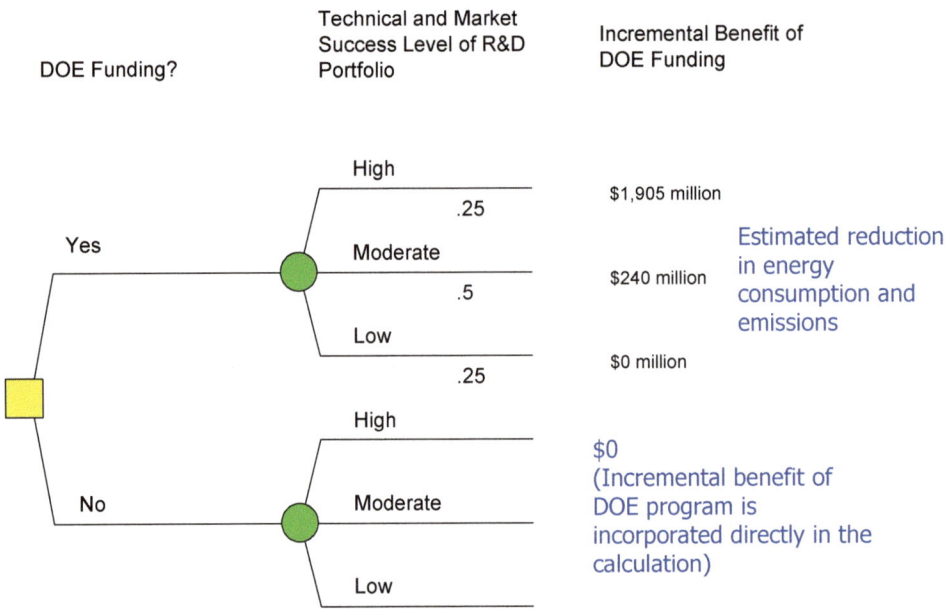

FIGURE M-3 Implied decision tree representing the panel's evaluation of the ITP–Chemicals portfolio. This tree is derived from the simulation of total portfolio benefit described in the report. There are several ways to derive a discrete distribution for a decision tree from a continuous distribution. For this figure, a well known shortcut for the bracket-mean approach was used, where the 10th, 50th, and 90th percentiles of the continuous distribution are selected to represent the full distribution and given weights of .25, .5, and .25 respectively.

Chemical Industrial Technologies Program. On the basis of its evaluation, the panel also concludes that the Chemical Industrial Technologies Program is addressing an important opportunity to produce energy savings in a major industrial segment by supporting early-stage research that industry is unlikely to support. However, to realize its potential benefits, the program must adapt to a seriously constrained budget and a changing domestic industrial environment. The following sections discuss more fully these conclusions about the program and the benefits estimation methodology.

Program Conclusions and Recommendations

As noted earlier in this report, the Chemical Industrial Technologies Program faces a very tight and rapidly declining budget. In addition, it seeks to accelerate energy savings in a dramatically changing U.S. chemical industry, notably one in which very few greenfield plants are being constructed in the United States but instead are going overseas to be closer to growth markets and/or raw materials. Rapidly increasing costs for energy and petroleum-derived raw materials could exacerbate these trends. DOE says it is moving toward the view that technologies that can be implemented as retrofits to existing plants will be favored so as to allow this country to directly reap most of the energy-saving benefits. Retrofits are subject to the same financial return tests as greenfield plants, with the possible exception that they are expected to generate earnings sooner after start-up since the primary objective is cost savings, not increased volume.

Against this background, and based on the experience of its members in managing and conducting research in the chemical industry, the panel wishes to underscore how essential it is for DOE management to focus its limited resources on the most promising opportunities available to it. The danger is more than just that of wasting funds on less promising projects. Equally if not more significant is the possibility of losing the benefits of high-priority projects because they were not pursued aggressively enough. This danger is particularly great for early-stage research of the type being funded by DOE, since such projects depend on a few creative and motivated individuals. If these persons cannot work on a project of importance to DOE, they will work on something else—with the result that the project may not be restarted when DOE has the funds to do so.

The panel recognizes that DOE has been pursuing projects in the Chemical Industrial Technologies Program that are of very high risk, but having high payoff. Faced with its severe budget constraint and the changing nature of the domestic industry, DOE should adhere to this fundamental strategy more closely than ever. Based on its members' experience with similar projects in industry, the panel suggests that DOE frequently review the projects in its portfolio to ensure that each is pursued just long enough—time, and money—to demonstrate feasibility. If it is judged at any time that the

		Global Scenario[a]		
		AEO Reference Case	High Oil and Gas Prices	Carbon Constrained
Program Risks	Technical Risks	Evaluated for each project in the portfolio. Most projects had technical risks associated with scale-up from laboratory-scale to commercial-scale applications. Estimated probability of technical success for the projects ranges from 5% to 50%.		
	Market Risks	Evaluated for each project in the portfolio. Estimated probability of market success for projects ranges from 10% to 70%.		
Program Benefits	Expected Economic Benefits	Economic benefits are calculated as the expected value of energy savings benefits from 2006 to 2030 resulting from the research portfolio, in 2003 dollars. Ranges shown are for the 10th and 90th percentiles:		
		$534 million at 3% ($0 to 1,550 million) $215 million at 7% ($0 to $640 million)	$950 million at 3% ($0 to $3,000 million) $390 million at 7% ($0 to $1,250 million)	$550 million at 3% ($0 to $1,600 million) $223 million at 7% ($0 to $700 million)
	Environmental Benefits	The same quantity of energy savings, and therefore the same environmental benefits, will be achieved under all three scenarios. Anticipated environmental benefits are estimated emission reductions from 2006 through 2030: CO: 15,100 (0 to 11,800) metric tons[b] CO_2: 1.7 (0-5.4) MMTCE SO_2: 9,200 (−2,000 to 30,000) metric tons[c] NO_x: 14,000 (0 to 42,000) metric tons Particulates: 180 (−50 to 600) metric tons VOCs: 340 (0 to 1,000) metric tons		
	Security Benefits	The same quantity of energy savings, and therefore the same security benefits, will be achieved under all three scenarios. Anticipated security benefits are the reduction in oil and natural gas consumption from 2006 through 2030: Natural gas: 89 (0-330) billion cubic feet Petroleum: 1.3 (0-0.01) million barrels[d]		

NOTE: Benefits are presented as expected values, with 10th and 90th percentile benefits in brackets. Economic benefits are shown as present values discounted at both 3% and 7% real; environmental and security benefits are discounted at 3%.

[a]The panel concluded that the scenarios would produce insignificantly small changes in the volumes of oil and gas saved; therefore, the physical quantities reported for environmental and security benefits are the same for all scenarios. Economic benefits differ because prices differ among the scenarios.

[b]CO reduction benefit derives primarily from a single project. This skews the distribution of benefits such that the expected value is higher than the 90th percentile.

[c]Several projects were estimated by DOE to result in increases of various types of emissions. For SO_2 and particulates, there were a sufficient number of these projects that the 10th percentile of the range of impacts represents an increase in emissions.

[d]Distribution on the reduction in petroleum usage is similarly skewed by a single project accounting for most of the reduction.

FIGURE M-4 Results matrix of the Panel on DOE's Industrial Technologies Program–Chemicals.

quality of the personnel and/or other resources being applied is not up to the task, or that progress is not up to expectations, the work should be terminated. The freed funds may then be applied elsewhere.

A model for the administration of projects of this kind might be the way Small Business Innovation Research grants (SBIR) are awarded and funded. Following this example, it might be appropriate to view the first year of funding of the chemical industrial projects as a first phase whose purpose is to define and assess the critical technology roadblocks that are potential showstoppers. In the second year-long phase, funds would be applied exclusively to address the most critical of these. Successful outcomes would trigger a request for further and often major increases in funding to demonstrate the approach. Sharing of funding and other resource commitments from industry should increase from phase to phase, reflecting growing confidence. The panel understands that DOE management is thinking along these lines.

Finally, the panel notes that a credible industry commitment to commercialization is crucial to the decision to move from the first phase to the second and third phases. In particular, the third phase must have industrial partners willing to commit considerable resources and have assurances from higher-level company management that there is a good probability of going forward to market. A marketing plan from the commercial partners should be available at this time. Absence of such a commitment should be cause for terminating the project.

Results Matrix

The panel believes it successfully applied the recommended methodology to this portfolio of small projects. This experience suggests that methodology could be applied to other situations in which a portfolio of diverse and technologically unrelated projects is to be evaluated. The panel made several judgments in applying the methodology, however, which should be considered when applying it to similar portfolio analyses. Specifically,

- The bottom-up analysis described above is more time consuming than a top-down approach, but the panel believes that it provides a more reliable estimate of expected benefits. The distribution of expected benefits at the program level can be inferred from the probability distributions developed by simulating the outcome of the program based on project-level risk estimates. Figure M-2 shows the result of this simulation for the Chemical Industrial Technologies Program.
- For early-stage research, as in the Chemical Industrial Technologies Program, a simplified calculation of the costs of commercialization is appropriate. The panel selected a simple payback model for this purpose. For portfolios comprising of nearer term research, direct calculation of commercialization costs may be more appropriate.[10]

The methodology developed for Phase Two spells out the information needed from DOE at the beginning of the project. Through its contractor, Energetics, the Chemical Industrial Technologies Program provided a one-page summary of the essential information about each project (see Figure M-5 for an example). This summary proved to be invaluable to the panel because it presented each project in a brief, consistent format, thus simplifying the review of 22 projects. The panel recommends that a similar summary be required for other projects being evaluated. It would be helpful, as well, to add to the summary sheet DOE's opinion of the technical and market risks and their distribution, the key drivers of uncertainty behind this risk assessment, and the metrics that would be useful for monitoring the progress of research in reducing risk.

Finally, although the Phase Two methodology suggests that benefits be evaluated every 3 years, a more frequent review of projects composed of early-stage research would be useful. This is because the projects are relatively short in duration and because less successful projects must be weeded out as quickly as possible.

Uncertainty About Portfolio Benefits

Figure M-2 illustrates the uncertainty about the total economic benefit of the research portfolio, based on a simulation of project-level risk evaluations. In the top chart, for any given portfolio value along the x-axis, the corresponding value along the y-axis shows the probability that the portfolio value will be equal to or less than that value. For example, there is about a 50 percent probability the value of the portfolio will be $250 million or less. The bottom chart displays a probability density function. The height of each bar represents the relative likelihood that the total portfolio value will lie in the range represented by the column. For example, it is about twice as likely that the total portfolio benefit lies between $0 and $150 million than between $150 million and $300 million.

Comments and Observations

The goal of the Chemical Industrial Technologies Program is to implement a successful DOE research program that helps the chemical industry to use 20 percent less energy in 2020 than in 2001. This translates into a reduction in energy use by the chemical industry of 1.3 quads per year in 2020 and proportional reduction in emissions. To achieve its goal the program has a portfolio of relatively small projects, all of which are competitively awarded and all of which involve 30-50 percent cost sharing. Twenty-two projects are currently being funded at a total of $7 million per year. If all projects were funded to completion and all were successful, DOE estimates they would achieve a saving of 0.303 quads per year, or 23 percent of the overall program goal for the Chemical Industrial Technologies Program.

The panel believes that the Chemical Industrial Technologies Program is seizing an important opportunity to produce energy savings in a major industrial segment by supporting early-stage research that industry is unlikely to support. However, to realize its potential benefits, the program must adapt to a seriously constrained budget and a changing domestic industrial environment. Cutting-edge research that can be done within the limited budget will probably continue to produce valuable but relatively small-scale advances. However, the program is also pursuing sweeping changes in process design in the hope that they can yield big energy savings.

The current portfolio of projects has an expected net economic benefit between $215 million and $534 million in the AEO Reference Case. Because the program is composed of early-stage research projects, the range of benefits is between $0 and $1.55 billion. Benefits in the High Oil and Gas Prices and Carbon Constrained scenarios are somewhat higher.

---

[10] It is important to note that the early-stage research considered here is nonetheless intended to produce discrete technologies to be brought to market. These are not curiosity-driven basic research projects, and the methodology used by the panel would not apply to such projects.

SUMMARY OF PROJECT IMPACTS
Chemical Reactions – New Process Chemistry/Synthesis

Project Title: Purification Process for PTA (Novel, Highly Efficient and Economic Purification Process Revolutionizing PTA Production)

Description: Purified terephthalic acid (PTA) is a chemical intermediate in the production of polyester. Two million tons are produced in the U.S. each year, for a variety of products including soft drink bottles, clothing and construction materials. The new process under development involves a two-step crystallization technique which can use less-pure feedstocks, can run at reduced temperature and pressure, eliminates use of hazardous chemical bromine, now deemed essential to processing, and requires 25% less capital investment than conventional technology, while reducing energy consumption.

Market Information

Market Segment	Ethylene- Polyester
Annual Market Production	2 Million Tons per year (2000)
Unit Size	Plant producing 770 million lb PTA per year
Number of Plants Operating	11
Market Annual Growth Rate	3%
Ultimate Potential Accessible Market	80%
Likely Technology Market Share	70%, 10 year market saturation curve
Commercial Prototype	2006
Year of Commercial Introduction	2008

Energy Consumption Data

Conventional Technology	Electricity: 0.055 billion kWh; Natural gas: 0.566 billion ft^3; feedstocks: 1.33 trillion Btu; waste: 0.579 trillion Btu
New Technology	Electricity: 0.0611 billion kWh; Natural gas: 0.704 billion ft^3; feedstocks: 1.20 trillion Btu; waste: 0.0174 trillion Btu

Technology Impacts

	2010	2015	2020	2025	2030
Market Penetration[a] (%)	3	21	71	96	99
Energy Savings/Year (trillion Btu)	.1	0.7	2.7	3.5	1.6
Energy Cost Savings/Year ($MM)	0.18	1.59	6.22	8.3	3.9

Notes
a percent of "likely technology market share" impacted, not the ultimate market.

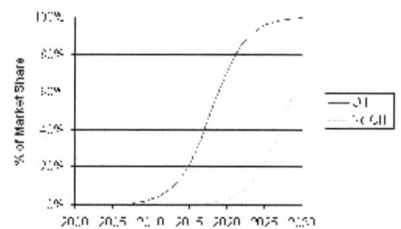

Market Penetration Curve for Purification Process for PTA (assumes 10 year acceleration of technology with Federal funds)

Impacts Summary for Industrial Technologies Program (ITP) Chemicals R&D Portfolio

FIGURE M-5 Sample project information sheet.

Technical and Market Risk Assessment

The panel assessed all of the projects in this program to be high risk and thus consistent with DOE's program strategy. Probability of technical success of individual projects ranged from 5 to 50 percent, with market success from 10 to 70 percent. Combining these assessments, the range for success for the entire portfolio is between 1 and 35 percent, with an average of 7.5 percent.

Benefits Estimation

The economic, environmental, and security benefits of the Chemical Industrial Technologies Program derive directly from the energy savings realizable from the projects. DOE estimates that the research will accelerate the development and implementation of the identified technology by 3 to 5 years. The panel agrees that in almost all cases, DOE support would be a significant accelerating factor. It calculated the expected total benefit of the portfolio by applying its probabilities of technical and market success to DOE's estimates of energy savings for each project in the portfolio and rolled up the individual project estimates into an expected value of gross benefits for the overall program. Net benefits are calculated by assuming that for any project that "succeeds," investments on the order of the net present value of the economic benefits for first 3 years will be required.

Program Observations

Based on the experience of its members in managing and conducting research in the chemical industry, the panel wishes to underscore how essential it is for DOE management to focus its limited resources on the most promising opportunities available to it. The danger is more than just the wasting of funds on less promising projects. Equally if not more significant is the possibility of losing the benefits of high-priority projects because they were not pursued aggressively enough. The panel suggests that DOE frequently review the projects in its existing portfolio to ensure that each project is pursued for as little time and money as it takes to demonstrate feasibility or infeasibility.

ATTACHMENT A
PANEL MEMBERS' BIOGRAPHIES

Robert W. Fri, *Chair*, is a visiting scholar and senior fellow emeritus at Resources for the Future, where he served as president from 1986 to 1995. From 1996 to 2001 he served as director of the National Museum of Natural History at the Smithsonian Institution. Before joining the Smithsonian, Mr. Fri served in both the public and private sectors, specializing in energy and environmental issues. In 1971 he became the first deputy administrator of the U.S. Environmental Protection Agency. In 1975, President Ford appointed him as the deputy administrator of the Energy Research and Development Administration. He served as acting administrator of both agencies for extended periods. From 1978 to 1986, Mr. Fri headed his own company, Energy Transition Corporation. He began his career with McKinsey & Company, where he was elected a principal. Mr. Fri is a senior advisor to private, public, and nonprofit organizations. He is a director of American Electric Power Company and of the Electric Power Research Institute and a trustee of Science Service, Inc. (publisher of *Science News* and organizer of the Intel Science Talent Search and International Science and Engineering Fair). He is a member of the National Petroleum Council and serves on the Advisory Council at the Marian E. Koshland Science Museum and on the Steering Committee at the Energy Future Coalition. In past years, he was a member of the President's Commission on Environmental Quality, the Secretary of Energy Advisory Board, and the University of Chicago board of governors for Argonne National Laboratory. He has chaired advisory committees of the National Research Council, the Carnegie Commission on Science, Technology and Government, the Electric Power Research Institute, and the Office of Technology Assessment. He served as chairman of the NRC Committee on Benefits of DOE's R&D in Energy Efficiency and Fossil Energy and of the Committee on Prospective Benefits of DOE's Energy Efficiency and Fossil Energy R&D Programs, Phase One. From 1978 to 1995 he was a director of Transco Energy Company, where he served as chair of the audit, compensation, and chief executive search committees. He is a member of Phi Beta Kappa and Sigma Xi and a national associate of the National Academies. He received his B.A. in physics from Rice University and his M.B.A. (with distinction) from Harvard University.

Anne Chaka is the group leader for computational chemistry at the National Institute of Standards and Technology in Gaithersburg, Maryland. She received her B.A. in chemistry from Oberlin College, her M.S. in clinical chemistry from Cleveland State University, and her Ph.D. in theoretical chemistry from Case Western Reserve University. In 1999-2000, she was Max Planck-Society Fellow at the Fritz-Haber-Institut in Berlin. She spent 10 years at the Lubrizol Corporation as head of the computational chemistry and physics program and previously was technical director of ICN Biomedicals, Inc., an analytical research chemist for Ferro Corporation, and a Cray programming consultant to Case Western Reserve University for the Ohio Supercomputer Center. Active areas of her research include metal oxide surface chemistry, atomistic descriptions of corrosion and materials, pericyclic reaction mechanisms, heterogeneous and homogeneous catalysis, thermochemistry, and oxidation.

Paul H. Kydd is the proprietor of Partnerships, a developer of lithium-ion batteries for applications in boats and automobiles. From 1999 to 2002 he was chairman and chief technology office of Parelec Inc., a firm that commercialized the Parmod technology for additive printing of electronic interconnects. He was president of Partnerships Limited Inc. from 1983 to 1999, a firm that developed chemical technology such as fuels and propellants. Previously, he was vice president for technology of Hydrocarbon Research, Inc., and oversaw approximately $20 million in synfuel and petrochemical process R&D. In addition, he was manager, chemical processes branch, GE Corporate R&D and combustion scientist, GE Research Laboratory. He has an A.B. from Princeton University and a Ph.D. from Harvard University, both in physical chemistry.

Alexander MacLachlan (NAE) retired at the end of 1993 from DuPont after more than 36 years of service. He was senior vice president for research and development and chief technical officer since 1986. In late 1994, he joined the U.S. Department of Energy (DOE) as deputy under secretary for technology partnerships and in 1995 was made deputy under secretary for R&D management. He left DOE in 1996 but remained on its Secretary of Energy Advisory Board, Laboratory Operations Board, Sandia President's Advisory Council, and the National Renewable Energy Laboratory's Advisory Council for several more years before resigning a few years ago. He has participated in several studies for the National Research Council, including *The Hydrogen Economy: Opportunities, Costs, Barriers, and R&D Needs, Containing* the Threat from Illegal Bombings: An Integrated National Strategy for Marking, Tagging, Rendering Inert, and Licensing Explosives and Their Precursors, Technology Commercialization: Russian Challenges, American Lessons, and Building an Effective Environmental Management Program: An Initial Assessment. Recently he was chair for the Committee for the Review of the Intelligent Vehicle Initiative. He is a member of Phi Beta Kappa and of the National Academy of Engineering. He is a graduate of Tufts University with a B.S. in chemistry and of MIT with a Ph.D. in physical organic chemistry. He currently serves as an outside director for the Millennium Cell Company, a company that strives to develop hydrogen source technology for fuel cells.

Gregory J. McRae is the Hoyt C. Hottel Professor of Chemical Engineering at MIT. His academic education includes a Ph.D. in engineering from the California Institute of Technology. Dr. McRae currently teaches undergraduate and graduate courses in process modeling, control, optimization, and computer-aided design. Another research focus is product and process design to avoid environmental problems and understanding the scientific aspects of problems involving pollutant transport and transformations in multimedia environments. His other interests include computational chemistry, process dynamics, turbulent fluid flow, computational fluid dynamics, reaction engineering, sensitivity/uncertainty analysis of complex systems, nonlinear parameter estimation, parallel computing, numerical analysis, and the design of cost-effective environmental controls. Dr. McRae is the recipient of numerous awards and prizes for his research in environmental and computational science, including the Presidential Young Investigator Award, the George Tallman Ladd Research Prize, the Forefronts of Computational Science Award, and an AAAS fellowship. He is a member of Sigma Xi, the American Chemical Society, and the American Institute of Chemical Engineers.

Michael L. Telson is presently serving as the director of National Laboratory Affairs for the University of California in its Washington Office of Federal Governmental Relations. He previously served as chief financial officer (CFO) of DOE from October of 1997 (after confirmation by the U.S. Senate) through May of 2001. He managed the relationship between DOE and the Office of Management and Budget (OMB), four congressional appropriations subcommittees, DOE's Inspector General, and the General Accounting Office (GAO). He reported directly to Secretaries Pena, Richardson, and Abraham, advising them on all financial matters, including the preparation and execution of DOE's nearly $20 billion annual budget, as well as reprogramming requests, in all of DOE's business lines, including national security, science, energy, and environmental quality. As CFO, Dr. Telson directed a staff of more than 200, also covering a number of other activities, including project management oversight; strategic planning and the Government Performance and Results Act (GPRA); privatization (including the sale of the Elk Hills Naval Petroleum Reserve, the initial public offering of stock in the U.S. Enrichment Corporation, and several environmental management privatization projects); payroll; and financial statement issues. Before the DOE, he served as a senior analyst on the staff of the Committee on the Budget, U.S. House of Representatives. He was responsible for reviewing energy, science, and space issues in the federal budget, including the programs of the DOE, the NSF, and NASA, government-wide R&D policy, and certain user fee programs (including FCC spectrum auction issues). He also served as staff economist to the House ad hoc Committee on Energy created to enact the 1978 National Energy Act. Dr. Telson is a member of Sigma Xi, Tau Beta Pi, and Eta Kappa Nu. He is a fellow of the AAAS, as well as of the APS, and received the Meritorious Service and Superior Performance awards from Energy Secretary Richardson and the Gold Medal for excellence from Energy Secretary Abraham. In 2002, he was named a senior fellow of the U.S. Association for Energy Economics. He holds B.S., M.S., E.E., and Ph.D. degrees in electrical engineering from MIT and an M.S. in management from the MIT Sloan School of Management.

William J. Ward (NAE) is a retired research engineer, GE Research and Development Center, which he joined in 1965. For 10 years he worked full-time in membrane gas separations. In subsequent years he worked part-time with GE and other colleagues on membranes. He did pioneering work on facilitated transport in immobilized liquid membranes and on ultrathin polymeric membranes. The latter resulted in a medical oxygen enrichment appliance. Dr. Ward was a manager from 1976 to 1979, after which he resumed full-time research in catalysis. His catalysis work in the 1980s provided new understanding of, and a much-improved catalyst for, the chemical reaction that is at the heart of the silicone polymer industry. From 1990 until 1995, Dr. Ward worked on understanding and improving the performance of polyurethane foam insulation and on eliminating chlorofluorocarbons as foam blowing agents. From 1996 through 1998 he was the technical leader of a team that made another major advance in the synthesis of silicone polymers. In his last 3 years at GE he was involved in a successful effort to develop a manufacturing process to produce a ceramic metal halide lamp. After retiring from GE in 2000, he has consulted for GE and other companies. Dr. Ward has 29 publications in refereed journals and 40 patents and is a member of the National Academy of Engineering. He holds a B.S. from Penn State and an M.S. and Ph.D. from the University of Illinois, all in chemical engineering.